P9-BIP-234

CELLULAR PHYSIOLOGY OF NERVE AND MUSCLE

The key to
short response
is key words & phrases

Cellular Physiology of Nerve and Muscle

Fourth Edition

Gary G. Matthews

Department of Neurobiology
State University of New York
at Stony Brook

Blackwell Publishing

© 2003 by Blackwell Science Ltd
a Blackwell Publishing company

BLACKWELL PUBLISHING
350 Main Street, Malden, MA 02148-5020, USA
9600 Garsington Road, Oxford OX4 2DQ, UK
550 Swanston Street, Carlton, Victoria 3053, Australia

The right of Gary G. Matthews to be identified as the Author of this Work has been asserted in accordance with the UK Copyright, Designs, and Patents Act 1988.

All rights reserved. No part of this publication may be reproduced, stored in a retrieval system, or transmitted, in any form or by any means, electronic, mechanical, photocopying, recording or otherwise, except as permitted by the UK Copyright, Designs, and Patents Act 1988, without the prior permission of the publisher.

First edition published 1986 by Blackwell Scientific Publications
Second edition published 1991
Third edition published 1998 by Blackwell Science, Inc.
Fourth edition published 2003 by Blackwell Science Ltd

6 2009

Library of Congress Cataloging-in-Publication Data

Matthews, Gary G., 1949 –
Cellular physiology of nerve and muscle / Gary G. Matthews.—4th ed.
p. cm.
Includes bibliographical references and index.
ISBN: 978-1-4051-0330-5
1. Neurons. 2. Muscle cells. 3. Nerves—Cytology. 4. Muscles—Cytology.
[DNLM: 1. Membrane Potentials—physiology. 2. Neurons—physiology.
3. Muscles—cytology. 4. Muscles—physiology. WL 102.5 M439c 2003]
I. Title.
QP363 .M38 2003
573.8'36—dc21 2002003951

A catalogue record for this title is available from the British Library.

Set in 11/12.5 pt Octavian
by Graphicraft Ltd, Hong Kong
Printed and bound in Singapore
by Utopia Press Pte Ltd

The publisher's policy is to use permanent paper from mills that operate a sustainable forestry policy, and which has been manufactured from pulp processed using acid-free and elementary chlorine-free practices. Furthermore, the publisher ensures that the text paper and cover board used have met acceptable environmental accreditation standards.

For further information on
Blackwell Publishing, visit our website:
www.blackwellpublishing.com

Contents

Preface to the Fourth Edition

The fourth edition of *Cellular Physiology of Nerve and Muscle* incorporates new material in several areas. An opening chapter has been added to introduce the basic characteristics of electrical signaling in the nervous system and to set the stage for the detailed topics covered in Part I. The coverage of synaptic transmission has been expanded to include synaptic plasticity, a topic requested by students and instructors alike. A new appendix has been included that covers the basic electrical properties of cells in greater detail for those who want a more quantitative treatment of this material.

Perhaps the most salient change is the artwork, with many new figures in this edition. As in previous editions, the goal of each figure is to clarify a single point of discussion, but I hope the new illustrations will also be more visually striking, while retaining their teaching purpose.

Students should also note that animations are available for selected figures, as indicated in the figure captions. The animations are available at www.blackwellscience.com by following the link for my general neurobiology text: *Neurobiology: Molecules, Cells, and Systems*.

Despite the numerous improvements in the fourth edition, the underlying core of the book remains the same: a step-by-step presentation of the physical and chemical principles necessary to understand electrical signaling in cells. This material is necessarily quantitative. However, I am confident that the approach taken here will allow students to arrive at a sophisticated understanding of how cells generate electrical signals and use them to communicate.

G.G.M.

Acknowledgments

Special thanks go to the following reviewers who offered their expert advice about the planned changes for the fourth edition. Their input was of great value.

Klaus W. Beyenbach, Cornell University
Scott Chandler, UCLA
Jon Johnson, University of Pittsburgh
Robert Paul Malchow, University of Illinois at Chicago
Stephen D. Meriney, University of Pittsburgh

Origin of Electrical Membrane Potential

part I

This book is about the physiological characteristics of nerve and muscle cells. As we shall see, the ability of these cells to generate and conduct electricity is fundamental to their functioning. Thus, to understand the physiology of nerve and muscle, we must understand the basic physical and chemical principles underlying the electrical behavior of cells.

Because an understanding of how electrical voltages and currents arise in cells is central to our goals in this book, Part I is devoted to this task. The discussion begins with the differences in composition of the fluids inside and outside cells and culminates in a quantitative understanding of how ionic gradients across the cell membrane give rise to a transmembrane voltage. This quantitative description sets the stage for the specific descriptions of nerve and muscle cells in Parts II and III of the book and is central to understanding how the nervous system functions as a transmitter of electrical signals.

as
an
aside,
what might
cause the
patellar reflex
<u>not</u> to occur?
Why might
that be a
bad thing?

Introduction to Electrical Signaling in the Nervous System

1

The Patellar Reflex as a Model for Neural Function

To set the stage for discussing the generation and transmission of signals in the nervous system, it will be useful to describe the characteristics of those signals using a simple example: the **patellar reflex**, also known as the knee-jerk reflex. Figure 1-1 shows the neural circuitry underlying the patellar reflex. Tapping the patellar tendon, which connects the knee cap (patella) to the bones of the lower leg, pulls the knee cap down and stretches the quadriceps muscle at the front of the thigh. Specialized nerve cells (**sensory neurons**) sense the stretch of the muscle and send a signal that travels along the thin fibers of the sensory

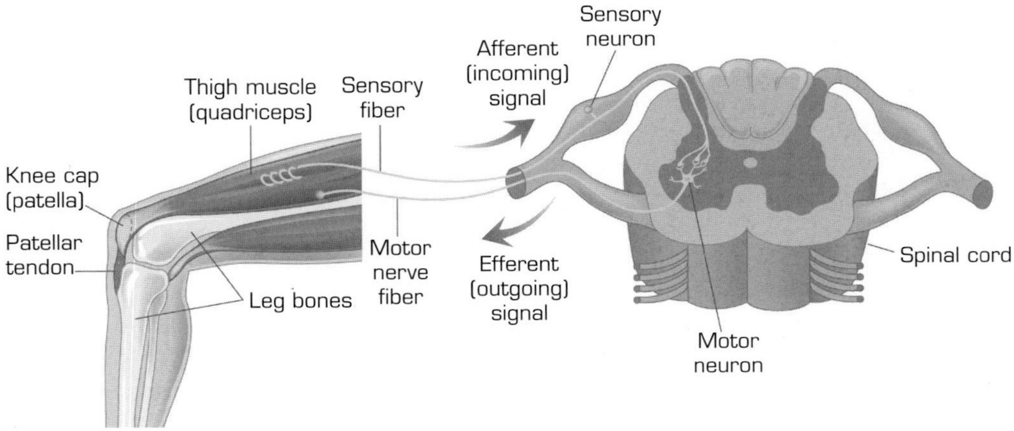

Figure 1-1 A schematic representation of the patellar reflex. The sensory neuron is activated by stretching the thigh muscle. The incoming (afferent) signal is carried to the spinal cord along the nerve fiber of the sensory neuron. In the spinal cord, the sensory neuron activates motor neurons, which in turn send outgoing (efferent) signals along the nerve back to the thigh muscle, causing it to contract.

neurons from the muscle to the spinal cord. In the spinal cord, the sensory signal is received by other neurons, called **motor neurons**. The motor neurons send nerve fibers back to the quadriceps muscle and command the muscle to contract, which causes the knee joint to extend.

The reflex loop exemplified by the patellar reflex embodies in a particularly simple way all of the general features that characterize the operation of the nervous system. A sensory stimulus (muscle stretch) is detected, the signal is transmitted rapidly over long distance (to and from the spinal cord), and the information is focally and specifically directed to appropriate targets (the quadriceps motor neurons, in the case of the sensory neurons, and the quadriceps muscle cells, in the case of the motor neurons). The sensory pathway, which carries information into the nervous system, is called the **afferent pathway**, and the motor output constitutes the **efferent pathway**. Much of the nervous system is devoted to processing afferent sensory information and then making the proper connections with efferent pathways to ensure that an appropriate response occurs. In the case of the patellar reflex, the reflex loop ensures that passive stretch of the muscle will be automatically opposed by an active contraction, so that muscle length remains constant.

The Cellular Organization of Neurons

Neurons are structurally complex cells, with long fibrous extensions that are specialized to receive and transmit information. This complexity can be appreciated by examining the structure of a motor neuron, shown schematically in Figure 1-2a. The cell body, or **soma**, of the motor neuron—where the nucleus resides—is only about 20–30 μm in diameter in the case of motor neurons involved in the patellar reflex. The soma is only a small part of the neuron, however, and it gives rise to a tangle of profusely branching processes called **dendrites**, which can spread out for several millimeters within the spinal cord. The dendrites are specialized to receive signals passed along as the result of the activity of other neurons, such as the sensory neurons of the patellar reflex, and to funnel those signals to the soma. The soma also gives rise to a thin fiber, the **axon**, that is specialized to transmit signals over long distances. In the case of the motor neuron in the patellar reflex, the axon extends all the way from the spinal cord to the quadriceps muscle, a distance of approximately 1 meter. As shown in Figure 1-2b, the sensory neuron of the patellar reflex is structurally simpler than the motor neuron. Its soma, which is located just outside the spinal cord in the **dorsal root ganglion**, gives rise to only a single nerve fiber, the axon. The axon splits into two branches shortly after it exits the dorsal root ganglion: one branch extends away from the spinal cord to contact the muscle cells of the quadriceps muscle, and the other branch passes into the spinal cord to contact the quadriceps motor neurons. The axon of the sensory neuron carries the signal generated by muscle stretch from the muscle into the spinal cord. Because the sensory

model: drg → sp. cord → mn ?? vs drg → sp. cord, muscle cell

what does it mean by drg branching to mn and muscle cells? shouldn't they be the same thing?

why does it branch?

(a) Motor neuron within spinal cord

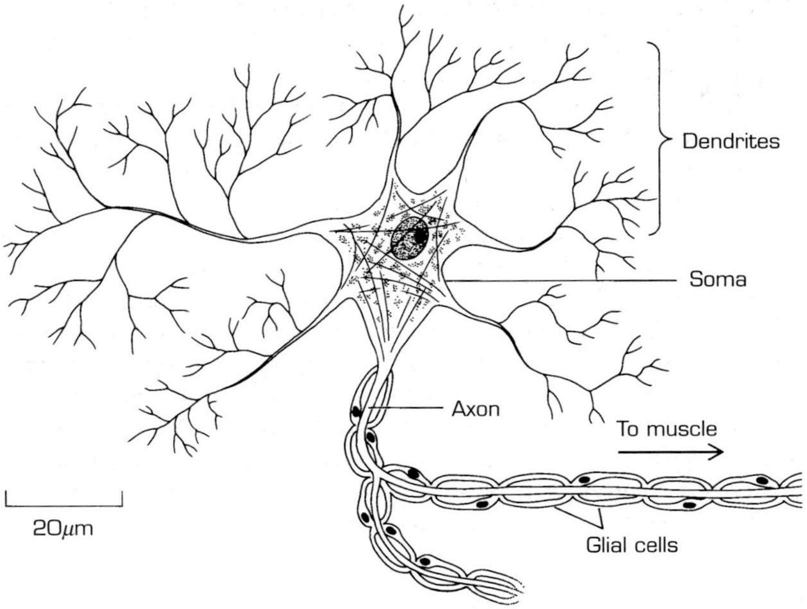

(b) Sensory neuron just outside spinal cord

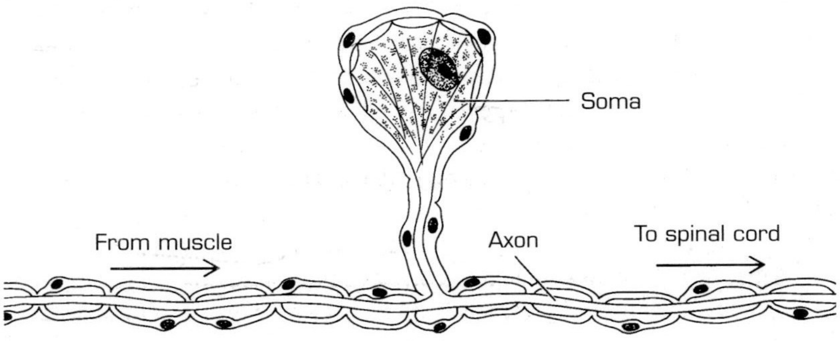

Figure 1-2 Structures of single neurons involved in the patellar reflex.

neuron receives its input signal from the sensory stimulus (muscle stretch) at the peripheral end of the axon instead of from other neurons, it lacks the dendrites seen in the motor neuron.

Electrical Signals in Neurons

To transmit information rapidly over long distances, neurons produce active electrical signals, which travel along the axons that make up the transmission paths. The electrical signal arises from changes in the electrical voltage difference across the cell membrane, which is called the **membrane potential**.

Although this transmembrane voltage is small—typically less than a tenth of a volt—it is central to the functioning of the nervous system. Information is transmitted and processed by neurons by means of changes in the membrane potential.

What does the electrical signal that carries the message along the sensory nerve fiber in the patellar reflex look like? To answer this question, we must measure the membrane potential of the sensory neuron by placing an ultrafine voltage-sensing probe, called an intracellular microelectrode, inside the sensory nerve fiber, as illustrated in Figure 1-3. A voltmeter is connected to measure the voltage difference between the tip of the intracellular microelectrode (point *a* in the figure) and a reference point in the extracellular space (point *b*). When the microelectrode is located outside the sensory neuron, both points *a* and *b* are in the extracellular space, and the voltmeter therefore records no voltage difference (Figure 1-3b). When the tip of the probe is inserted inside the sensory neuron, however, the voltmeter measures an electrical potential between points *a* and *b*, representing the voltage difference between the inside and the outside of the neuron—that is, the membrane potential of the neuron. As shown in Figure 1-3b, the inside of the sensory nerve fiber is negative with respect to the outside by about seventy-thousandths of a volt (1 millivolt, abbreviated mV, equals one-thousandth of a volt). Because the potential outside the cell is our reference point and the inside is negative with respect to the outside, the membrane potential is represented as a negative number, i.e., −70 mV.

As long as the sensory neuron is not stimulated by stretching the muscle, the membrane potential remains constant at this resting value. For this reason, the unstimulated membrane potential is known as the **resting potential** of the cell. When the muscle is stretched, however, the membrane potential of the sensory neuron undergoes a dramatic change, as shown in Figure 1-3b. After a delay that depends on the distance of the recording site from the muscle, the membrane potential suddenly moves in the positive direction, transiently reverses sign for a brief period, and then returns to the resting negative level. This transient jump in membrane potential is the **action potential**—the long-distance signal that carries information in the nervous system.

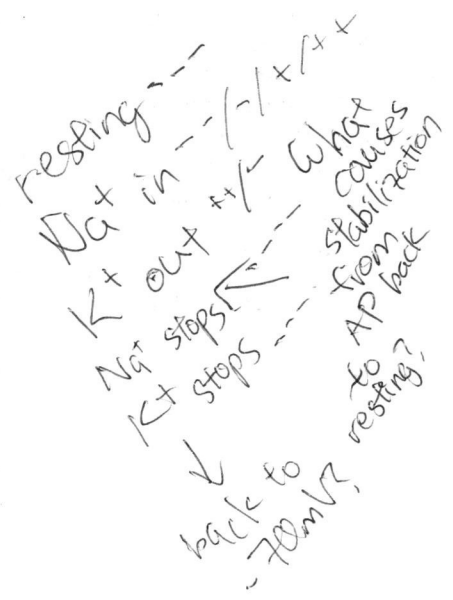

Transmission between Neurons

What happens when the action potential reaches the end of the neuron, and the signal must be transmitted to the next cell? In the patellar reflex, signals are relayed from one cell to another at two locations: from the sensory neuron to the motor neuron in the spinal cord, and from the motor neuron to the muscle cells in the quadriceps muscle. The point of contact where signals are transmitted from one neuron to another is called a **synapse**. In the patellar reflex, both the synapse between the sensory neuron and the motor neuron and the synapse between the motor neuron and the muscle cells are chemical synapses, in which

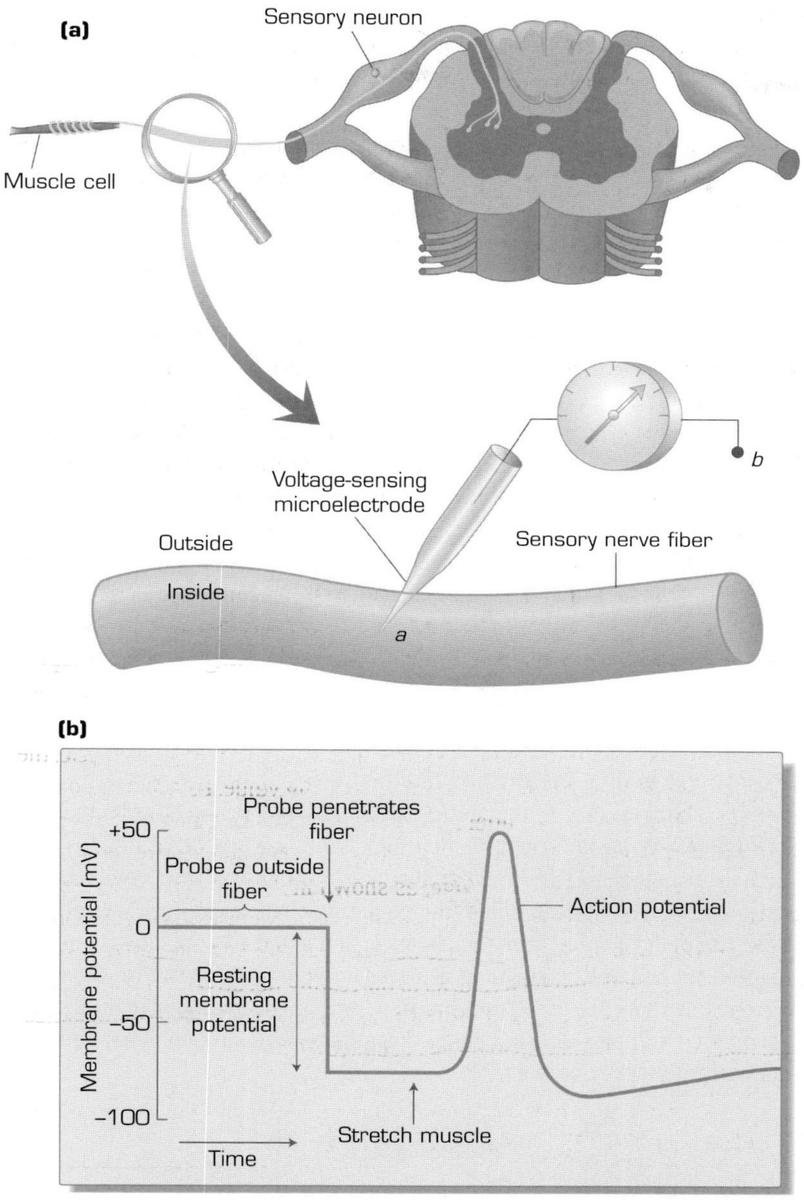

Figure 1-3 Recording the action potential in the nerve fiber of the sensory neuron in the patellar stretch reflex. (a) A diagram of the recording configuration. A tiny microelectrode is inserted into the sensory nerve fiber, and a voltmeter is connected to measure the voltage difference (E) between the inside (a) and the outside (b) of the nerve fiber. (b) When the microelectrode penetrates the fiber, the resting membrane potential of the nerve fiber is measured. When the sensory neuron is activated by stretching the muscle, an action potential occurs and is recorded as a rapid shift in the recorded membrane potential of the sensory nerve fiber.

an action potential in the input cell (the presynaptic cell) causes it to release a chemical substance, called a **neurotransmitter**. The molecules of neurotransmitter then diffuse through the extracellular space and change the membrane potential of the target cell (the postsynaptic cell). The change in membrane potential of the target then affects the firing of action potentials

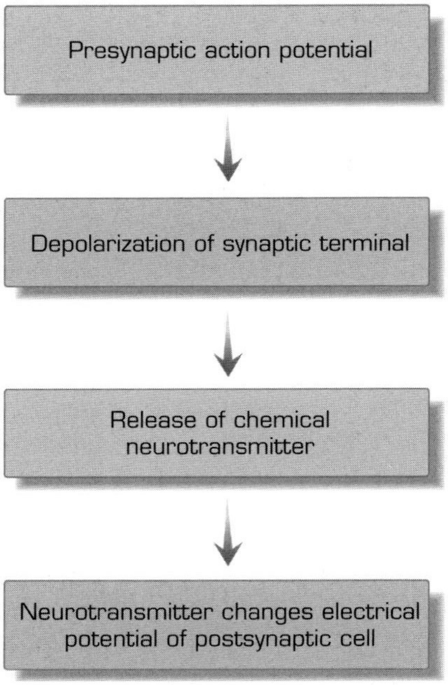

Figure 1-4 Chemical transmission mediates synaptic communication between cells in the patellar reflex. The flow diagram shows the sequence of events involved in the release of chemical neurotransmitter from the synaptic terminal.

by the postsynaptic cell. This sequence of events during synaptic transmission is summarized in Figure 1-4.

Because signaling both within and between cells in the nervous system involves changes in the membrane potential, the brain is essentially an electrochemical organ. Therefore, to understand how the brain functions, we must first understand the electrochemical mechanisms that give rise to a transmembrane voltage in cells. The remaining chapters in Part I are devoted to the task of developing the basic chemical and physical principles required to comprehend how cells communicate in the nervous system. In Part II, we will then consider how these electrochemical principles are exploited in the nervous system for both long-distance communication via action potentials and local communication at synapses.

Composition of Intracellular and Extracellular Fluids

2

When we think of biological molecules, we normally think of all the special molecules that are unique to living organisms, such as proteins and nucleic acids: enzymes, DNA, RNA, and so on. These are the substances that allow life to occur and that give living things their special characteristics. Yet, if we were to dissociate a human body into its component molecules and sort them by type, we would find that these special molecules are only a small minority of the total. Of all the molecules in a human body, only about 0.25% fall within the category of these special biological molecules. Most of the molecules are far more ordinary. In fact, the most common molecule in the body is water. Excluding nonessential body fat, water makes up about 75% of the weight of a human body. Because water is a comparatively light molecule, especially when compared with massive protein molecules, this 75% of body weight translates into a staggering number of molecules of water. Thus, water molecules account for about 99% of all molecules in the body. The remaining 0.75% consists of other simple inorganic substances, mostly sodium, potassium, and chloride ions. In the first part of this book we will be concerned in large part with the mundane majority of molecules, the 99.75% made up of water and inorganic ions.

Why should we study these mundane molecules? Many enzymatic reactions involving the more glamorous organic molecules require the participation of inorganic cofactors, and most biochemical reactions within cells occur among substances that are dissolved in water. Nevertheless, most inorganic molecules in the body never participate in any biochemical reactions. In spite of this, a sufficient reason to study these inorganic substances is that cells could not exist and life as we know it would not be possible if cells did not possess mechanisms to control the distribution of water and ions across their membranes. The purpose of this chapter is to see why that is true and to understand the physical principles that underlie the ability of cells to maintain their integrity in a hostile physicochemical environment.

Intracellular and Extracellular Fluids

The water in the body can be divided into two compartments: intracellular and extracellular fluid. About 55% of the water is inside cells, and the remainder is outside. The extracellular fluid, or ECF, can in turn be subdivided into plasma, lymphatic fluid, and interstitial fluid, but for now we can lump all the ECF together into one compartment. Similarly there are subcompartments within cells, but it will suffice for now to treat cells as uniform bags of fluid. The wall that separates the intracellular and extracellular fluid compartments is the outer cell membrane, also called the **plasma membrane** of the cell.

Both organic and inorganic substances are dissolved in the intracellular and extracellular water, but the compositions of the two fluid compartments differ. Table 2-1 shows simplified compositions of ECF and intracellular fluid (ICF) for a typical mammalian cell. The compositions shown in the table are simplified by including only those substances that are important in governing the basic osmotic and electrical properties of cells. Many other kinds of inorganic and organic solutes beyond those shown in the table are present in both the ECF and ICF, and many of them have important physiological roles in other contexts. For the present, however, they can be ignored.

The principal cation (positively charged ion) outside the cell is sodium, although there is also a small amount of potassium, which will be important to consider when we discuss the origin of the membrane potential of cells. Inside cells, the situation is reversed, with a small amount of sodium and potassium being the principal cation. Negatively charged chloride ions, which are present at a high concentration in ECF, are relatively scarce in ICF. The major anion (negatively charged ion) inside cells is actually a class of molecules that bear a net negative charge. These intracellular anions, which we will abbreviate A^-, include protein molecules, acidic amino acids like aspartate and glutamate, and inorganic ions like sulfate and phosphate. For the purposes of this

Table 2-1 Simplified compositions of intracellular and extracellular fluids for a typical mammalian cell.

	Internal concentration (m*M*)	External concentration (m*M*)	Can it cross plasma membrane?
K^+	125	5	Y
Na^+	12	120	N*
Cl^-	5	125	Y
A^-	108	0	N
H_2O	55,000	55,000	Y

Membrane potential = −60 to −100 mV

*As we will see in Chapter 3, this "No" is not as simple as it first appears.

book, the anions of this class outside cells can be ignored, and we will simplify the situation by assuming that the sole extracellular anion is chloride.

It will also be important to consider the concentration of water on the two sides of the membrane, which is also shown in Table 2-1. It may seem odd to speak of the "concentration" of the solvent in ECF and ICF. However, as we shall see when we consider the maintenance of cell volume, the concentration of water must be the same inside and outside the cell, or water will move across the membrane and cell volume will change.

Another important consideration will be whether a particular substance can cross the plasma membrane—that is, whether the membrane is permeable to that substance. The plasma membrane is permeable to water, potassium, and chloride, but is effectively impermeable to sodium (however, we will reconsider the sodium permeability later). Of course, if the membrane is to do its job properly, it must keep the organic anions inside the cell; otherwise, all of a cell's essential biochemical machinery would simply diffuse away into the ECF. Thus, the membrane is impermeable to A^-.

As described in Chapter 1, there is an electrical voltage across the plasma membrane, with the inside of the cell being more negative than the outside. The voltage difference is usually about 60–100 millivolts (mV), and is referred to as the **membrane potential** of the cell. By convention, the potential outside the cell is called zero; therefore, the typical value of the membrane potential (abbreviated E_m) is −60 to −100 mV, as shown in Table 2-1. A major concern of the first section of this book will be the origin of this electrical membrane potential. In later sections, we will discuss how the membrane potential influences the movement of charged particles across the cell membrane and how the electrical energy stored in the membrane potential can be tapped to generate signals that can be passed from one cell to another in the nervous system.

The Structure of the Plasma Membrane

Before we consider the mechanisms that allow cells to maintain the differences in ECF and ICF shown in Table 2-1, it will be helpful to look at the structure of the outer membrane of the cell, the plasma membrane. The control mechanisms responsible for the differences between ICF and ECF reside within the plasma membrane, which forms the barrier between the intracellular and extracellular compartments.

It has long been known that the contents of a cell will leak out if the cell is damaged by being poked or prodded with a glass probe. Also, some dyes will not enter cells when dissolved in the ECF, and the same dyes will not leak out when injected inside cells. These observations, first made in the nineteenth century, led to the idea that there is a selectively permeable barrier—the plasma membrane—separating the intracellular and extracellular fluids.

contents leak if cell prodded vs. some dyes are impermeable to membrane ↓ idea of selective perm. barrier

The first systematic observations of the kinds of molecules that would enter cells and the kinds that were excluded were made by Overton in the early part of the twentieth century. He found that, in general, substances that are highly soluble in lipids enter cells more easily than substances that are less soluble in lipids. Lipids are molecules that are not soluble in water or other polar solvents, but are soluble in oil or other nonpolar solvents. Thus, Overton suggested that the plasma membrane of a cell is made of lipids and that substances can cross the membrane if they can dissolve in the membrane lipids.

There were some exceptions to the general lipid solubility rule. Electrically charged substances, like potassium and chloride ions, are almost totally insoluble in lipids, yet they manage to cross the plasma membrane. Other substances, such as urea, entered cells more easily than expected from their lipid solubility alone. To take account of these exceptions, Overton suggested that the lipid membrane is shot through with tiny holes or pores that allow highly water soluble (hydrophilic) substances, such as ions, to cross the membrane. Only hydrophilic substances that are small enough to fit through these small aqueous pores can cross the membrane. Larger molecules like proteins and amino acids cannot fit through the pores and thus cannot cross the membrane without the help of special transport mechanisms.

The molecules of the lipid skin of cell membranes appear to be arranged in a layer only two molecules thick. Evidence for this arrangement was obtained from experiments in which the lipids were chemically extracted from the plasma membranes of cells and spread out on a trough of water in such a way that they formed a film only one molecule thick. When the area of this monolayer "oil slick" was measured, it was found to be about twice the total surface area of the intact cells from which the lipids were obtained. This suggests that the membrane of the intact cells was two molecules thick. Such a membrane is called a lipid bilayer membrane.

The bilayer arrangement of the cell membrane makes chemical sense when we consider the characteristics of the particular lipid molecules found in the plasma membrane. The cell lipids are largely phospholipids, which are molecules that have both a polar region that is hydrophilic and a nonpolar region that is hydrophobic. When surrounded by water, these lipid molecules tend to aggregate, with the hydrophilic regions oriented outward toward the surrounding water and the hydrophobic regions pointed inward toward each other. When spread out in a sheet with water on each side of the sheet, the phospholipids can maintain their preferred state by forming a bimolecular sandwich, with the hydrophilic parts on the outside toward the water, and the hydrophobic parts in the middle, pointed toward each other. This bilayer model for the cell plasma membrane is illustrated in Figure 2-1.

Figure 2-1 also shows another important characteristic of cell membranes. They contain not only lipid molecules but also protein molecules. Some proteins are attached to the inner or outer surface of the cell membrane, and others penetrate all the way through the membrane so that they form a bridge from one side to the other. Some of these transmembrane proteins form the aqueous pores, or channels, that allow ions and other small hydrophilic molecules to cross the membrane. If we separate membranes from the rest of the cell and

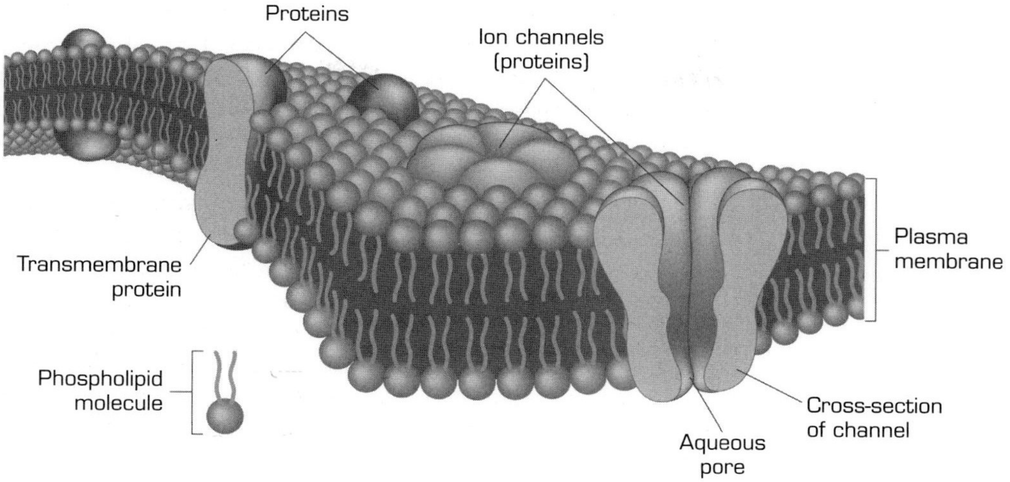

Figure 2-1 A schematic diagram of a section of the plasma membrane. The backbone of the membrane is a sheet of lipid molecules two molecules thick. Inserted into this sheet are various types of protein molecules. Some protein molecules extend all the way across the sheet, from the inner to the outer face. These transmembrane proteins sometimes form aqueous pores or channels through which small hydrophilic molecules, such as ions, can cross the membrane. The diagram shows two such channels; one is cut in cross-section to reveal the interior of the pore.

analyze their composition, we find that, by weight, only about one-third of the membrane material is lipid; most of the rest is protein. Thus, the lipids form the backbone of the membrane, but proteins are an important part of the picture. We will see later that the proteins are very important in controlling the movement of substances, particularly ions, across the cell membrane.

We can get an idea of the importance of membrane proteins for life by examining how much of the entire genome of a simple organism is taken up by genes encoding membrane proteins. One of the smallest genomes of any free-living organism is that of *Mycoplasma genitalium*, a microbe whose genome can be regarded as close to the minimum required for an independent, cellular life form. The DNA of *M. genitalium* has been completely sequenced, revealing a total of 482 individual genes. Of this total, 140 genes, or about 30%, code for membrane proteins. Thus, *M. genitalium* expends a large fraction of its total available DNA for the membrane proteins that sit at the interface between the microbe and its external environment. This points out the central role of these proteins in the maintenance of cellular life.

Anatomical evidence also supports the model shown in Figure 2-1. The cell membrane is much too thin to be seen with the light microscope. In fact, it is almost too thin to be seen with the electron microscope. However, with an electron microscope it is possible to see at the outer boundary of a cell a three-layered (trilaminar) profile like a railroad track, with a light central region separating two darker bands. Figure 2-2 is an example of an electron micrograph showing the plasma membranes of two cells lying in close contact. The interpretation of the trilaminar profile is that the two dark bands represent the polar heads of the membrane phospholipids and protein molecules on the inner and outer surfaces of the membrane and that the lighter region between the two dark bands represents the nonpolar tails of the lipid molecules. The total thickness of the sandwich is about 7.5 nm. The lighter-colored "fuzz" surrounding the trilaminar profiles of the two cell membranes in Figure 2-2 consists in part of portions of membrane-associated protein molecules extending out into the intracellular and extracellular spaces. The two cells shown in Figure 2-2 are nerve cells (neurons) in the brain, and the region of close contact is a specialized junction, called a synapse, where electrical activity is relayed from one nerve cell to another. The synapse is the basic mechanism of information transfer in the brain, and one of our major goals in this book is to understand how synapses work.

By using a special form of microscopy called freeze-fracture electron microscopy, it is possible to visualize more clearly the protein molecules that are embedded in the plasma membrane. A schematic representation of the freeze-fracture technique is shown in Figure 2-3. A small sample of the tissue to be examined is frozen in liquid nitrogen, and then a thin sliver of the frozen tissue is shaved off with a sharp knife. Because the tissue is frozen, however, the sliver is not so much sliced off as broken off from the sample. In some cases, like that shown in Figure 2-3, the line of fracture runs between the two lipid layers of the membrane bilayer, leaving holes where protein molecules are ripped out of the lipid monolayer and protrusions where membrane

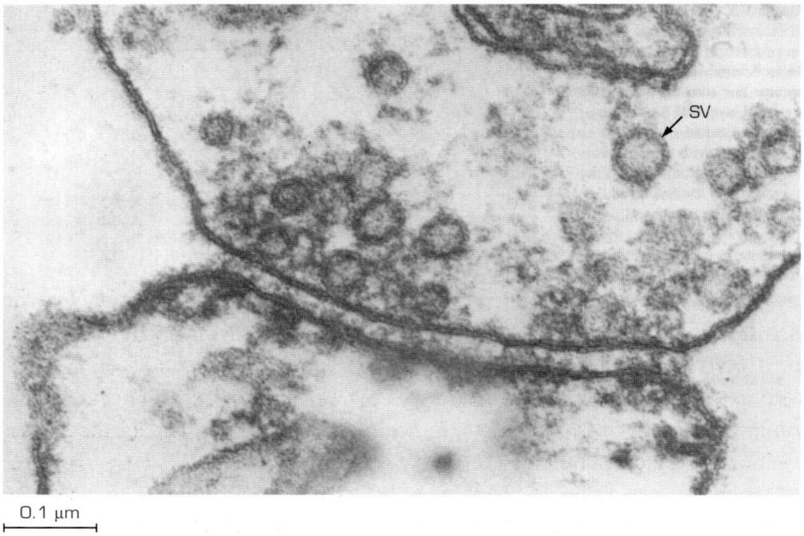

Figure 2-2 High-power electron micrograph of the plasma membranes of two neighboring cells. Note the two dark bands separated by a light region at the outer surface of each cell. The two cells are nerve cells from the brain, and the point of close contact between them is a **synapse**, the point of information transfer in the nervous system. Note also the membrane-bound intracellular structures (labeled SV), called **synaptic vesicles**, inside one of the cells; the vesicle membranes also have the trilaminar profile seen in the plasma membranes. We will learn more about synaptic vesicles and synapses in Chapters 8 and 9. (Courtesy of A. L. deBlas of the University of Connecticut.)

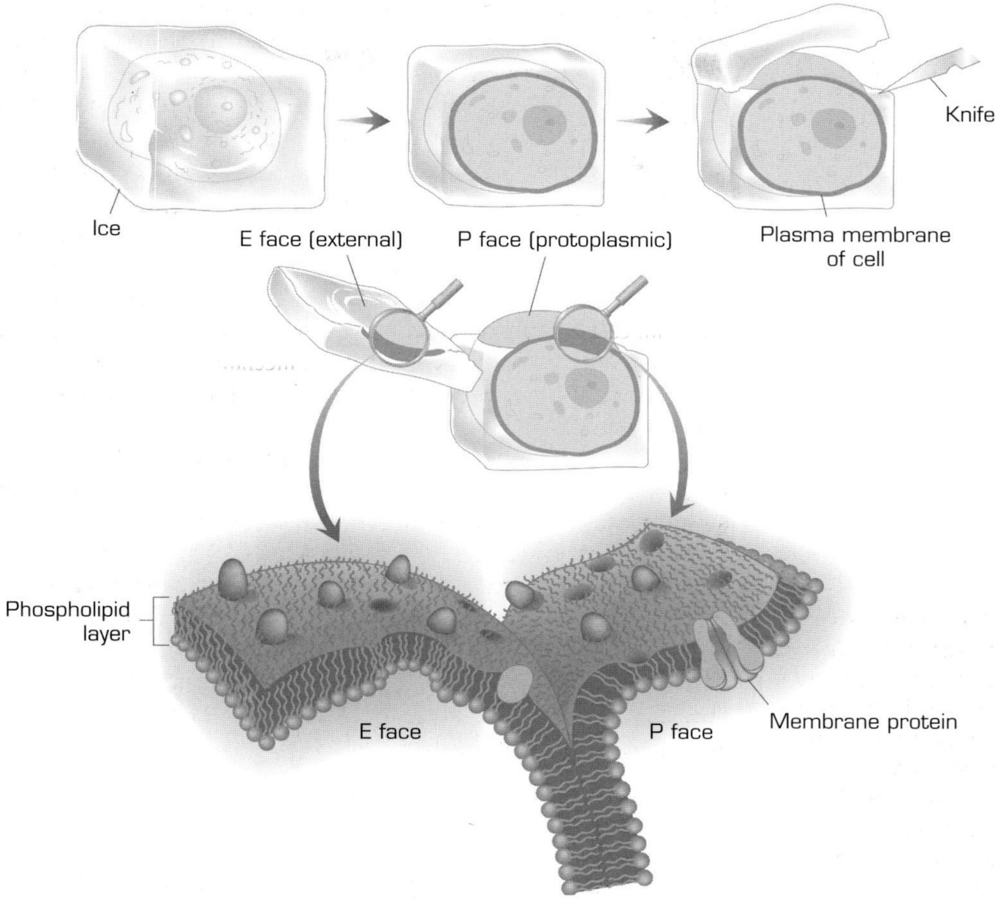

Figure 2-3 Schematic illustration of the freeze-fracture procedure for electron microscopy. When a fracture line runs between the two lipid layers of the plasma membrane, some membrane proteins stay with one monolayer, others with the other layer. When the fractured surface is then examined with the electron microscope, the remaining proteins appear as protruding bumps in the surface.

proteins are ripped out of the opposing monolayer and come along with the shaved sliver. An example of such a freeze-fracture sample viewed through the electron microscope is shown in Figure 2-4. The membrane proteins appear as small bumps in the otherwise smooth surface of the plasma membrane, like grains of sand sprinkled on a freshly painted surface. In the discussion of the transmission of signals at synapses in Chapter 8, we will see other examples of freeze-fracture electron micrographs and see how they can provide important evidence about the physiological functioning of cells.

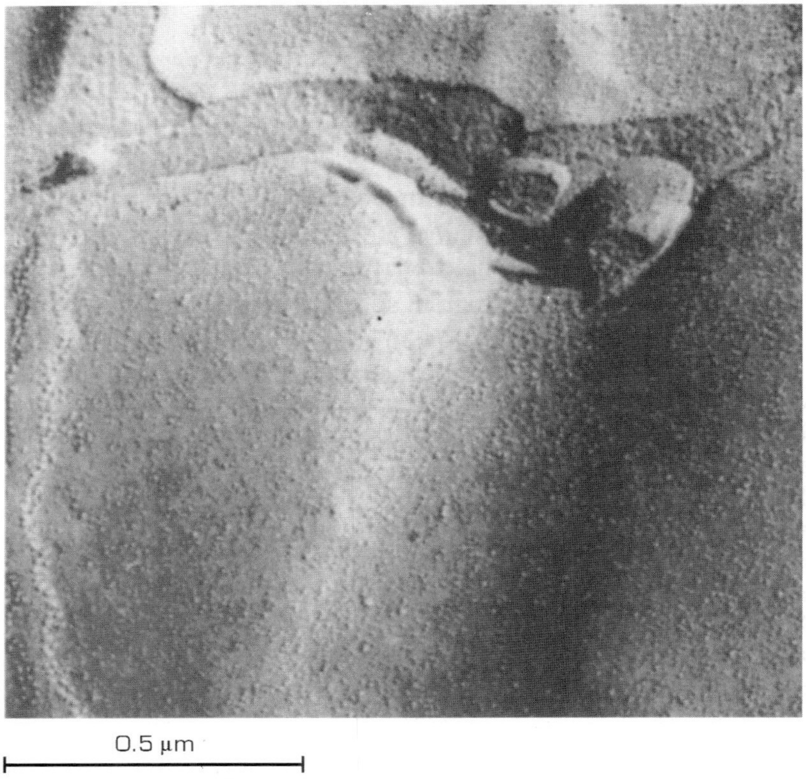

Figure 2-4 Example of a fractured membrane surface containing protein molecules, viewed through the electron microscope. The membrane surface shown is that of the presynaptic nerve terminal at the nerve–muscle junction, which will be discussed in detail in Chapter 8. The protein molecules are the small bumps scattered about on the planar surface of the membrane. (Reproduced from C.-P. Ko, Regeneration of the active zone at the frog neuromuscular junction. *Journal of Cell Biology* 1984;98:1685–1695; by copyright permission of the Rockefeller University Press.)

Summary

The most common molecules in the body are water and simple inorganic molecules—mainly sodium, potassium, and chloride ions. The water in the body can be divided into two compartments: the intracellular and extracellular fluids. The barrier between those two compartments is the plasma membrane of the cell, which is a phospholipid bilayer with protein molecules inserted into it. The extracellular fluid is high in sodium and chloride, but low in potassium, while the intracellular fluid is low in sodium and chloride, but high in potassium. This difference is maintained and regulated by control mechanisms residing in the plasma membrane, which acts as a selectively permeable barrier permitting some substances to cross but excluding others.

Maintenance of Cell Volume

3

At an early stage of evolution, before the development of cells, life might well have been nothing more than a loose confederation of enzyme systems and self-replicating molecules. A major problem faced by such acellular systems must have been how to keep their constituent parts from simply diffusing away into the surrounding murk. The solution to this problem was the development of a cell membrane that was impermeable to the organic molecules. This was the origin of cellular life. However, the cell membrane, while solving one problem, brought with it a new problem: how to achieve osmotic balance. To see how this problem arises, it will be useful to begin with a review of solutions, osmolarity, and osmosis. We will then turn to an analysis of the cellular mechanisms used to deal with problems of osmotic balance.

Molarity, Molality, and Diffusion of Water

Examine the situation illustrated in Figure 3-1. We take 1 liter of pure water and dissolve some sugar in it. The dissolved sugar molecules take up some space that was formerly occupied by water molecules, and thus the volume of the solution increases. Recall that the concentration of a substance is defined as the number of molecules of that substance per unit volume of solution. In Figure 3-1, this means that the concentration of water in the sugar–water solution is lower than it was in the pure water before the sugar was dissolved. This is because the total volume increased after the sugar was added, but the total

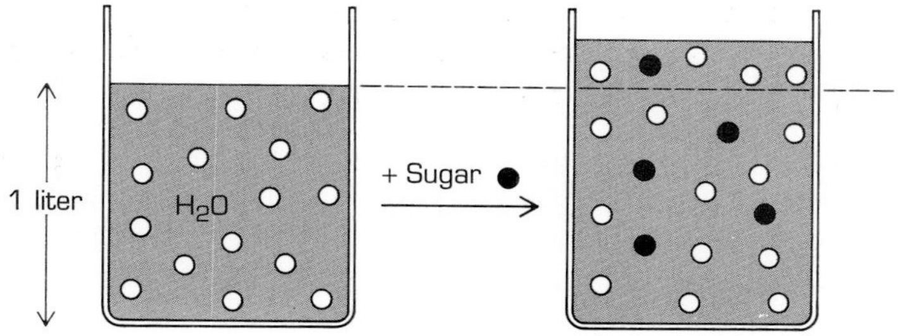

Figure 3-1 When sugar molecules (filled circles) are dissolved in a liter of water, the resulting solution occupies a volume greater than a liter. This is because the sugar molecules have taken up some space formerly occupied by water molecules (open circles). Therefore, the concentration of water (number of molecules of water per unit volume) is lower in the sugar–water solution.

number of water molecules present is the same before and after dissolving the sugar in the water.

To compare the concentrations of water in solutions containing different concentrations of dissolved substances, we will use the concept of **osmolarity**. A solution containing 1 mole of dissolved particles per liter of solution (a 1 molar, or 1 *M*, solution) is said to have an osmolarity of 1 osmolar (1 Osm), and a 1 millimolar (1 mM) solution has an osmolarity of 1 milliosmolar (1 mOsm). *The higher the osmolarity of a solution, the lower the concentration of water.* For practical purposes in biological solutions, it doesn't matter what the dissolved particle is; that is, the concentration of water is effectively the same in a solution of 0.1 Osm glucose, 0.1 Osm sucrose, or 0.1 Osm urea. To be strictly correct in discussing the concentration of water in various solutions, we would have to speak of the **molality**, rather than the molarity, of the solutions. Whereas molarity is defined as moles of solute per liter of solution, molality is defined as moles of solute per kilogram of solvent. This definition means that molality takes into account the fact that solutes having a higher molecular weight displace more water per mole of solute than do solutes with a lower molecular weight. That is, a liter of solution containing 1 mole of a large molecule, like a protein, would contain less water (and hence fewer grams of water) than a liter of solution containing 1 mole of a small molecule, like urea. Thus, the molality of the protein solution would be higher than the molality of the urea solution, even though both solutions have the same molarity (1 *M*). For our purposes, however, it will be adequate to treat molarity and osmolarity as equivalent to molality and osmolality.

It is important in determining the osmolarity of a solution to take into account how many dissolved particles result from each molecule of the dissolved substance. Glucose, sucrose, and urea molecules don't dissociate when they dissolve, and thus a 0.1 *M* glucose solution is a 0.1 Osm solution. A solution of sodium chloride, however, contains two dissolved particles—a sodium and a chloride ion—from each molecule of salt that goes into solution. Thus, a 0.1 *M* NaCl solution is a 0.2 Osm solution. To be strictly correct, we would have to take into account interactions among the ions in a solution, so that the effective osmolarity might be less than we would expect from assuming that all dissolved particles behave independently. But for dilute solutions like those we usually encounter in cell biology, such interactions are weak and can be safely ignored. Thus, for practical purposes we will assume that all dissolved particles act independently in determining the total osmolarity of a solution. Under this assumption, then, solutions containing 300 m*M* glucose, 150 m*M* NaCl, 100 m*M* NaCl + 100 m*M* glucose, or 75 m*M* NaCl + 75 m*M* KCl would all have the same total osmolarity—300 mOsm.

When solutions of different osmolarity are placed in contact through a barrier that allows water to move across, water will diffuse across the barrier down its concentration gradient (that is, from the lower osmolar solution to the higher). This movement of water down its concentration gradient is called osmosis. Consider the example shown in Figure 3-2a, which shows a container

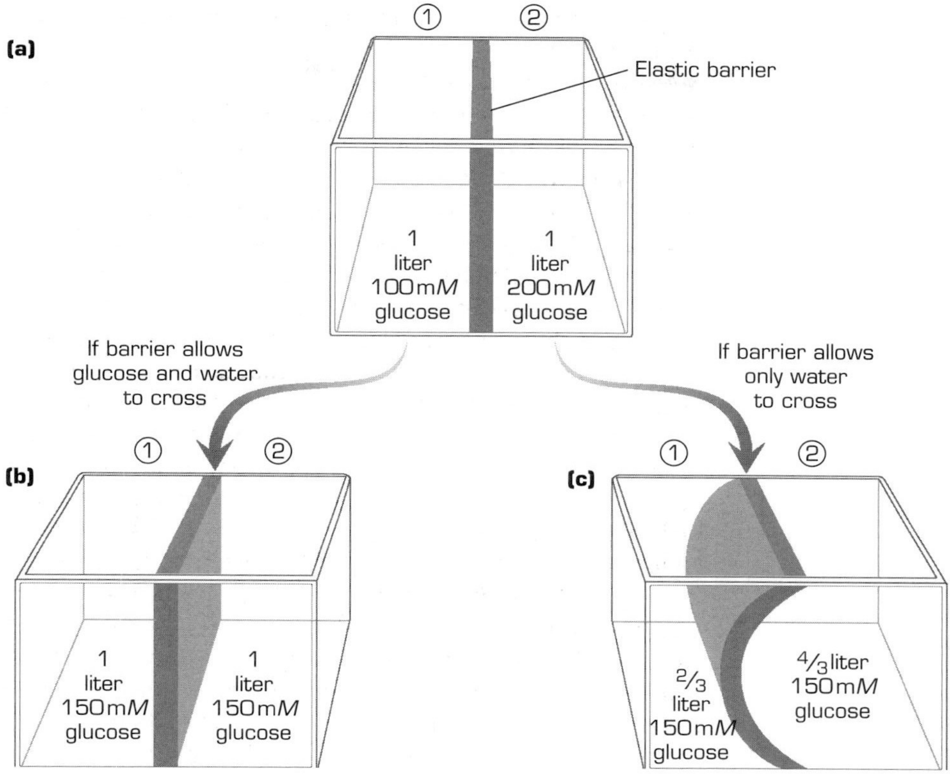

Figure 3-2 The effect of properties of the barrier separating two different glucose solutions on final volumes of the solutions. The starting conditions are shown in (a). (b) If the barrier allows both glucose and water to cross, the volumes of the two solutions do not change when equilibrium is reached. (c) If the barrier allows only water to cross, osmolarities of the two solutions are the same at equilibrium, but the final volumes differ.

divided into two equal compartments that are filled with glucose solutions. Imagine that the barrier dividing the container is made of an elastic material, so that it can stretch freely. If the barrier allows both water and glucose to cross, then water will move from side 1 to side 2, down its concentration gradient, and glucose will move from side 2 to side 1. The movement of water and glucose will continue until their concentrations on the two sides of the barrier are equal. Thus, side 1 gains glucose and loses water, and side 2 loses glucose and gains water until the glucose concentration on both sides is 150 m*M*. There will be no net change in the volume of solution on either side of the barrier, as shown in Figure 3-2b.

If the barrier in Figure 3-2a allows water but not glucose to cross, however, the outcome will be quite different from that shown in Figure 3-2b. Once again, water will move down its concentration gradient from side 1 to side 2. In this case, though, the loss of water will not be compensated by a gain of glucose. As water continues to leave side 1 and accumulates on side 2, the volume of side 2 will increase and the volume of side 1 will decrease. The accumulating water will exert a pressure on the elastic barrier, causing it to expand to the left to accommodate the volume changes (as shown in Figure 3-2c). The resulting volume changes will increase the osmolarity of side 1 and decrease the osmolarity of side 2, and this process will continue until the osmolarities of the two sides are equal—150 mOsm. In order to prevent the changes in volume, we would have to exert a pressure against the elastic barrier from side 1 to keep it from stretching. This pressure would be equal to the pressure moving water down its concentration gradient and would provide a measure of the osmotic pressure across the barrier.

Osmotic Balance and Cell Volume

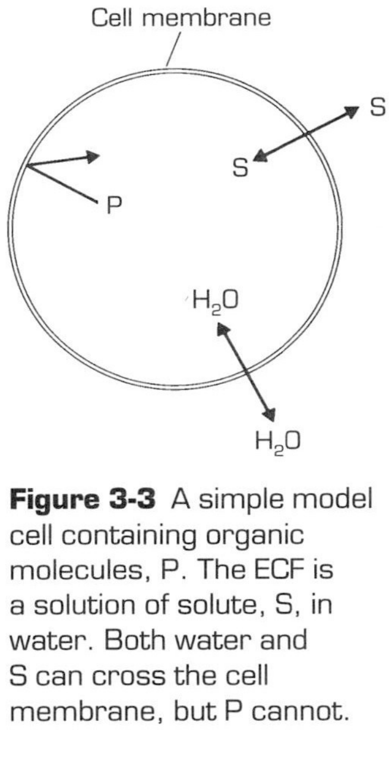

Figure 3-3 A simple model cell containing organic molecules, P. The ECF is a solution of solute, S, in water. Both water and S can cross the cell membrane, but P cannot.

Return now to the hypothetical primitive cell, early after the development of a cell membrane. In order for the cell membrane to do its job, it must be impermeable to the organic molecules inside the cell. But if the compositions of the extracellular and intracellular fluids are the same, with the exception of the internal organic molecules, the cell faces an imbalance of water on the two sides of the membrane. This situation is shown schematically in Figure 3-3. Here, the solutes that are in common in ICF and ECF are grouped together and symbolized by S. The extra solute inside the cell—the organic molecules (symbolized by P, for protein)—cause the concentration of water inside the cell to be less than it is outside. Put another way, the total osmolarity inside the cell is greater than it is outside the cell. There are two solutes inside, S and P, and only one outside. Water will therefore enter the cell and will continue to enter until the osmolarity on the two sides of the membrane is the same. Because the volume of the sea is essentially infinite relative to the volume of a cell and can thus be treated as constant, this end point could be reached only when the internal concentration of organic solutes is zero. This would require the volume of the cell to be infinite. Real cell membranes are not infinitely elastic, and thus water will enter the cell, causing it to swell, until the membrane ruptures and the cell bursts.

It will be convenient to summarize this situation in equation form. *If a substance is at diffusion equilibrium across a cell membrane, there is no net movement of that substance across the membrane.* For any solute, S, that can cross the cell membrane, this diffusion equilibrium will be reached when

$$[S]_i = [S]_o \qquad (3\text{-}1)$$

The square brackets indicate the concentration of a substance, and the subscripts i and o refer to the inside and outside of the cell. Thus, in order for water to be at equilibrium, we would expect that

$$[S]_i + [P]_i = [S]_o \qquad (3\text{-}2)$$

which is the same as saying that at equilibrium, the total osmolarity inside the cell must be the same as the total osmolarity outside the cell. For the cell of Figure 3-3, diffusion equilibrium will be reached only when the concentrations of all substances that can cross the membrane (in this case, S and water) are the same inside and outside the cell. This would require that Equations (3-1) and (3-2) be true simultaneously, which can occur only if $[P]_i$ is zero.

Answers to the Problem of Osmotic Balance

What solutions exist to this apparently fatal problem? There are three basic strategies that have developed in different types of cells. First, the problem could be eliminated by making the cell membrane impermeable to water. This turns out to be quite difficult to do and is not a commonly found solution to the problem of osmotic balance. However, certain kinds of epithelial cells have achieved very low permeability to water. A second strategy is commonly found and was likely the first solution to the problem. Here, the basic idea is to use brute force: build an inelastic wall around the cell membrane to physically prevent the cell from swelling. This is the solution used by bacteria and plants. The third strategy is that found in animal cells: achieve osmotic balance by making the cell membrane impermeable to selected extracellular solutes. This solution to the problem of osmotic balance works by balancing the concentration of nonpermeating molecules inside the cell with the same concentration of nonpermeating solutes outside the cell.

To see how the third strategy works, it will be useful to work through some examples using a simplified model animal cell whose membrane is permeable to water. Suppose the model cell contains only one solute: nonpermeating protein molecules, P, dissolved in water at a concentration of 0.25 *M*. We will then perform a series of experiments on this model cell by placing it in various extracellular fluids and deducing what would happen to its volume in each case. Assume that the initial volume of the cell is one-billionth of a liter (1 nanoliter, or 1 nl) and that the volume of the ECF in each case is infinite. This latter assumption means that the concentration of extracellular solutes does not change during the experiments, because the infinite extracellular volume provides an infinite reservoir of both water and external solutes.

The first experiment will be to place the cell in a 0.25 *M* solution of sucrose, which does not cross cell membranes. This is shown in Figure 3-4a. In this situation, only water can cross the cell membrane. For water to be at equilibrium, the internal osmolarity must equal the external osmolarity, or:

$$[P]_i = [\text{sucrose}]_o \qquad (3\text{-}3)$$

Because the internal and external osmolarities are both 0.25 Osm, this condition is met. Thus, there will be no net diffusion of water, and cell volume will not change.

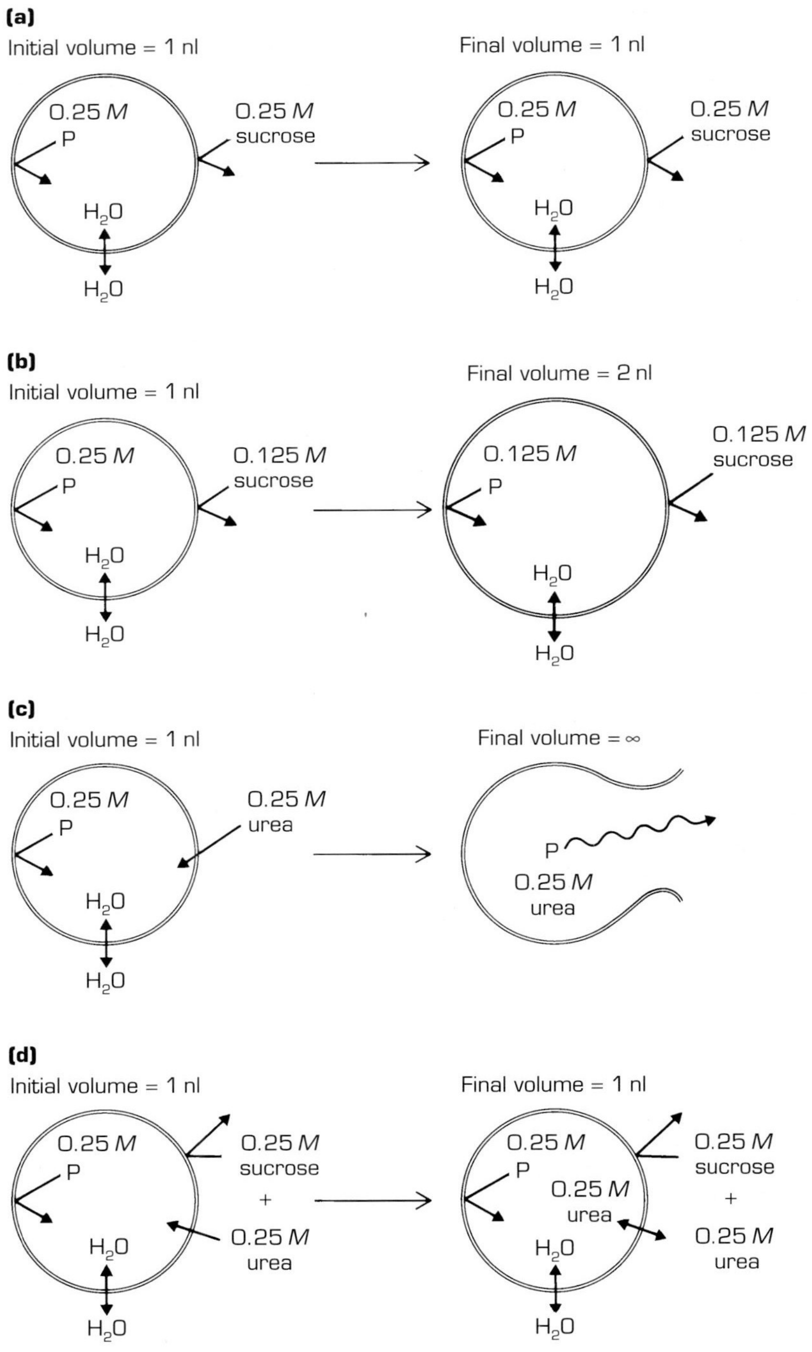

Figure 3-4 Effects of various extracellular fluids on the volume of a simple model. (a) The ECF contains an impermeant solute (sucrose), and the osmolarity is the same as that inside the cell. (b) The ECF contains an impermeant solute, and the osmolarity is lower than that inside the cell. (c) The ECF contains a permeant solute (urea) and external and internal osmolarities are equal. (d) The ECF contains a mixture of permeant and impermeant solutes.

In the second example, shown in Figure 3-4b, the cell is placed in 0.125 *M* sucrose rather than 0.25 *M* sucrose. Again, only water can cross the membrane, and Equation (3-3) must be satisfied for equilibrium to be reached. In 0.125 *M* sucrose, however, the internal osmolarity (0.25 Osm) is greater than the external (0.125 Osm), and water will enter the cell until internal osmolarity falls to 0.125 *M*. This will happen when the cell volume is twice normal, that is, 2 nl. What would the equilibrium cell volume be if we placed the cell in 0.5 *M* sucrose rather than 0.125 *M*?

The point of the previous two examples is that *water will be at equilibrium if the concentration of impermeant extracellular solute is the same as the concentration of impermeant internal solute.* To see that the external solute must not be able to cross the cell membrane, consider the example shown in Figure 3-4c. In this case, the model cell is placed in 0.25 *M* urea, rather than sucrose. Unlike sucrose, urea can cross the cell membrane, and thus we must take into account both urea and water in determining diffusion equilibrium. In equation form, equilibrium will be reached when these two relations hold:

$$[\text{urea}]_i = [\text{urea}]_o \qquad (3\text{-}4)$$

$$[\text{urea}]_i + [\text{P}]_i = [\text{urea}]_o \qquad (3\text{-}5)$$

Here, Equation (3-4) specifies diffusion equilibrium for urea, and Equation (3-5) applies to diffusion equilibrium for water. Because the external volume is infinite, $[\text{urea}]_o$ will be 0.25 *M* at equilibrium, and according to Equation (3-4) $[\text{urea}]_i$ must also be 0.25 *M* at equilibrium. Together, Equations (3-4) and (3-5) require that $[\text{P}]_i$ must be zero at equilibrium. Thus, the equilibrium volume is infinite, and the cell will swell until it bursts. Qualitatively, when the cell is first placed in 0.25 *M* urea, there will be no net movement of water across the membrane because internal and external osmolarities are both 0.25 Osm. But as urea enters the cell down its concentration gradient, internal osmolarity rises as urea accumulates. Water will then begin to enter the cell down its concentration gradient. The cell begins to swell and continues to do so until it bursts. Thus, an extracellular solute that can cross the cell membrane cannot help a cell achieve osmotic balance.

An interesting example is shown in Figure 3-4d. In this experiment, the model cell is placed in mixture of 0.25 *M* urea and 0.25 *M* sucrose. The equilibrium for urea will once again be given by Equation (3-4), and water will be at equilibrium when

$$[\text{urea}]_i + [\text{P}]_i = [\text{urea}]_o + [\text{sucrose}]_o \qquad (3\text{-}6)$$

Both Equation (3-4) and Equation (3-6) will be satisfied when $[\text{P}]_i = 0.25\ M$, which is the initial condition. Therefore, in this example, the cell volume at diffusion equilibrium will be the normal volume, 1 nl. The point is that even if some extracellular solutes can cross the cell membrane, the presence of a nonpermeating external solute at the same concentration as the nonpermeating internal solute allows the cell to achieve diffusion equilibrium for water and thus to maintain its volume. This is the strategy taken by animal cells to avoid bursting. As shown in Table 2-1, the impermeant extracellular solute in the case of real cells is sodium.

In all the examples of osmotic equilibrium we just worked through, the answer was arrived at using just one rule: *For each permeating substance (including water), the inside concentration must equal the outside concentration at equilibrium.*

Tonicity

In the examples in Figure 3-4, 0.25 *M* sucrose and 0.25 *M* urea had the same osmolarity: 0.25 Osm. But the two solutions had dramatically different effects on cell volume. In 0.25 *M* sucrose, cell volume didn't change, while in 0.25 *M* urea the cell exploded. To take into account the differing biological effects of solutions of the same osmolarity, we will use the concept of **tonicity**. An **isotonic** solution has no final effect on cell volume; a solution that causes cells to swell at equilibrium is called a **hypotonic** solution; and a solution that causes cells to shrink at equilibrium is called a **hypertonic** solution. Thus, the 0.25 *M* sucrose solution was isotonic, and the 0.25 *M* urea solution was hypotonic. Note that an isotonic solution must have the same osmolarity as the fluid inside the cell, but that having the same osmolarity as the ICF does not guarantee that an external fluid is isotonic.

Time-course of Volume Changes

So far in the discussion of maintenance of cell volume, we have considered only the final, equilibrium effect of a solution on cell volume and have ignored any transient effects that may occur. To see such transient effects, consider what happens to the model cell immediately after it is placed in the solution in Figure 3-4d, 0.25 *M* urea + 0.25 *M* sucrose. This is summarized in Figure 3-5. At the start, the osmolarity outside (0.5 Osm) is greater than the osmolarity inside (0.25 Osm), and water will initially leave the cell as it diffuses down its concentration gradient. Urea, however, begins to diffuse into the cell down its

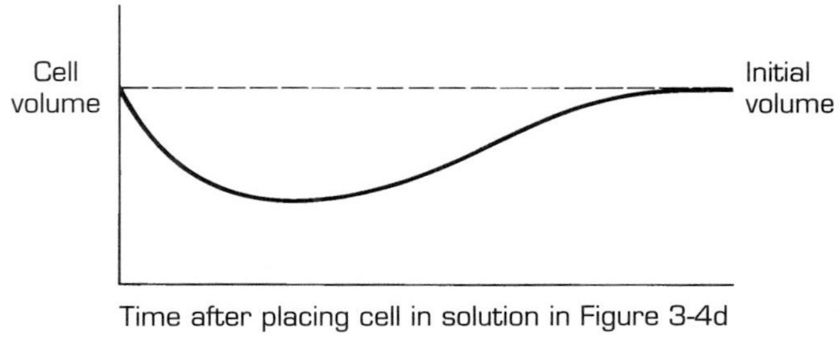

Figure 3-5 Time-course of cell volume when the model cell is placed in the solution used in Figure 3-4d.

concentration gradient. Thus, the internal osmolarity begins to rise as a result of the increasing $[\text{urea}]_i$ and the loss of intracellular water. The leakage of water out of the cell slows down and finally ceases altogether when $[\text{P}]_i + [\text{urea}]_i = 0.5\ M$; that is, at the point when internal and external osmolarities are equal. At this point, however, $[\text{P}]_i$ is higher than its initial value (0.25 *M*) because of the reduction in cell volume, and $[\text{urea}]_i$ is thus less than 0.25 *M*. Urea therefore continues to enter the cell to reach its own diffusion equilibrium, and the internal osmolarity rises above 0.5 Osm, so that water enters the cell and volume begins to increase. This situation continues until the final equilibrium state governed by Equations (3-4) and (3-6) is reached. What would you expect the time-course of cell volume to be if the model cell were placed in an infinite volume of a solution of 0.5 *M* urea?

Summary

If animal cells are to survive, it is essential that they regulate the movement of water across the plasma membrane. Given that proteins and other organic constituents of the ICF cannot be allowed to cross the membrane, diffusion of water becomes a problem. Animal cells have solved this problem by excluding a compensating extracellular solute, sodium ions. We'll discuss in more detail later exactly how they go about excluding Na^+.

Diffusion equilibrium is reached when internal and external concentrations are equal for all substances that can cross the membrane. For uncharged substances, such as those we have considered in our examples so far, we do not have to consider the influence of electrical force on the equilibrium state. However, the solutes of the ICF and ECF of real cells bear a net electrical charge. In the next chapter, we will consider what role electric fields play in the movements of these charged substances across the membranes of animal cells.

4 Membrane Potential: Ionic Equilibrium

The central topics in Chapter 3 were the factors that influence the distribution of water across the plasma membrane and the strategies by which cells can attain osmotic equilibrium. For clarity, all the examples so far have used only uncharged particles; however, a glance at Table 2-1 in Chapter 2 shows that all the solutes of both ICF and ECF are electrically charged. For charged particles, movement across the membrane will be determined not only by their concentration gradients, but also by the electrical potential across the membrane. This chapter will consider how cells can achieve equilibrium in the situation where both diffusional and electrical forces must be taken into account.

To illustrate the important principles that apply to ionic equilibrium, it will be useful to work through a series of examples that are increasingly complex and increasingly similar to the situation in real animal cells. At the end of the series of examples, we will see how a model cell, with internal and external compositions like those given in Table 2-1, could be in electrical and chemical equilibrium. However, we will also see that this equilibrium model of the electrochemical state of cells does not apply to real animal cells. Instead, real cells must expend energy to maintain the distribution of ions across the plasma membrane.

Diffusion Potential

In solution, positively charged particles accumulate around a wire connected to the negative pole—the cathode—of a battery, whereas negatively charged particles are attracted to a wire connected to the positive pole—the anode. This observation gives rise to the names **cation** (attracted to the cathode) for positively charged ions and **anion** (attracted to the anode) for negatively charged ions. The battery sets up a gradient of electrical potential (a **voltage** gradient) in the solution, and the movement of the ions in the solution is influenced by that voltage gradient. Thus, the distribution of ions in a solution depends on the presence of an electric field in that solution. The other side of the coin is that a differential distribution of ions in a solution *gives rise to* a voltage gradient in the solution. As an example of how an electrical potential can arise from spatial

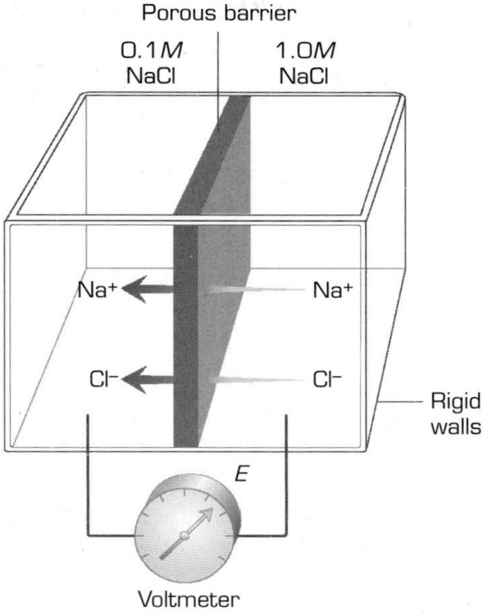

Figure 4-1 Schematic diagram of an apparatus for measuring the diffusion potential. A voltmeter measures the electrical voltage difference across the barrier separating the two salt solutions.

differences in the distribution of ions, we will consider the origin of **diffusion potentials**.

Diffusion potentials arise in the situation where two or more ions are moving down a concentration gradient. Examine the situation illustrated in Figure 4-1, which shows a rigid container divided into two compartments by a porous barrier. In the left compartment we place a 0.1 *M* NaCl solution and in the right compartment a 1.0 *M* NaCl solution. The porous barrier allows Na^+, Cl^-, and water to cross, but because of the rigid walls the compartment volume is not free to change and water cannot move. Thus, osmotic factors can be neglected for the moment. However, both Na^+ and Cl^- will move down their concentration gradients from right to left until their concentrations are equal in both compartments. In aqueous solution, Na^+ and Cl^- do not move at the same rate; Cl^- is more mobile and moves from right to left more quickly than Na^+. This is because ions dissolved in water carry with them a loosely associated "cloud" of water molecules, and Na^+ must drag along a larger cloud than Cl^-, causing it to move more slowly.

In Figure 4-1, then, the concentration of Cl^- on the left side will rise faster than the concentration of Na^+. In other words, there will be more negative than positive charges in the left compartment, and a voltmeter connected between the two sides would record a voltage difference, E, across the barrier, with the left compartment being negative with respect to the right compartment. This voltage difference is the diffusion potential. Notice that the electrical potential across the barrier tends to retard movement of Cl^- and speed up movement of Na^+ because the excess negative charges on the left repel Cl^- and attract Na^+. The diffusion potential will continue to build up until the electrical effect on the ions exactly counteracts the greater mobility of Cl^-, and the two ions cross the barrier at the same rate.

Another name for voltage is electromotive force. This name emphasizes the fact that voltage is the driving force for the movement of electrical charges through space; without a voltage gradient there is no net movement of charged particles. Thus, voltage can be thought of as a pressure driving charges in a particular direction, just as the pressure in the water pipe drives water out through your tap when you open the valve. Unlike the pressure in a hydraulic system, however, a voltage gradient can move charges in two opposing directions, depending on the polarity of the charge. Thus, the negative pole of a battery simultaneously attracts positively charged particles and repels negatively charged particles.

Equilibrium Potential

The Nernst Equation

The diffusion potential example of Figure 4-1 does not describe an equilibrium condition, but rather a transient situation that occurs only as long as there is a net diffusion of ions across the barrier. Equilibrium would be achieved in Figure 4-1 only when $[Na^+]$ and $[Cl^-]$ are the same in compartments 1 and 2. At that point, there would be no concentrational force to support net diffusion of either Na^+ or Cl^- across the membrane and there would be no electrical potential across the barrier. Under what conditions might there be a steady electrical potential at equilibrium? To see this, consider a small modification to the previous example, shown in Figure 4-2. In the new example, everything is as before, except that the barrier between the two compartments of the box is selectively permeable to Cl^-: Na^+ cannot cross. Once again, we assume that the box has rigid walls so that we can neglect movement of water for the present.

The analysis of the situation in Figure 4-2 is similar to that of the diffusion potential, except that now the "mobility" of Na^+ is reduced effectively to zero by the permeability characteristics of the barrier. Chloride ions will move down their concentration gradient from compartment 1 to compartment 2, but now no positive charges accompany them and negative charges will quickly build up in compartment 2. Thus, the voltmeter will record an electrical potential across the barrier, with side 2 being negative with respect to side 1. Because only Cl^- can cross the barrier, equilibrium will be reached when there is no further net movement of chloride across the barrier. This happens when the electrical force driving Cl^- out of compartment 2 exactly balances the concentrational force driving Cl^- out of compartment 1. Thus, at equilibrium a chloride ion moves from side 1 to side 2 down its concentration gradient for every chloride ion that moves from side 2 to side 1 down its electrical gradient. There will be no further change in $[Cl^-]$ in the two compartments, and no further change in the electrical potential, once this equilibrium has been reached.

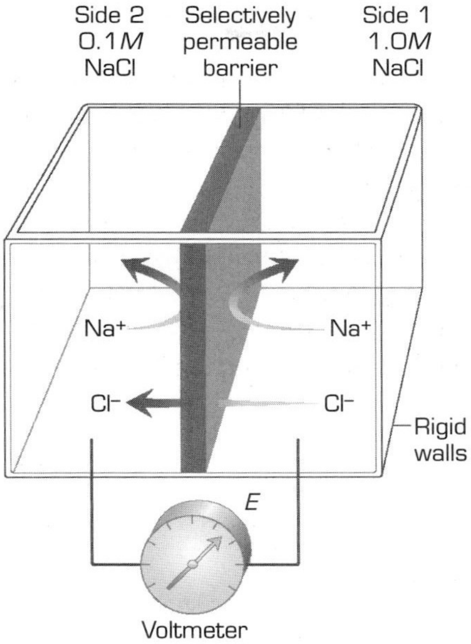

Figure 4-2 Schematic diagram of an apparatus for measuring the equilibrium, or Nernst, potential for a permeant ion. At equilibrium, a steady electrical potential (the equilibrium potential) is measured across the selectively permeable barrier separating the two salt solutions.

Equilibrium for an ion is determined not only by concentrational forces but also by electrical forces. Movement of an ion across a cell membrane is determined both by the concentration gradient for that ion across the membrane and by the electrical potential difference across the membrane. We will use these ideas extensively in this book, so the remainder of this chapter will be spent examining how these principles apply in simple model situations and in real cells.

What would be the measured value of the voltage across the barrier at equilibrium in Figure 4-2? This is a quantitative question, and the answer is provided by Equation (4-1), which is called the **Nernst equation** after the physical chemist who derived it. The Nernst equation for Figure 4-2 can be written as

$$E_{\mathrm{Cl}} = \left(\frac{RT}{ZF}\right) \ln \left(\frac{[\mathrm{Cl}^-]_1}{[\mathrm{Cl}^-]_2}\right) \tag{4-1}$$

Here, E_{Cl} is the voltage difference between sides 1 and 2 at equilibrium, R is the gas constant, T is the absolute temperature, Z is the valence of the ion in question (−1 for chloride), F is Faraday's constant, ln is the symbol for the natural, or base e, logarithm, and $[\mathrm{Cl}^-]_1$ and $[\mathrm{Cl}^-]_2$ are the chloride concentrations in compartments 1 and 2.

The value of electrical potential given by Equation (4-1) is called the **equilibrium potential**, or **Nernst potential**, for the ion in question. For example, in Figure 4-2 the permeant ion is chloride and the electrical potential, E_{Cl}, across the barrier is called the chloride equilibrium potential. If the barrier in

Figure 4-2 allowed Na^+ to cross rather than Cl^-, Equation (4-1) would again apply, except that $[Na^+]_1$ and $[Na^+]_2$ would be used instead of $[Cl^-]$, and the valence would be +1 instead of −1. If sodium were the permeant ion, the resulting potential, E_{Na}, would be the sodium equilibrium potential. *The Nernst equation applies only to one ion at a time and only to ions that can cross the barrier.*

A derivation of Equation (4-1) is given in Appendix A. The Nernst equation comes from the realization that at equilibrium the total change in energy encountered by an ion in crossing the barrier must be zero. If the change in energy were not zero, there would be a net force driving the ion in one direction or the other, and the ion would not be at equilibrium. There are two important sources of energy change involved in crossing the barrier shown in Figure 4-2: the electric field and the concentration gradient. Nernst arrived at his equation by setting the sum of the concentrational and electrical energy changes across the barrier to zero.

In biology, we usually work with a simplified form of Equation (4-1):

$$E_{Cl} = \left(\frac{58\ \text{mV}}{Z}\right) \log\left(\frac{[Cl^-]_1}{[Cl^-]_2}\right) \qquad (4\text{-}2)$$

The simplification arises from converting from base e to base 10 logarithms, evaluating (RT/F) at standard room temperature (20°C), and expressing the result in millivolts (mV). That is where the constant 58 mV comes from in Equation (4-2). From the simplified Nernst equation, it can be seen that E_{Cl} in Figure 4-2 would be −58 mV. That is, in crossing the barrier from side 1 to side 2, we would encounter a potential change of 58 mV, with side 2 being negative with respect to side 1. This is as expected from the fact that chloride ions, and therefore negative charges, are accumulating on side 2. If the barrier were selectively permeable to Na^+ rather than Cl^-, the voltage across the barrier would be given by E_{Na}, which would be +58 mV given the values in Figure 4-2. What would be the equilibrium potential for chloride in Figure 4-2 if the concentration of NaCl was 1.0 *M* on both sides of the barrier? (Hint: in that case the concentration gradient would be zero.)

The Principle of Electrical Neutrality

In arriving at −58 mV for the chloride equilibrium potential in Figure 4-2, we used 1.0 *M* and 0.1 *M* for $[Cl^-]_1$ and $[Cl^-]_2$. These are the initial concentrations in the two compartments, even though in our qualitative analysis we said that Cl^- moved from compartment 1 to 2, producing an excess of negative charge in compartment 2 and giving rise to the electrical potential. This would seem to suggest that $[Cl^-]$ changes from its initial value, invalidating our sample calculation. It is legitimate to use initial concentrations, however, because the increment in the electrical gradient caused by the movement of a single charged particle from compartment 1 to 2 is very much larger than the decrement in concentration gradient resulting from movement of that same particle. Thus, *only a very small number of charges need accumulate in order to counter even a large concentration gradient*.

In Figure 4-2, for example, it is possible to calculate that if the volume of each compartment were 1 ml and if the barrier between compartments were 1 cm^2 of the same material as found in cell membranes, it would require less than one-billionth of the chloride ions of side 1 to move to side 2 in order to reach the equilibrium potential of −58 mV. (The basis of this calculation is explained below.) Clearly, such a small change in concentration would produce an insignificant difference in the result calculated according to Equation (3-2), and we can safely ignore the movement of chloride necessary to achieve equilibrium.

This leads to an important principle that will be useful in the examples following in this chapter. This principle, called the **principle of electrical neutrality**, states that *under biological conditions, the bulk concentration of cations within any compartment must be equal to the bulk concentration of anions in that compartment*. This is an acceptable approximation because the number of charges necessary to reach transmembrane potentials of the magnitude encountered in biology is insignificant compared with the total numbers of cations and anions in the intracellular and extracellular fluids.

The Cell Membrane as an Electrical Capacitor

This section explains how we were able to calculate the number of charges necessary to produce the equilibrium membrane potential of −58 mV in the preceding section. The calculation was made by treating the barrier between the two compartments as an electrical capacitor, which is a charge-storing device consisting of two conducting plates separated by an insulating barrier. In Figure 4-2, the two conducting plates are the salt solutions in the two compartments, and the barrier is the insulator. In a real cell, the ICF and ECF are the conductors, and the lipid bilayer of the plasma membrane is the insulating barrier. When a capacitor is hooked up to a battery as shown in Figure 4-3, the voltage of the battery causes electrons to be removed from one conducting plate and to accumulate on the other plate. This will continue until the resulting

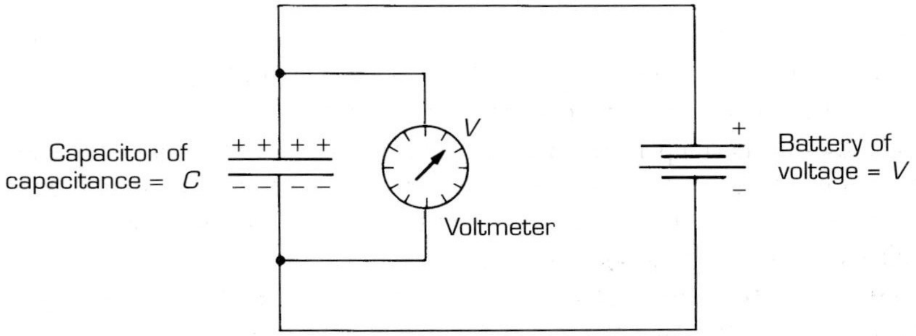

Figure 4-3 When a battery is connected to a capacitor, charge accumulates on the capacitor until the voltage across the capacitor is equal to the voltage of the battery.

voltage gradient across the capacitor is equal to the voltage of the battery. Basic physics tells us that the amount of charge, q, stored on the capacitor at that time will be given by $q = CV$, where V is the voltage of the battery and C is the **capacitance** of the capacitor. A capacitor's capacitance is directly proportional to the area of the plates (bigger plates can store more charge) and inversely proportional to the distance separating the two plates. Capacitance also depends on the characteristics of the insulating material between the plates; in the case of cells, that insulating material is the lipid plasma membrane. The unit of capacitance is the farad (F): a 1 F capacitor can store 1 coulomb of charge when hooked up to a 1 V battery. Biological membranes, like the plasma membrane, have a capacitance of 10^{-6} F (that is, 1 microfarad, or μF) per cm^2 of membrane area.

If the barrier in Figure 4-2 were 1 cm^2 of cell membrane, it would therefore have a capacitance of 10^{-6} F. From $q = CV$, it follows that an equilibrium potential of −58 mV would store 5.8×10^{-8} coulomb of charge on the barrier. Note that the charge on the membrane barrier in Figure 4-2 is carried by ions, not by electrons as in Figure 4-3. Thus, to know the total number of excess anions on side 2 of the barrier at equilibrium, we must convert from coulombs of charge to moles of ion. This can be done by dividing the number of coulombs on the barrier by Faraday's constant (approximately 10^5 coulombs per mole of monovalent ion), yielding 5.8×10^{-13} mole or about 3.5×10^{11} chloride ions moving from side 1 to side 2 in Figure 4-2. If the volume of each compartment were 1 ml, then side 2 would contain about 6×10^{20} chloride and sodium ions. These leads to the conclusion stated in the previous section that less than one-billionth of the chloride ions in side 1 cross to side 2 to produce the equilibrium voltage across the barrier.

Incorporating Osmotic Balance

The example shown in Figure 4-2 illustrates how ionic equilibrium can be reached and how the Nernst equation can be used to calculate the value of the membrane potential at equilibrium. However, the simple situation in the example is not very similar to the situation in real animal cells. For one thing, animal cells are not enclosed in a box with rigid walls, and thus osmotic balance must be taken into account. An example of how equilibrium can be reached when water balance must be considered is shown in Figure 4-4a. In this example the rigid walls are removed, so that osmotic balance must be achieved in order to reach equilibrium. In addition, an impermeant intracellular solute, P, has been added. For now, P has no charge; the effect of adding a charge on the intracellular organic solute will be considered later.

In Figure 4-4a, it is assumed that the model cell contains 50 mM Na^+ and 100 mM P. What must the concentrations of the other intracellular and extracellular solutes be in order for the model cell to be at equilibrium? The principal of electrical neutrality tells us that for practical purposes, the concentrations of

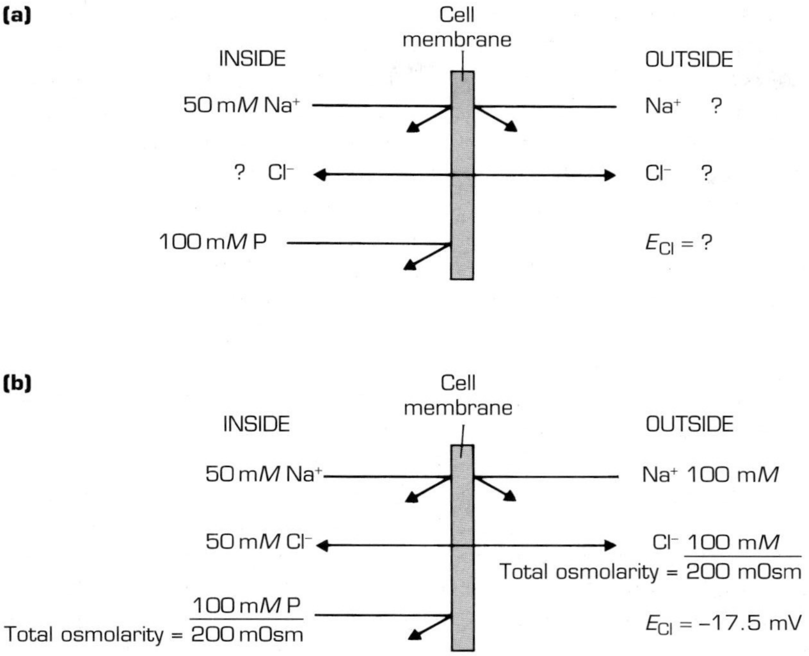

Figure 4-4 A model cell in which both osmotic and electrical factors must be considered at equilibrium. (a) The starting conditions, with initial values of some parameters provided. (b) The values of all parameters required for the cell to be at equilibrium.

cations and anions within any compartment are equal. Thus, because P is assumed to have no charge, $[Cl^-]_i = [Na^+]_i = 50\ mM$. For osmotic balance, the external osmolarity must equal the internal osmolarity, which is 200 mOsm. The principal of electrical neutrality again requires that $[Na^+]_o = [Cl^-]_o$. This requirement, together with the requirement for osmotic balance, can be satisfied if $[Na^+]_o = [Cl^-]_o = 100$ mM. The model cell of Figure 4-4a can therefore be at equilibrium if the concentrations of intracellular and extracellular solutes are as shown in Figure 4-4b. At this equilibrium, the voltage across the membrane of the model cell (the membrane potential, E_m) would be given by the Nernst equation for chloride:

$$E_m = E_{Cl} = -58\ \text{mV}\ \log\left(\frac{[Cl^-]_o}{[Cl^-]_i}\right) = -17.5\ \text{mV}$$

Donnan Equilibrium

The example of Figure 4-4b shows how we could construct a model cell that is simultaneously at osmotic and ionic equilibrium. However, the situation in Figure 4-4b is not very much like that in real animal cells. A major difference is that the principal internal cation in real cells is K^+, not Na^+. Also, there is some potassium in the ECF, and the cell membrane is permeable to K^+ as well as Cl^-.

In this situation, there are two ions that can cross the membrane: K^+ and Cl^-. If equilibrium is to be reached, the electrical potential across the cell membrane must simultaneously balance the concentration gradients for both K^+ and Cl^-. Because the membrane potential can have only one value, this equilibrium condition will be satisfied only when the equilibrium potentials for Cl^- and K^+ are equal. In equation form, this condition can be written as:

$$E_K = 58\text{ mV}\log\left(\frac{[K^+]_o}{[K^+]_i}\right) = E_{Cl} = -58\text{ mV}\log\left(\frac{[Cl^-]_o}{[Cl^-]_i}\right)$$

Here, the minus sign on the far right arises from the fact that the valence of chloride is -1. Canceling 58 mV from the above relation leaves

$$\log\left(\frac{[K^+]_o}{[K^+]_i}\right) = -\log\left(\frac{[Cl^-]_o}{[Cl^-]_i}\right) \tag{4-3}$$

The minus sign on the right side can be moved inside the parentheses of the logarithm to yield $\log([Cl^-]_i/[Cl^-]_o)$. Thus, equilibrium will be reached when

$$\left(\frac{[K^+]_o}{[K^+]_i}\right) = \left(\frac{[Cl^-]_i}{[Cl^-]_o}\right) \tag{4-4}$$

This equilibrium condition is called the **Donnan** or **Gibbs–Donnan equilibrium**, and it specifies the conditions that must be met in order for two ions that can cross a cell membrane to be simultaneously at equilibrium. Equation (4-4) is usually written in a slightly rearranged form as the product of concentrations:

$$[K^+]_o[Cl^-]_o = [K^+]_i[Cl^-]_i \tag{4-5}$$

In words, for a Donnan equilibrium to hold, the product of the concentrations of the permeant ions outside the cell must be equal to the product of the concentrations of those two ions inside the cell.

To see how the Donnan equilibrium might apply in an animal cell, consider the example shown in Figure 4-5a. Here a model cell containing K^+, Cl^-, and P is placed in ECF containing Na^+, K^+, and Cl^-. As an exercise, we will calculate the values of all concentrations at equilibrium assuming that $[Na^+]_o$ is 120 mM and $[K^+]_o$ is 5 mM. From the principal of electrical neutrality, $[Cl^-]_o$ must be 125 mM. Also, because P is assumed for the present to be uncharged, the principle of electrical neutrality requires that $[K^+]_i$ must equal $[Cl^-]_i$. Because two ions—K^+ and Cl^-—can cross the membrane, the defining relation for a Donnan equilibrium shown in Equation (3-5) must be obeyed. Thus, if the model cell of Figure 4-5a is to be at equilibrium, $[K^+]_i[Cl^-]_i$ must equal $[K^+]_o[Cl^-]_o$, which is 5×125, or 625 mM^2. Because $[K^+]_i = [Cl^-]_i$, the Donnan condition reduces to $[K^+]_i^2 = 625$ mM^2; thus, $[K^+]_i$ and $[Cl^-]_i$ must be 25 mM

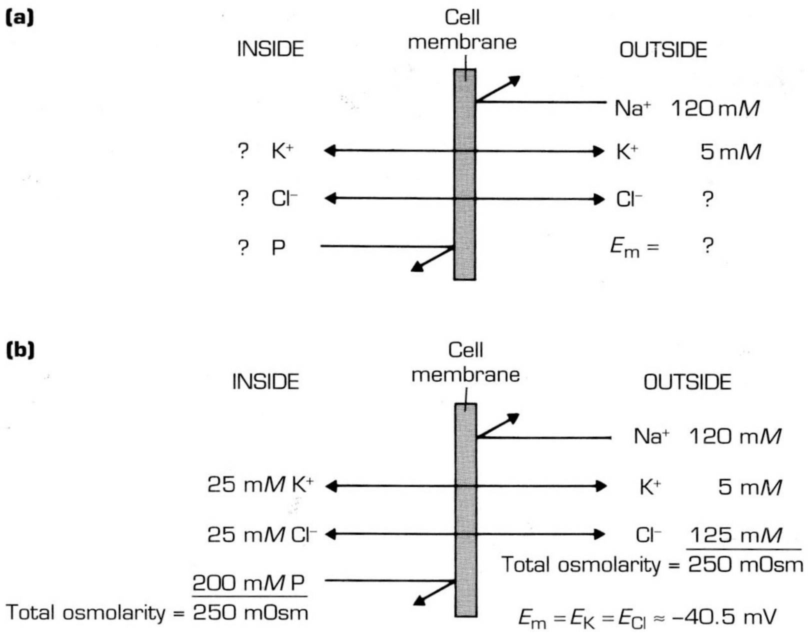

Figure 4-5 An example of a model cell at Donnan equilibrium. The cell membrane is permeable to both potassium and chloride. (a) The starting conditions, with initial values of some parameters provided. (b) The values of all parameters required for the cell to be at equilibrium.

at equilibrium. For osmotic balance, the internal osmolarity must equal the external osmolarity, which is 250 mOsm. This requires that $[P]_i$ must be 200 m*M* for the model cell to be at equilibrium. The results of this example are summarized in Figure 4-5b, which represents a model cell at equilibrium. What would be the membrane potential of this equilibrated model cell? The Nernst equation—Equation (4-2)—tells us that the membrane potential for a cell at equilibrium with $[K^+]_o = 5$ m*M* and $[K^+]_i = 25$ m*M* is about −40.5 mV, inside negative. You should satisfy yourself that the Nernst equation for chloride yields the same value for membrane potential.

A Model Cell that Looks Like a Real Animal Cell

The model cell of Figure 4-5b still lacks many features of real animal cells. For instance, as Table 2-1 shows, the internal organic molecules are charged, and this charge must be considered in the balance between cations and anions required by the principle of electrical neutrality. Recall that the category of internal anions, A^-, actually represents a diverse group of molecules, including proteins, charged amino acids, and sulfate and phosphate ions. Some of these bear a single negative charge, others two, and some even three net negative charges. Taken as a group, however, the average charge per molecule is slightly greater than −1.2. Thus, the internal impermeant anions can be represented as $A^{1.2-}$.

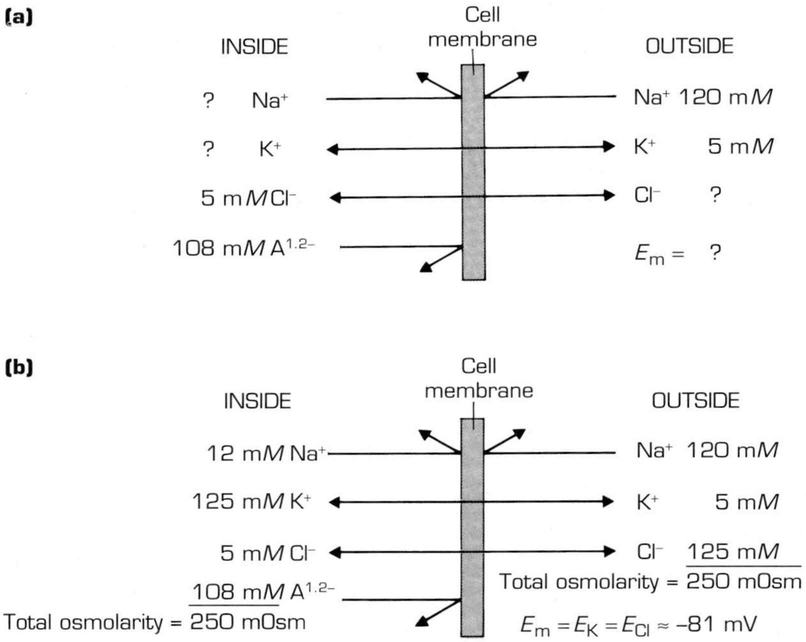

Figure 4-6 An example of a realistic model cell that is at both electrical and osmotic equilibrium. The compositions of ECF and ICF for this equilibrated model cell are the same as for a typical mammalian cell (see Table 2-1). (a) The starting conditions, with initial values of some parameters provided. (b) The values of all parameters required for the cell to be at equilibrium.

In addition, the model cell of Figure 4-5b lacked Na^+ inside the cell, while real ICF does contain a small amount of sodium. Addition of these complicating factors leads to the model cell of Figure 4-6a, which now contains all the constituents shown in Table 2-1. If the cell of Figure 4-6a is to be at equilibrium, what concentrations of the various ions in ECF and ICF would be required, and what would be the transmembrane potential? To begin, we will take some values from Table 2-1 and determine what the remaining parameters must be for the cell to be at equilibrium. Assume that $[K^+]_o = 5$ mM, $[Na^+]_o = 120$ mM, $[Cl^-]_i = 5$ mM, and $[A^{1.2-}]_i = 108$ mM. (Actually, it is not necessary to assume the concentration of A; it could be calculated from the other parameters. For mathematical simplicity, however, we will assume that it is known from the start.) Because Cl^- is the sole external anion, the principle of electrical neutrality requires that $[Cl^-]_o$ be 125 mM. Both K^+ and Cl^- can cross the membrane, so that the conditions for a Donnan equilibrium—Equation (4-5)—must be satisfied. This requires that $[K^+]_i = 125$ mM. The equilibrated value of $[Na^+]_i$ can then be obtained from the requirements for osmotic balance; $[Na^+]_i$ must be 12 mM if internal and external osmolarities are to be equal. From the Nernst equation for either Cl^- or K^+, the membrane potential at equilibrium can be determined to be about −81 mV.

The equilibrium values for this model cell are shown in Figure 4-6b. Note that the concentrations of all intracellular and extracellular solutes are the same for the model cell and for real mammalian cells (Table 2-1). The values in Figure 4-6b were arrived at by assuming that the cell was in equilibrium, and this implies that the real cell, which has the same ECF and ICF, is also at equilibrium. Thus, the model cell, and by extension the real cell, will remain in the state summarized in Figure 4-6b without expending any metabolic energy at all. From this viewpoint, the animal cell is a beautiful example of efficiency, existing at perfect equilibrium, both ionic and osmotic, in harmony with its electrochemical environment. The problem, however, is that the model cell is not an accurate representation of the situation in real animal cells: *real cells are not at equilibrium and must expend metabolic energy to maintain the status quo.*

The Sodium Pump

For some time, the model in Figure 4-6b was thought to be an accurate description of real animal cells. The difficulty with this scheme arose when it became apparent that real cells are permeable to sodium, while the model cell is assumed to be impermeable to sodium. Permeability to sodium, however, would be catastrophic for the model cell. If sodium can cross the membrane, then all extracellular solutes can cross the membrane. Recall from Chapter 3, however, what happens to cells that are placed in ECF containing only permeant solutes (like the urea example in Figure 3-4c): the cell swells and bursts. The cornerstone of the strategy employed by animal cells to achieve osmotic balance is that the cell membrane must exclude an extracellular solute to balance the impermeant organic solutes inside the cell. Sodium ions played that role for the model cell of Figure 4-6b.

How can the permeability of the plasma membrane to sodium be reconciled with the requirement for osmotic balance? An answer to this question was suggested by the experiments that demonstrated the sodium permeability of the cell membrane in the first place. In these experiments, red blood cells were incubated in an external medium containing radioactive sodium ions. When the cells were removed from the radioactive medium and washed thoroughly, it was found that they remained radioactive, indicating that the cells had taken up some of the radioactive sodium. This showed that the plasma membrane was permeable to sodium. In addition, it was found that the radioactive cells slowly lost their radioactive sodium when incubated in normal ECF. This latter observation was surprising because both the concentration gradient and the electrical gradient for sodium are directed inward; neither would tend to move sodium out of the cell. Further, the rate of this loss of radioactive sodium from the cell interior was slowed dramatically by cooling the cells, indicating that a source of energy other than simple diffusion was being tapped to actively "pump" sodium out of the cell against its concentrational and electrical gradients. It turns out that this energy source is metabolic energy in the form of the high-energy phosphate compound adenosine triphosphate (ATP).

This active pumping of sodium out of the cell effectively prevents sodium from accumulating intracellularly as it leaks in down its concentration and electrical gradients. Thus, even though sodium can cross the membrane, it is actively extruded at a rate sufficiently high to counterbalance the inward leak. The net result is that *sodium behaves osmotically as though it cannot cross the membrane*. Note however that this mechanism is fundamentally different from the situation in the model cell of Figure 4-6b. The model was in *equilibrium* and required no energy input to maintain itself. By contrast, real animal cells are in a finely balanced *steady state*, in which there is no net movement of ions across the cell membrane, but which requires the expenditure of metabolic energy.

Metabolic inhibitors, such as cyanide or dinitrophenol, prevent the pumping of sodium out of the cell and cause cells to gain sodium and swell. If ATP is added, the pump can operate once again and the accumulated sodium will be extruded. Similarly, other manipulations that reduce the rate of ATP production, like cooling, cause sodium accumulation and increased cell volume. Experiments of this type demonstrated the role of ATP in the active extrusion of sodium and the maintenance of cell volume. The mechanism of the sodium pump has been studied biochemically. The pump itself is a particular kind of membrane-associated protein molecule that can bind both sodium ions and ATP at the intracellular face of the membrane. The protein then acts as an enzyme to cleave one of the high-energy phosphate bonds of the ATP molecule, using the released energy to drive the bound sodium out across the membrane by a process that is not yet completely understood.

The action of the sodium pump also requires potassium ions in the ECF. Binding of K^+ to a part of the protein on the outer surface of the cell membrane is required for the protein to return to the configuration in which it can again bind another ATP and sodium ions at the inner surface of the membrane. The potassium bound on the outside is released again on the inside of the cell, so that the protein molecule acts as a bidirectional pump carrying sodium out across the membrane and potassium in. Thus, the sodium pump is more correctly referred to as the sodium–potassium pump, and can be thought of as a shuttle carrying Na^+ out across the membrane, releasing it in the ECF, then carrying K^+ in across the membrane and releasing it in the ICF. Because the pump molecule splits ATP and binds both sodium and potassium ions, biochemists refer to this membrane-associated enzyme as a Na^+/K^+ ATPase.

Summary

The movement of charged substances across the plasma membrane is governed not only by the concentration gradient across the membrane but also by the electrical potential across the membrane. Equilibrium for an ion across the membrane is reached when the electrical gradient exactly balances the concentration gradient for that ion. The equation that expresses this equilibrium condition quantitatively is the Nernst equation, which gives the value of membrane potential that will exactly balance a given concentration gradient.

If more than one ion can cross the cell membrane, both can be at equilibrium only if the Nernst, or equilibrium, potentials for both ions are the same. This requirement leads to the defining properties of the Donnan, or Gibbs–Donnan, equilibrium, which applies simultaneously to two permeant ions. By working through a series of examples, we saw how it is possible to build a model cell that is at equilibrium and that has ICF, ECF, and membrane potential like that of real animal cells.

Real cells, however, were found to be permeable to sodium ions. This removed an important cornerstone of the equilibrated model cell, and forced a change in viewpoint about the relation between animal cells and their environment. Real cells must expend metabolic energy, in the form of ATP, in order to "pump" sodium out against its concentration and electrical gradients and thus to maintain osmotic balance. In the next chapter, we will consider what effect the sodium permeability of the plasma membrane might have on the electrical membrane potential. We will see how the membrane potential depends not only on the concentrations of ions on the two sides of the membrane, as in the Nernst equation, but also on the relative permeability of the membrane to those ions.

5 Membrane Potential: Ionic Steady State

In Chapter 4, we learned that in a Donnan equilibrium, two permeant ions can be at equilibrium provided the membrane potential is simultaneously equal to the Nernst potentials for both ions. However, real animal cells are permeable to sodium, and thus there are three major ions—potassium, chloride, and sodium—that can cross the plasma membrane. This chapter will be concerned with the effect of sodium permeability on membrane potential and with the quantitative relation between ion permeabilities and ion concentrations on the one hand and electrical membrane potential on the other.

Equilibrium Potentials for Sodium, Potassium, and Chloride

If the permeability of the cell membrane to sodium is not zero, then the resting membrane potential of the cell must have a contribution from Na^+ as well as from K^+ and Cl^-. This is true even though the sodium pump eventually removes any sodium that leaks into the cell. There are two reasons for this. First, recall that electrical force per particle is much stronger than concentrational force per particle; therefore, even a tiny trickle of sodium that would cause a negligible change in internal concentration could produce large changes in membrane potential. Because the sodium pump responds only to changes in the bulk concentration of sodium inside the cell, it could not detect and respond to the tiny changes that would occur for even large changes in membrane potential. Second, even though sodium that leaks in is eventually pumped out, the efflux of sodium through the pump is coupled with an influx of potassium. Thus, there is a net transfer of positive charge into the cell associated with leakage of sodium.

Application of the Nernst equation to the concentrations of sodium, potassium, and chloride in the ICF and ECF of a typical mammalian cell (Table 2-1) shows that the membrane potential cannot possibly be simultaneously at the equilibrium potentials of all three ions. As we calculated in Chapter 4, $E_K = E_{Cl}$ = about −80 mV (actually a bit greater than −81 mV, given the values in Table 2-1). But with $[Na^+]_o = 120$ mM and $[Na^+]_i = 12$ mM, E_{Na} would be

+58 mV. The membrane potential, E_m, cannot simultaneously be at −80 mV and +58 mV. The actual value of membrane potential will fall somewhere between these two extreme values. If the sodium permeability of the membrane were in fact zero, E_m would be determined solely by E_K and E_{Cl} and would be −80 mV. Conversely, if chloride and potassium permeability were zero, E_m would be determined only by sodium and would lie at E_{Na}, +58 mV. Because the permeabilities of all three ions are nonzero, there will be a struggle between Na^+ on the one hand, tending to make E_m equal +58 mV, and K^+ and Cl^- on the other, tending to make E_m equal −80 mV. Two factors determine where E_m will actually fall: (1) ion concentrations, which determine the equilibrium potentials for the ions; and (2) relative ion permeabilities, which determine the relative importance of a particular ion in governing where E_m lies. Before expressing these relations quantitatively, it will be useful to consider the mechanism of ionic permeability in more detail.

Ion Channels in the Plasma Membrane

The permeability of a membrane to a particular ion is a measure of the ease with which that ion can cross the membrane. It is a property of the membrane itself. Recall that ions cannot cross membranes through the lipid portion of the membrane; they must cross through aqueous pores or channels in the membrane. Thus, the ionic permeability of a membrane is determined by the properties of the ionic pores or channels in the membrane. The total permeability of a membrane to a particular ion is governed by the total number of membrane channels that allow that ion to cross and by the ease with which the ion can go through a single channel. Ion channels are protein molecules that are associated with the membrane, and thus an important function of membrane proteins is the regulation of ionic permeability of the cell membrane. In later chapters, we will discuss how specialized channels modulate ionic permeability in response to chemical or electrical signals and the role of such changes in permeability in the processing of signals in the nervous system.

Not all membrane channels allow all ions to cross with equal ease. Some channels allow only cations through, others only anions. Some channels are even more selective, allowing only K^+ through but not Na^+, or vice versa. Thus, it is possible for a membrane to have very different permeabilities to different ions, depending on the number of channels for each ion.

Membrane Potential and Ionic Permeability

As an example of how the actual value of membrane potential depends on the relative permeabilities of the competing ions, consider the situation illustrated in Figure 5-1. This model cell is much more permeable to K^+ than to Na^+. In other words, there are many channels that allow K^+ to cross the membrane but

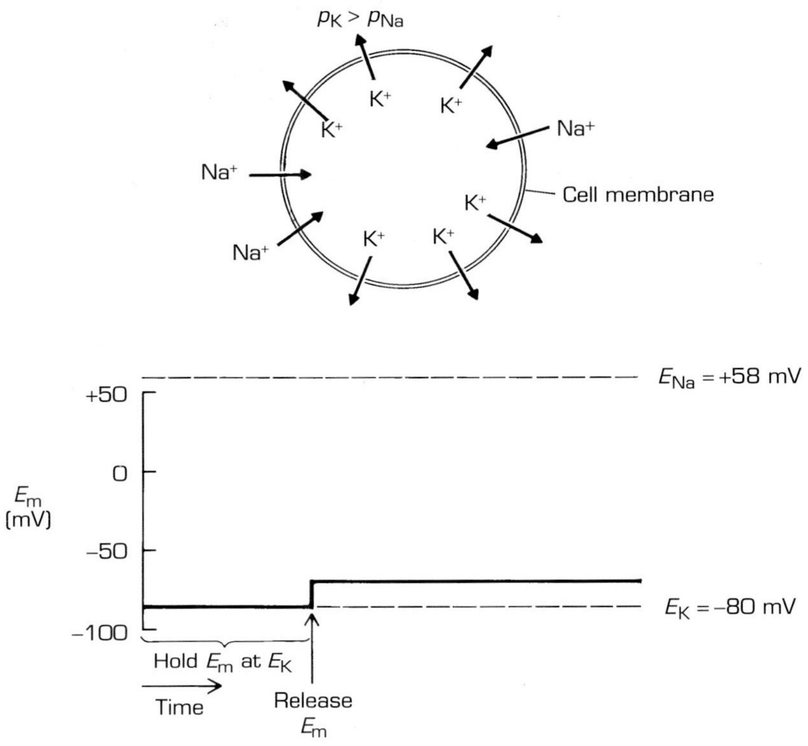

Figure 5-1 The resting membrane potential of a cell that is more permeable to potassium than to sodium. At the upward arrow, an apparatus that artificially holds the membrane potential at E_K abruptly switched off, and E_m is allowed to seek its own resting level.

only a few that allow Na^+ to cross. Imagine that initially we connect the cell to an apparatus that artificially maintains the resting membrane potential at E_K, so that $E_m = E_K = -80$ mV. (This could be accomplished experimentally using a voltage clamp apparatus, as described in Chapter 7.) What will happen to E_m when we switch off the apparatus and allow E_m to take on any value it wishes? In order to determine what will happen, it is necessary to keep in mind one important principle: *if the membrane potential is not equal to the equilibrium potential for an ion, that ion will move across the membrane in such a way as to force* E_m *toward the equilibrium potential for that ion*. For example, Figure 5-2 illustrates the movement of K^+ across a cell membrane in response to changes in E_m. In this example, a cell is connected to an apparatus that allows us to set the membrane potential to any value we choose. Initially, we set E_m to E_K. Recall from Chapter 4 that when $E_m = E_K$ there is a balance between the electrical force driving K^+ into the cell and the concentrational force driving K^+ out of the cell. At time = a, however, we suddenly make the interior of the cell less negative, reducing the electrical potential across the cell membrane and therefore decreasing the electrical force driving K^+ into the cell. Such a reduction in the electrical potential across the membrane is called a depolarization of the membrane. The electrical force will then be weaker than the oppositely directed concentrational force, and there will be a net movement of K^+ out of the cell.

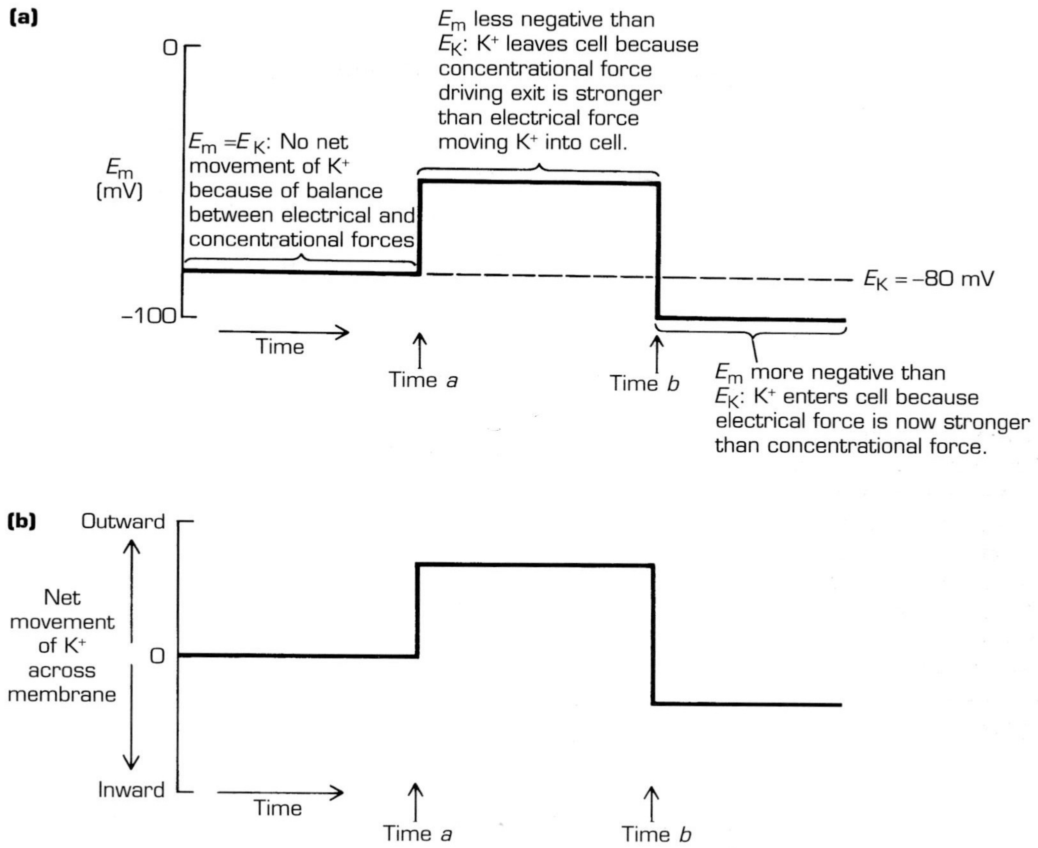

Figure 5-2 Effect of changes in membrane potential on the movement of potassium ions across the plasma membrane. (a) The membrane potential is artificially manipulated with respect to E_K, as indicated. (b) In response to the changes in membrane potential, potassium ions move across the membrane in a direction governed by the difference between E_m and E_K.

Note that this movement is in the proper direction to make E_m move back toward E_K; that is, to make the interior of the cell more negative because of the efflux of positive charge. At time = *b*, we suddenly make E_m more negative than E_K; that is, we hyperpolarize the membrane. Now the electrical force will be stronger than the concentrational force and there will be a net movement of K^+ into the cell. Again, this is in the proper direction to make E_m move toward E_K, in this case by adding positive charge to the interior of the cell.

Return now to Figure 5-1. We would expect that Na^+, which has an equilibrium potential of +58 mV, will enter the cell. That is, Na^+ will bring positive charge into the cell, and when we switch off the apparatus forcing E_m to remain at E_K, this influx of sodium ions will cause the membrane potential to become more positive (that is, move toward E_{Na}). As E_m moves toward E_{Na}, however, it

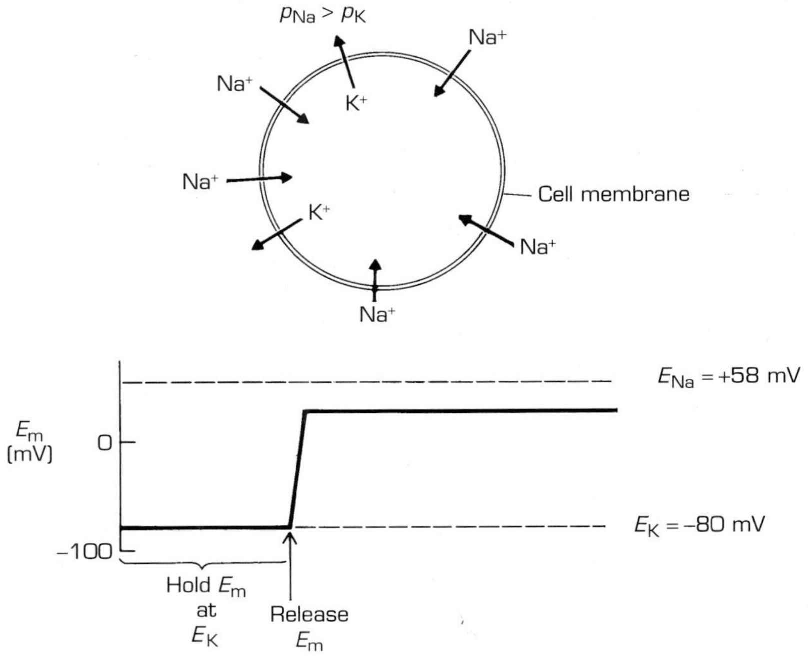

Figure 5-3 The resting membrane potential of a cell that is more permeable to sodium than to potassium. As in Figure 5-1, an apparatus holding E_m at E_K is abruptly turned off at the upward arrow.

will no longer be equal to E_K, and K^+ will move out of the cell in response to the resulting imbalance between the potassium concentrational force and electrical force. Thus, there will be a struggle between K^+ efflux forcing E_m toward E_K and Na^+ influx forcing E_m toward E_{Na}. Because K^+ permeability is much higher than Na^+ permeability, potassium ions can move out readily to counteract the electrical effect of the trickle of sodium ions into the cell. Thus, in this situation, the balance between the movement of Na^+ into the cell and the exit of K^+ from the cell would be struck relatively close to E_K.

Figure 5-3 shows a different situation. In this case, everything is as before except that the sodium permeability is much greater than the potassium permeability. That is, there are more channels that allow Na^+ across than allow K^+ across. Once again, we start with $E_m = E_K = -80$ mV and then allow E_m to seek its own value. Sodium, with $E_{Na} = +58$ mV, enters the cell down its electrical and concentration gradients. The resulting accumulation of positive charge again causes the cell to depolarize, as before. Now, however, potassium cannot move out as readily as sodium can move in, and the influx of sodium will not be balanced as readily by efflux of potassium. Thus, E_m will move farther from E_K and will reach a steady value closer to E_{Na} than to E_K.

The point of the previous two examples is that the value of membrane potential will be governed by the relative permeabilities of the permeant ions. *If a cell membrane is highly permeable to an ion, that ion can respond readily to deviations away from its equilibrium potential and E_m will tend to be near that equilibrium potential.*

The Goldman Equation

The examples discussed so far have been concerned with the qualitative relation between membrane potential and relative ionic permeabilities. The equation that gives the quantitative relation between E_m on the one hand and ion concentrations and permeabilities on the other is the **Goldman equation**, which is also called the **constant-field equation**. For a cell that is permeable to potassium, sodium, and chloride, the Goldman equation can be written as:

$$E_m = \frac{RT}{F} \ln\left(\frac{p_K[K^+]_o + p_{Na}[Na^+]_o + p_{Cl}[Cl^-]_i}{p_K[K^+]_i + p_{Na}[Na^+]_i + p_{Cl}[Cl^-]_o}\right) \tag{5-1}$$

This equation is similar to the Nernst equation (see Chapter 4), except that it simultaneously takes into account the contributions of all permeant ions. Some information about the derivation of the Goldman equation can be found in Appendix B. Note that the concentration of each ion on the right side of the equation is scaled according to its permeability, p. Thus, if the cell is highly permeable to potassium, for example, the potassium term on the right will dominate and E_m will be near the Nernst potential for potassium. Note also that if p_{Na} and p_{Cl} were zero, the Goldman equation would reduce to the Nernst equation for potassium, and E_m would be exactly equal to E_K, as we would expect if the only permeant ion were potassium.

Because it is easier to measure relative ion permeabilities than it is to measure absolute permeabilities, the Goldman equation is often written in a slightly different form:

$$E_m = 58 \text{ mV} \log\left(\frac{[K^+]_o + b[Na^+]_o + c[Cl^-]_i}{[K^+]_i + b[Na^+]_i + c[Cl^-]_o}\right) \tag{5-2}$$

In this case, the permeabilities have been expressed relative to the permeability of the membrane to potassium. Thus, $b = p_{Na}/p_K$, and $c = p_{Cl}/p_K$. We have also evaluated RT/F at room temperature, converted from ln to log, and expressed the result in millivolts.

For most nerve cells, the Goldman equation can be simplified even further: the chloride term on the right can be dropped altogether. This approximation is valid because the contribution of chloride to the resting membrane potential is insignificant in most nerve cells. In this case, the Goldman equation becomes

$$E_m = 58 \text{ mV} \log\left(\frac{[K^+]_o + b[Na^+]_o}{[K^+]_i + b[Na^+]_i}\right) \tag{5-3}$$

This is the form typically encountered in neurophysiology. In nerve cells, the ratio of sodium to potassium permeability, b, is commonly about 0.02, although this value may vary somewhat from one type of cell to another. That is, p_K is about 50 times higher than p_{Na}. Thus, Equation (5-3) tells us that E_m would be about −71 mV for a cell with $[K^+]_i = 125$ m*M*, $[K^+]_o = 5$ m*M*, $[Na^+]_i = 12$ m*M*, $[Na^+]_o = 120$ m*M*, and $b = 0.02$. What would E_m be for the same cell if b were 1.0 (that is, if $p_{Na} = p_K$) instead of 0.02?

The Goldman equation tells us quantitatively what we would expect qualitatively. If p_K is 50 times higher that p_{Na}, we would expect E_m to be nearer to E_K than to E_{Na}. Indeed, Equation (5-3) yields $E_m = -71$ mV, which is much nearer to E_K (−80 mV) than to E_{Na} (+58 mV). The difference between E_m and E_K reflects the steady influx of sodium ions carrying positive charge into the cell and maintaining a depolarization from E_K.

The applicability of the Goldman equation to a real cell can be tested experimentally by varying the concentration of potassium in the ECF and measuring the resulting changes in membrane potential. If membrane potential were determined solely by the distribution of potassium ions across the cell membrane—that is, if the factor b in Equation (5-3) were zero—we know that E_m would be determined by the potassium equilibrium potential. In this situation, a plot of measured membrane potential against log $[K^+]_o$ would yield a straight line with a slope of 58 mV per tenfold change in $[K^+]_o$. This straight line would merely be a plot of the E_m calculated from the Nernst equation at different values for external potassium concentration, and it is shown by the dashed line in Figure 5-4. Look, however, at the actual data from a real experiment in Figure 5-4. These data show the measured values of E_m of a nerve fiber observed at a number of different external potassium concentrations. The data do not follow the line expected from the Nernst equation, but instead fall along the solid line. That line was drawn according to the form of the Goldman

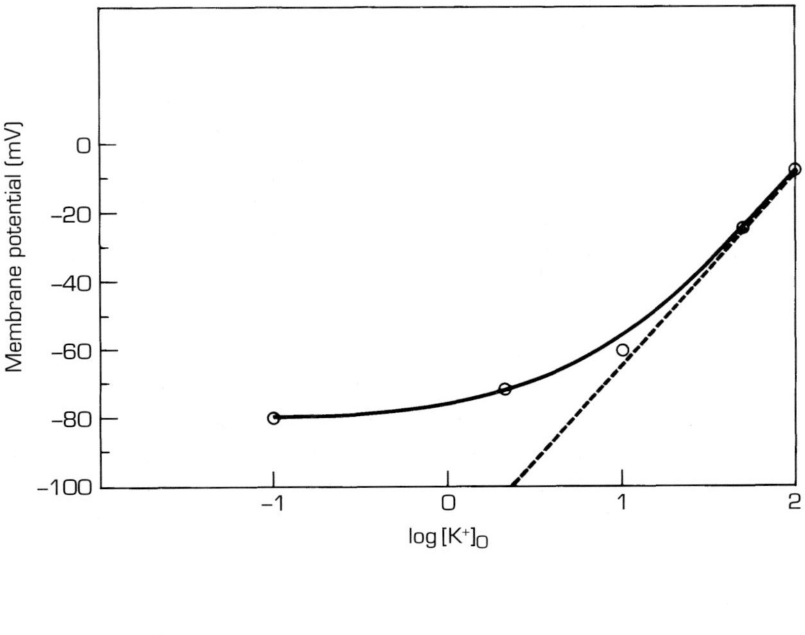

Figure 5-4 Experimentally determined relation between external potassium concentration and resting membrane potential of an axon in the spinal cord of the lamprey. The circles show the measured value of membrane potential at five different values of $[K^+]_o$. The dashed line gives the potassium equilibrium potential calculated from the Nernst equation. The solid line shows the prediction from the Goldman equation with internal and external sodium and potassium concentrations appropriate for the lamprey nervous system.

equation given in equation (5-3), and this experiment demonstrates that the real value of membrane potential in the nerve fiber is determined jointly by potassium and sodium ions. Experiments of this type by Hodgkin and Katz in 1949 first demonstrated the role of sodium ions in the resting membrane potential of real cells.

Equation (5-3) is a reasonable approximation to Equation (5-2) only if p_{Cl}/p_K is negligible. To determine if it is valid to ignore the contribution of chloride—that is, to use Equation (5-3)—experiments like that summarized in Figure 5-4 can be performed in which the concentration of chloride in the ECF is varied rather than the concentration of potassium. When that was done on the type of nerve cell used in the experiment of Figure 5-4, it was found that a tenfold reduction of $[Cl^-]_o$ caused only a 2 mV change in the resting membrane potential. Thus, for that type of cell, membrane potential is relatively unaffected by chloride concentration, and Equation (5-3) is valid. This is also true for other nerve cells. It is important to emphasize, however, that the membranes of other kinds of cells, such as muscle cells, have larger chloride permeability; therefore, the membrane potential of those cells would be more strongly dependent on external chloride concentration. This has been demonstrated experimentally for muscle cells by Hodgkin and Horowicz.

Ionic Steady State

The Goldman equation represents the actual situation in animal cells. The membrane potential of the cell takes on a steady value that reflects a fine balance between competing influences. It is important to keep in mind that neither sodium ions nor potassium ions are at equilibrium at that steady value of potential: sodium ions are continually leaking into the cell and potassium ions are continually leaking out. If this were allowed to continue, the concentration gradients for sodium and potassium would eventually run down and the membrane potential would decline to zero as the ion gradients collapsed. It is like a flashlight that has been left on: the batteries slowly discharge.

To prevent the intracellular accumulation of sodium and loss of potassium, the cell must expend energy to restore the ion gradients. Here again is an important role for the sodium pump. Metabolic energy stored in ATP is used to extrude the sodium that leaks in and to regain the potassium that was lost. In this way, the batteries are recharged using metabolic energy. Viewed in this light, we can see that the steady membrane potential of a cell represents chemical energy that has been converted into a different form and stored in the ion gradients across the cell membrane. In Part II of this book, beginning with Chapter 6, we will see how some cells, most notably the cells that make up the nervous system, are able to tap this stored energy to generate signals that can carry information and allow animals to move about and function in their environment.

The Chloride Pump

Because the resting membrane potential of a cell is not at either the sodium or potassium equilibrium potentials, there is a continuous net flux of sodium across the membrane. As we have just seen, metabolic energy must be expended in order to maintain the ion gradients for sodium and potassium. What about chloride? The equilibrium potential for chloride given the internal and external concentrations shown in Table 2-1 would be about −80 mV, but the resting membrane potential is about −71 mV. Thus, we would expect that there would be a steady influx of chloride into the cell because of this imbalance between the electrical and concentration gradients for chloride. Eventually, this influx would raise the internal chloride concentration to the point where the new chloride equilibrium potential would be −71 mV, the same as the resting membrane potential. At that point the concentration gradient for chloride would be reduced sufficiently to come into balance with the resting membrane potential. We can calculate from the Nernst equation that chloride would have to rise to about 7.5 m*M* from its usual 5 m*M* in order for this new equilibrium state to be established.

In some cells, this does indeed appear to happen: chloride reaches a new equilibrium governed by the resting membrane potential of the cell. (The cell would also gain the same small amount of potassium; because there is so much potassium inside, a change of a few millimolar in potassium concentration makes very little change in the potassium equilibrium potential, however.) In other cells, however, the chloride equilibrium potential remains different from the resting membrane potential, just as the sodium and potassium equilibrium potentials remain different from E_m. The only way this nonequilibrium condition can be maintained is by expending energy to keep the internal chloride constant—that is, there must also be a chloride pump similar in function to the sodium–potassium pump. In most cells, the chloride pump moves chloride ions out of the cell, so that the chloride equilibrium potential remains more negative than the resting membrane potential. In a few cases, however, an inwardly directed chloride pump has been discovered. Less is known about the molecular machinery of the chloride pump than that of the sodium–potassium pump. It is thought to involve an ATPase in some instances, so that the energy released by hydrolysis of ATP is the immediate driving energy for the pumping. In other cases, the pump may use energy stored in gradients of other ions to drive the movement of chloride.

Electrical Current and the Movement of Ions Across Membranes

An electrical current is the movement of charge through space. In a wire like that carrying electricity in your house, the electrical current is a flow of electrons; in a solution of ions, however, a flow of current is carried by movement of ions. That is, in a solution, the charges that move during an electrical current flow are the charges on the ions in solution. Thus, the movement of ions through space—such as from the outside of a cell to the inside of a cell—constitutes an electrical current, just as the movement of electrons through a wire constitutes an electrical current.

By thinking of ion flows as electrical currents, we can get a different perspective on the factors governing the steady-state membrane potential of cells. We have seen that at the steady-state value of membrane potential, there is a steady influx of sodium ions into the cell and a steady efflux of potassium ions out of the cell. This means that there is a steady electrical current, carried by sodium ions, flowing across the cell membrane in one direction and another current, carried by potassium ions, flowing across the membrane in the opposite direction. By convention, it is assumed that electrical current flows from the plus to the minus terminal of a battery; that is, we talk about currents in a wire as though the current is carried by positive charges. By extension, this convention means that the sodium current is an *inward* membrane current (the transfer of positive charge from the outside to the inside of the membrane), and the potassium current is an *outward* membrane current.

As we saw in our discussion of the Goldman equation above, a steady value of membrane potential will be achieved when the influx of sodium is exactly balanced by the efflux of potassium. In electrical terms, this means that in the steady state the sodium current, i_{Na}, is equal and opposite to the potassium current i_K. In equation form, this can be written

$$i_K + i_{Na} = 0 \qquad (5\text{-}4)$$

Thus, at the steady state the net membrane current is zero. This makes electrical sense, if we keep in mind that the cell membrane can be treated as an electrical capacitor (see Chapter 4). If the sum of i_{Na} and i_K were not zero, there would be a net flow of current across the membrane. Thus, there would be a movement of charge onto (or from) the membrane capacitor. Any such movement of charge would change the voltage across the capacitor (the membrane potential); that is, from the relation $q = CV$, if q changes and C remains constant then V must of necessity change. Equation (5-4), then, is a requirement of the steady-state condition; if the equation is not true, the membrane potential cannot be at a steady level.

In cells in which there is an appreciable flow of chloride ions across the membrane, Equation (5-4) must be expanded to include the chloride current, i_{Cl}:

$$i_K + i_{Na} + i_{Cl} = 0 \qquad (5\text{-}5)$$

Equation (5-5) is, in fact, the starting point in the derivation of the Goldman equation (see Appendix B). Note that because of the negative charge of chloride and because of the electrical convention for the direction of current flow, an outward movement of chloride ions is actually an inward membrane current.

Factors Affecting Ion Current Across a Cell Membrane

What factors govern the amount of current carried across the membrane by a particular ion? We would expect that one important factor would be the difference between the equilibrium potential for the ion and the actual membrane potential. As an example, consider the movement of potassium ions across the membrane. We know that if $E_m = E_K$, there is a balance between the electrical and concentrational forces for potassium and there is no net movement of potassium across the membrane. In this situation, then, $i_K = 0$. As shown in Figure 5-2, if E_m does not equal E_K, the resulting imbalance in electrical and concentrational forces will drive a net movement of potassium across the membrane. The larger the difference between E_m and E_K, the larger the imbalance between the electrical and concentration gradients and the larger the net movement of potassium. Thus, i_K depends on $E_m - E_K$. This difference is called the **driving force** for membrane current carried by an ion.

We would also expect that the permeability of the membrane to an ion would be an important determinant of the amount of membrane current carried by that ion. If the permeability is high, the ion current at a particular value of driving force will be higher than if the permeability were low. Thus, because p_K is much greater than p_{Na}, the potassium current resulting from a 10 mV difference between E_m and E_K will be much larger than the sodium current resulting from a 10 mV difference between E_m and E_{Na}. This is, in electrical terms, the reason that the steady-state membrane potential of a cell lies close to E_K rather than E_{Na}: in order for Equation (5-4) to be obeyed, the driving force for sodium entry $(E_m - E_{Na})$ must be much greater than the driving force for potassium exit $(E_m - E_K)$.

Membrane Permeability vs. Membrane Conductance

To place the discussion in the preceding section on more quantitative ground, it will be necessary to introduce a new concept that is closely related to membrane permeability: **membrane conductance**. The conductance of a membrane to an ion is an index of the ability of that ion to carry current across the membrane: the higher the conductance, the greater the ion current for a given driving force. Conductance is analogous to the reciprocal of the resistance of an electrical circuit to current flow: the higher the resistance of a circuit, the lower the amount of current that flows in response to a particular voltage. This behavior of electrical circuits can be conveniently summarized by Ohm's law: $i = V/R$. Here, i is the current flowing through a resistor, R, in the presence of a voltage gradient, V. The equivalent form for the flow of an ion current across a membrane is, using potassium as an example:

$$i_K = g_K(E_m - E_K) \tag{5-6}$$

where g_K is the conductance of the membrane to potassium ions. The unit of electrical conductance is the Siemen, abbreviated S; a 1 V battery will drive 1 ampere of current through a 1 S conductance. Similar equations can be written for sodium and chloride:

$$i_{Na} = g_{Na}(E_m - E_{Na}) \tag{5-7}$$

$$i_{Cl} = g_{Cl}(E_m - E_{Cl}) \tag{5-8}$$

Note that for the usual values of E_m (−71 mV), E_K (−80 mV), and E_{Na} (+58 mV), the potassium current is a positive number and the sodium current is a negative number, as required by the fact that the two currents flow in opposite directions across the membrane. By convention in neurophysiology, an outward membrane current (such as i_K, at the steady-state E_m) is positive and an inward current (such as i_{Na}, at the steady-state E_m) is negative.

The membrane conductance to an ion is closely related to the membrane permeability to that ion, but the two are not identical. The membrane current carried by a particular ion, and hence the membrane conductance to that ion, is proportional to the rate at which ions are crossing the membrane (that is, the ion flux). That rate depends not only on the permeability of the membrane to the ion, but also on the number of available ions in the solution. As an example, imagine a cell membrane with many potassium channels (Figure 5-5). The permeability of this membrane to potassium is thus high. If there are few potassium ions in solution, on the one hand, the chance is small that a K^+ will encounter a channel and cross the membrane. In this case, the potassium current will be low and the conductance of the membrane to K^+ will be low even though the permeability is high. On the other hand, if there are many potassium ions available to cross the membrane (Figure 5-5b), the chance that

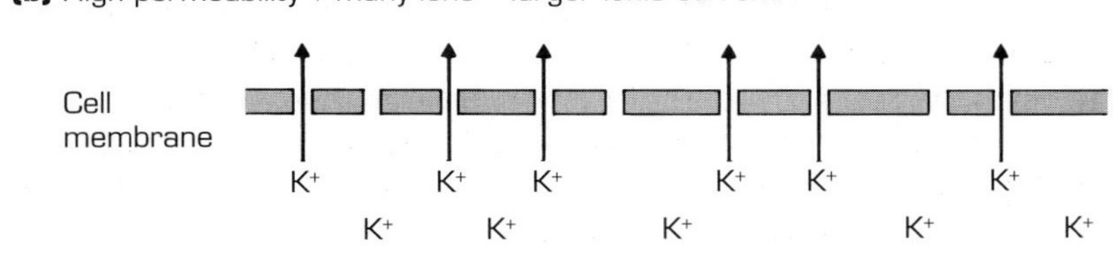

Figure 5-5 Illustration of the difference between permeability and conductance. (a) A cell membrane is highly permeable to potassium, but there is little potassium in solution. Therefore, the ionic current carried by potassium ions is small and the membrane conductance to potassium is small. (b) The same cell membrane in the presence of higher potassium concentration has a larger potassium conductance because the potassium current is larger. The permeability, however, is the same as in (a).

a K^+ will encounter a channel is high, and the rate of K^+ flow across the membrane will be high. The permeability remains fixed but the ionic conductance increases when more potassium ions are available. The point is that the potassium conductance of the membrane depends on the concentration of potassium at the membrane. For the most part, however, a change in permeability of a membrane to an ion produces a corresponding change in the conductance of the membrane to that ion. Thus, when we are dealing with changes in membrane conductance—as in the next chapter—we can treat a conductance change as a direct index of the underlying permeability change.

Behavior of Single Ion Channels

At this point, it is worthwhile considering the properties of the ion current flowing through an individual ion channel. We have already seen that the total membrane permeability of a cell to a particular ion depends on the number of channels the cell has that allow the ion to cross. But in addition, the total permeability will also depend on how readily ions go through a single channel. In Equations (5-6) through (5-8), we showed that the ion current across a membrane is equal to the product of the electrical driving force and the membrane conductance. Similar considerations apply to the ion current flowing through a single open ion channel, and we can write (for a potassium channel, for example):

$$i_S = g_S(E_m - E_K) \qquad (5\text{-}9)$$

where i_S is the single-channel current and g_S is the single-channel conductance for the potassium channel in question. Analogous equations could be written for single sodium or chloride channels.

What would we expect to see if we could measure directly the electrical current flowing through a single ion channel in a cell membrane? Up to now, we have treated ion channels as simple open pores or holes that allow ions to cross the membrane. But real ion channels show somewhat more complex behavior: the protein molecule that makes up the channel can apparently exist in two conformational states, one in which the pore is open and ions are free to move through it, and one in which the pore is closed and ions are not allowed through. (Actually, channels frequently show more than two functional states, but for our purposes in this book, we can treat channels as being either open or closed.) Thus, channels behave as though access to the pore is controlled by a "gate" that can be open or closed; for this reason, we refer to the opening and closing of the channel as **channel gating**. The electrical behavior of such a gated ion channel is illustrated in Figure 5-6, which shows the electrical current we would measure through a small patch of cell membrane containing a single potassium channel. We find that when the channel is in the closed state, there is no current across the membrane patch because potassium ions have no path across the membrane; however, when the channel protein abruptly undergoes

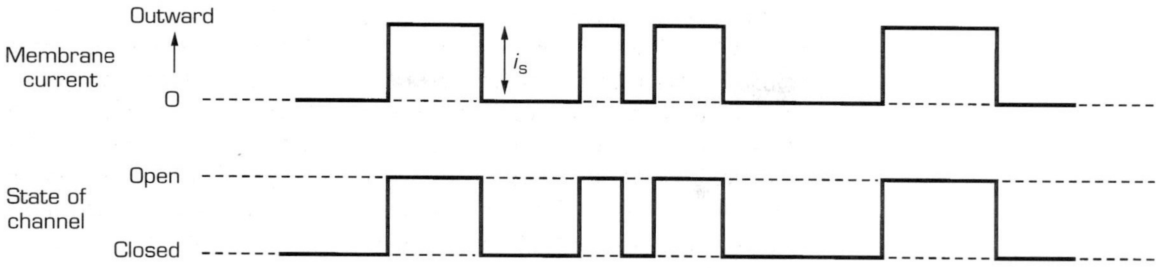

Figure 5-6 The electrical current flowing through a single potassium channel. The bottom trace shows the state of the channel (either open or closed), and the top trace shows the resulting ionic current through the channel. At the beginning of the trace, the channel is closed and so there is no ionic current flowing. When the channel opens, potassium ions begin to exit the cell through the channel, carrying an outward membrane current. The magnitude of the current (i_S) is given by Equation (5-9).

a transition to the open state, an outward current will suddenly appear as potassium ions begin to exit the cell through the open pore. When the channel makes a transition back to the closed state, the current will abruptly disappear again. As we will see in the next section of this book, the control of ion channel gating, and thus of the ionic conductance of the cell membrane, is an important way in which biological signals are passed both within a cell and between cells.

The amplitude of the current that flows through the open channel will be given by Equation (5-9); that is, the single-channel current will depend on the electrical driving force and on the single-channel conductance. How big is the single-channel current in real life? That depends on exactly what particular kind of ion channel we are talking about, because there is considerable variation in single-channel conductance among the various kinds of channel we would encounter in a cell membrane (however, all of the individual channels of a particular kind would have the same single-channel conductance). But a value of about 20 pS might be considered typical (pS is the abbreviation for picoSiemen, or 10^{-12} Siemen). If the conductance is 20 pS and the driving force is 50 mV, then Equation (5-9) tells us that the single-channel current would be 10^{-12} A (or 1 pA; this corresponds to about 6 million monovalent ions per second). A small current indeed! Nevertheless, it has proved possible, using a measurement technique called the patch clamp, to measure directly the electrical current flowing through a single open ion channel. This technique, invented by Erwin Neher and Bert Sakmann, will be discussed in more detail in Chapter 8. The ability to make such measurements from single channels has revolutionized the study of ion channels and made possible a great deal of what we know about how channels work.

What is the relationship between the total conductance of the cell membrane to an ion (Equations (5-6), (5-7), and (5-8)) and the single-channel conductance? If a cell has only one type of potassium channel with single-channel conductance g_S, then the total membrane conductance to potassium, g_K, would be given by:

$$g_K = N g_S P_o \tag{5-10}$$

where N is the number of potassium channels in the entire cell membrane and P_o is the average proportion of time that an individual channel is in the open state. You can see that if the individual channels are always closed, then P_o is zero and g_K would also be zero. Conversely, if individual channels are always open, then P_o is 1 and g_K will simply be the sum of all the individual single-channel conductances (i.e., $N \times g_S$).

Summary

In real cells, the resting membrane potential is the point at which sodium influx is exactly balanced by potassium efflux. This point depends on the relative membrane permeabilities to sodium and potassium; in most cells p_K is much higher than p_{Na} and the balance is struck close to E_K. The Goldman equation gives the quantitative expression of the relation between membrane potential on the one hand and ion concentrations and permeabilities on the other. Because the steady-state membrane potential lies between the equilibrium potentials for sodium and potassium, there is a constant exchange of intracellular potassium for sodium. This would lead to progressive decline of the ion gradients across the membrane if it were not for the action of the sodium–potassium pump. Thus, metabolic energy, in the form of ATP used by the pump, is required for the long-term maintenance of the sodium and potassium gradients. In the absence of chloride pumping, the chloride equilibrium potential will change to come into line with the value of membrane potential established by sodium and potassium. In some cells, however, a chloride pump maintains the internal chloride concentration in a nonequilibrium state, just as the sodium–potassium pump maintains internal sodium and potassium concentrations at nonequilibrium values.

The steady fluxes of potassium and sodium ions constitute electrical currents across the cell membrane, and at the steady-state E_m these currents cancel each other so that the net membrane current is zero. The membrane current carried by a particular ion is given by an ionic form of Ohm's law—that is, by the product of the driving force for that ion and the membrane conductance to that ion. The driving force is the difference between the actual value of membrane potential and the equilibrium potential for that ion. Conductance is a measure of the ability of the ion to carry electrical current across the membrane, and it is closely related to the membrane permeability to the ion.

Individual ion channels behave as though access to the pore through which ions can cross the membrane is controlled by a gate that may be open or closed. When the gate is open, the channel conducts and electrical current flows across the membrane; when the gate is closed, there is no current flow. The current through a single open channel is again given by the ionic form of Ohm's law—that is, the driving force multiplied by the single-channel conductance.

Cellular Physiology of Nerve Cells

part II

Part I focused on general properties that are shared by all cells. Every cell must achieve osmotic balance, and all cells have an electrical membrane potential. Part II considers properties that are peculiar to particular kinds of cells: those that are capable of modulating their membrane potential in response to stimulation from the environment. These cells are called excitable cells because they can generate active electrical responses that serve as signals or triggers for other events. The most notable examples of excitable cells are the cells of the nervous system, which are called neurons.

The nervous system must receive information from the environment, transmit and analyze that information, and coordinate an appropriate action in response. The signals passed along in the nervous system are electrical signals, produced by modulating the membrane potential. Part II describes these electrical signals, including how the signals arise, how they propagate, and how the signals are passed along from one neuron to another. We will see that simple modifications of the scheme for the origin of the membrane potential, presented in Chapter 5, can explain how neurons carry out their vital signaling functions.

Generation of Nerve Action Potential

6

This chapter examines the mechanism of the action potential, the signal that carries messages over long distances along axons in the nervous system. We begin here with a descriptive introduction to the action potential and its mechanism. Then, Chapter 7 presents in more advanced form the physiological experiments that first established the mechanism of the action potential.

The Action Potential

Ionic Permeability and Membrane Potential

In Chapter 5, we learned that membrane potential is governed by the relative permeability of the cell membrane to sodium and potassium, as specified by the Goldman equation. If sodium permeability is greater than potassium permeability, the membrane potential will be closer to E_{Na} than to E_K. Conversely, if potassium permeability is greater than sodium permeability, E_m will be closer to E_K. Until now, we have treated ionic permeability as a fixed characteristic of the cell membrane. However, the ionic permeability of the plasma membrane of excitable cells can vary. Specifically, a transient, dramatic increase in sodium permeability underlies the generation of the basic signal of the nervous system, the action potential.

Measuring the Long-distance Signal in Neurons

What kind of signal carries the message along the sensory neuron in the patellar reflex? As described in Chapter 1, the signals in the nervous system are electrical signals, and to monitor these signals it is necessary to measure the changes in electrical potential associated with the activation of the reflex. This can be done by placing an intracellular microelectrode inside the sensory axon to measure the electrical membrane potential of the neuron. A diagram illustrating this kind of experiment is shown in Figure 6-1. A voltmeter is connected to measure the voltage difference between point *a*, at the tip of the microelectrode, and point *b*, a reference point in the ECF. As shown in Figure 6-1b, when

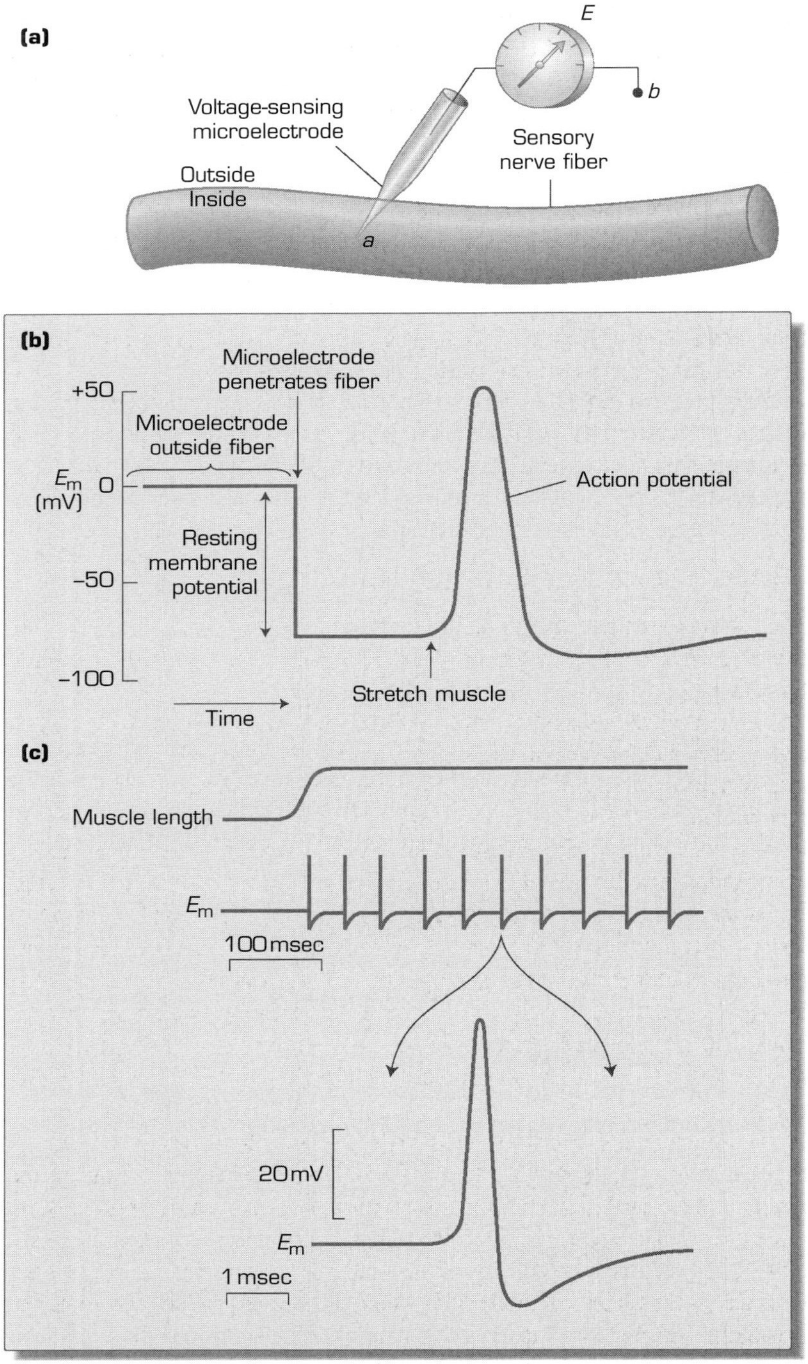

Figure 6-1 An example of an action potential in a neuron. (a) An experimental arrangement for recording the membrane potential of a nerve cell fiber. (b) Resting membrane potential and an action potential recorded by a microelectrode inside the sensory neuron of the patellar reflex loop. (c) A series of action potentials in a single stretch-receptor sensory fiber during stretch of the muscle. The lower trace shows a single action potential on an expanded time scale to illustrate its waveform in more detail.

the microelectrode is outside the sensory axon, both the microelectrode and the reference point are in the ECF, and the voltmeter records no voltage difference. When the electrode is inserted into the sensory fiber, however, it measures the voltage difference between the inside and outside of the neuron, the membrane potential. As expected from the discussion in Chapter 5, the membrane potential of the sensory fiber is about −70 mV.

When the muscle is stretched (Figure 6-1b), the membrane potential in the sensory fiber undergoes a dramatic series of rapid changes. After a small delay, the membrane potential suddenly jumps transiently in a positive direction (a depolarization) and actually reverses in sign for a brief period. When the potential returns toward its resting value, it may transiently become more negative than its normal resting value. The transient jump in potential is called an action potential, which is the long-distance signal of the nervous system. If the stretch is sufficiently strong, it might elicit a series of several action potentials, each with the same shape and amplitude, as illustrated in Figure 6-1c.

Characteristics of the Action Potential

The action potential has several important characteristics that will be explained in terms of the underlying ionic permeability changes. These include the following:

1. Action potentials are triggered by depolarization. The stimulus that initiates an action potential in a neuron is a reduction in the membrane potential—that is, depolarization. Normally, depolarization is produced by some external stimulus, such as the stretching of the muscle in the case of the sensory neuron in the patellar reflex, or by the action of another neuron, as in the transmission of excitation from the sensory neuron to the motor neuron in the patellar reflex.

2. A threshold level of depolarization must be reached in order to trigger an action potential. A small depolarization from the normal resting membrane potential will not produce an action potential. Typically, the membrane must be depolarized by about 10–20 mV in order to trigger an action potential. Thus, if a neuron has a resting membrane potential of about −70 mV, the membrane potential must be reduced to −60 to −50 mV to trigger an action potential.

3. Action potentials are all-or-none events. Once a stimulus is strong enough to reach threshold, the amplitude of the action potential is independent of the strength of the stimulus. The event either goes to completion (if depolarization is above threshold) or doesn't occur at all (if the depolarization is below threshold). In this manner, triggering an action potential is like firing a gun: the speed with which the bullet leaves the barrel is independent of whether the trigger was pulled softly or forcefully.

4. An action potential propagates without decrement throughout a neuron, but at a relatively slow speed. If we record simultaneously from the sensory fiber in the patellar reflex near the muscle and near the spinal cord, we would find that the action potential at the two locations has the same amplitude and form. Thus, as the signal travels from the muscle—where it originated—to the spinal cord, its amplitude remains unchanged. However, there would be a significant delay of about 0.1 sec between the occurrence of the action potential near the muscle and its arrival at the spinal cord. The conduction speed of an action potential in a typical mammalian nerve fiber is about 10–20 m/sec, although speeds as high as 100 m/sec have been observed.

5. At the peak of the action potential, the membrane potential reverses sign, becoming inside positive. As shown in Figure 6-1, the membrane potential during an action potential transiently overshoots zero, and the inside of the cell becomes positive with respect to the outside for a brief time. This phase is called the overshoot of the action potential. When the action potential repolarizes toward the normal resting membrane potential, it transiently becomes more negative than normal. This phase is called the undershoot of the action potential.

6. After a neuron fires an action potential, there is a brief period, called the absolute refractory period, during which it is impossible to trigger another action potential. The absolute refractory period varies somewhat from one neuron to another, but it usually lasts about 1 msec. The refractory period limits the maximum firing rate of a neuron to about 1000 action potentials per second.

The goal of the remainder of this chapter is to explain all of these characteristics of the nerve action potential in terms of the underlying changes in the ionic permeability of the cell membrane and the resulting movements of ions.

Initiation and Propagation of Action Potentials

Some of the fundamental properties of action potentials can be studied experimentally using an apparatus like that diagrammed in Figure 6-2a. Imagine that a long section of a single axon is removed and arranged in the apparatus so that intracellular probes can be placed inside the fiber at two points, *a* and *b*, which are 10 cm apart. The probe at *a* is set up to pass positive or negative charge into the fiber and to record the resulting change in membrane potential, while the probe at *b* records membrane potential only. The effect of injecting negative charge at a constant rate at *a* is shown in Figure 6-2b. The extra negative charges make the interior of the fiber more negative, and the membrane potential increases; that is, the membrane is hyperpolarized. At the same time, the probe at *b* records no change in membrane potential at all, because the plasma membrane is leaky to charge. In Chapter 3, we discussed the cell membrane as an electrical capacitor. In addition, the membrane behaves like an electrical resistor; that is, there is a direct path through which ionic current may flow across the membrane. As we saw in Chapter 5, that current path is through the ion channels that are inserted into the lipid bilayer of the plasma membrane. Thus, the charges injected at *a* do not travel very far down the fiber

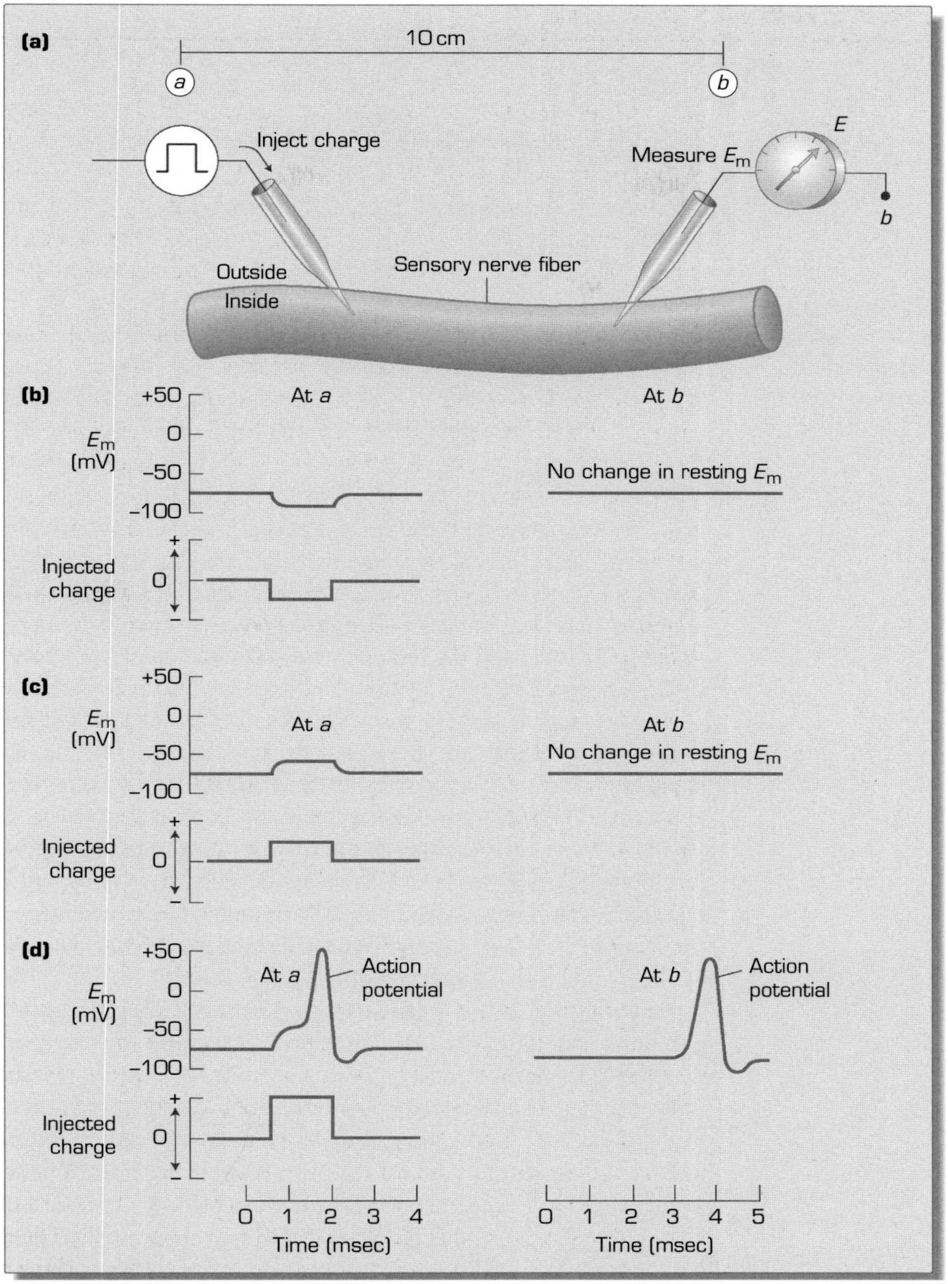

Figure 6-2 The generation and propagation of an action potential in a nerve fiber. (a) Apparatus for recording electrical activity of a segment of a sensory nerve fiber. The probes at points *a* and *b* allow recording of membrane potential, and the probe at *a* also allows injection of electrical current into the fiber. (b) Injecting negative charges at *a* causes hyperpolarization at *a*. All injected charges leak out across the membrane before reaching *b*, and no change in membrane potential is recorded at *b*. (c) Injection of a small amount of positive charge produces a depolarization at *a* that does not reach *b*. (d) If a stronger depolarization is induced at *a*, an action potential is generated. The action potential propagates without decrement along the fiber and is recorded at full amplitude at *b*.

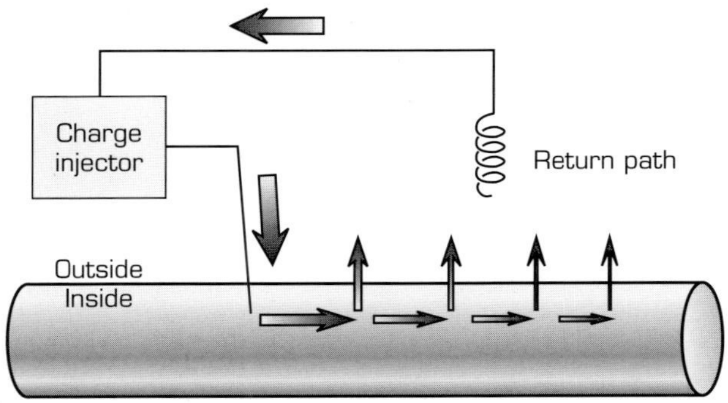

Figure 6-3 A schematic representation of the decay of injected current in an axon with distance from the site of current injection.

before leaking out of the cell across the plasma membrane. None of the charges reaches *b*, and so there is no change in membrane potential at *b*. When we stop injecting negative charges at *a*, all the injected charge leaks out of the cell, and the membrane potential returns to its normal resting value. The electrical properties of cells and the response to charge injection are described in more detail in Appendix C.

Another way of looking at the situation in Figure 6-2b is in terms of the flow of electrical current. The negative charges injected into the cell at a constant rate constitute an electrical current originating from the experimental apparatus. The return path for the current to the apparatus lies in the ECF, so that in order to complete the circuit the current must exit across the plasma membrane. Two paths are available for the current at the point where it is injected: it can flow across the membrane immediately or it can move down the axon to flow out through a more distant segment of axon membrane. This situation is illustrated in Figure 6-3 (also see Appendix C). The injected current will thus divide, some taking one path and some the other. The proportion of current taking each path depends on the relative resistances of the two paths: more current will flow down the path with less resistance. With each increment in distance along the axon, that fraction of the injected current that flowed down the axon again faces two paths; it can continue down the interior of the axon or it can cross the membrane at that point. The current will again divide, and some fraction of the remaining injected current will continue down the nerve fiber. This process will continue until all the injected current has crossed the membrane, and no current is left to flow further down the interior of the axon. At that point, the injected current will not influence the membrane potential because there will be no remaining injected current. Thus, the change in membrane potential produced by current injection (Figure 6-2a) decays with distance from the injection site. The greatest effect occurs at the injection site, and there is progressively less effect as injected current is progressively lost across the plasma membrane. Appendix C presents a quantitative discussion of this decay of voltage with distance along a nerve fiber. The cell membrane is not a particularly good insulator (it has a low resistance to current flow compared, for example, with the insulator surrounding the electrical wires in your house), and the ICF inside the axon is not a particularly good conductor (its resistance to current flow is high compared with that of a copper wire). This set of circumstances favors the rapid decay of injected current with distance. In real axons, the hyperpolarization produced by current injected at a point decays by about 95% within 1–2 mm of the injection site.

Let's return now to the experiment shown in Figure 6-2. The effect of injecting positive charges into the axon is shown in Figure 6-2c. If the number of positive charges injected is small, the effect is simply the reverse of the effect of injecting negative charges; the membrane depolarizes while the charges are injected, but the effect does not reach *b*. When charge injection ceases, the extra positive charges leak out of the fiber, and membrane potential returns to normal. If the rate of injection of positive charge is increased, as in Figure 6-2d, the depolarization is larger. If the depolarization is sufficiently large, an all-or-none action potential, like that recorded when the muscle was stretched (Figure 6-1), is triggered at *a*. Now, the probe at *b* records a replica of the action potential at *a*, except that there is a time delay between the occurrence of the action potential at *a* and its arrival at *b*. Thus, action potentials are triggered by depolarization, not by hyperpolarization (characteristic 1, above), the depolarization must be large enough to exceed a threshold value (characteristic 2), and the action potential travels without decrement throughout the nerve fiber (characteristic 4). What ionic properties of the neuron membrane can explain these properties?

Changes in Relative Sodium Permeability During an Action Potential

The key to understanding the origin of the action potential lies in the discussion in Chapter 5 of the factors that influence the steady-state membrane potential of a cell. Recall that the resting E_m for a neuron will lie somewhere between E_K and E_{Na}. According to the Goldman equation, the exact point at which it lies will be determined by the ratio p_{Na}/p_K. As we saw in Chapter 5, p_{Na}/p_K of a resting neuron is about 0.02, and E_m is near E_K.

What would happen to E_m if sodium permeability suddenly increased dramatically? The effect of such an increase in p_{Na} is diagrammed in Figure 6-4. In the example, p_{Na} undergoes an abrupt thousandfold increase, so that $p_{Na}/p_K = 20$ instead of 0.02. According to the Goldman equation, E_m would then swing from about −70 mV to about +50 mV, near E_{Na}. When p_{Na}/p_K returns to 0.02, E_m will return to its usual value near E_K. Note that the swing in membrane potential in Figure 6-4 reproduces qualitatively the change in potential during an action potential. Indeed, it is a transient increase in sodium permeability, as in Figure 6-4, that is responsible for the swing in membrane polarization from near E_K to near E_{Na} and back during an action potential.

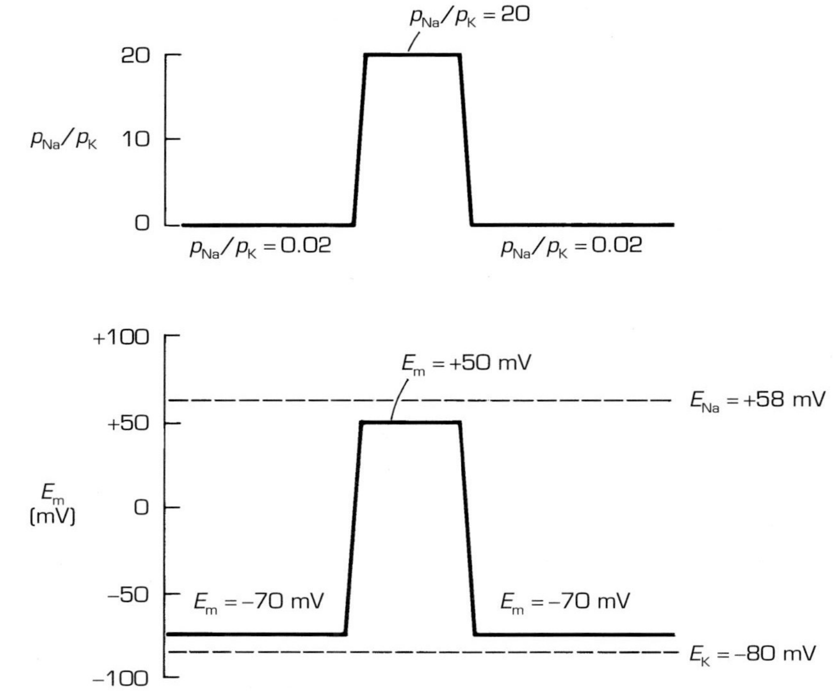

Figure 6-4 The relation between relative sodium permeability and membrane potential. When the ratio of sodium to potassium permeability (upper trace) is changed, the position of E_m relative to E_K and E_{Na} changes accordingly.

Voltage-dependent Sodium Channels of the Neuron Membrane

Recall that ions must cross the membrane through transmembrane pores or channels. A dramatic increase in sodium permeability like that shown in Figure 6-4 requires a dramatic increase in the number of membrane channels that allow sodium ions to enter the cell. Thus, the resting p_{Na} of the membrane of an excitable cell is only a small fraction of what it could be because most membrane sodium channels are closed at rest. What stimulus causes these hidden channels to open and produces the positive swing of E_m during an action potential? It turns out that the conducting state of sodium channels of excitable cells depends on membrane potential. When E_m is at the usual resting level or more negative, these sodium channels are closed, Na^+ cannot flow through them, and p_{Na} is low. These channels open, however, when the membrane is depolarized. The stimulus for opening of the voltage-dependent sodium channels of excitable cells is a reduction of the membrane potential.

Because the voltage-dependent sodium channels respond to depolarization, the response of the membrane to depolarization is regenerative, or explosive. This is illustrated in Figure 6-5. When the membrane is depolarized, p_{Na} increases, allowing sodium ions to carry positive charge into the cell. This depolarizes the cell further, causing a greater increase in p_{Na} and more depolarization. Such a process is inherently explosive and tends to continue until all sodium channels are open and the membrane potential has been driven up to

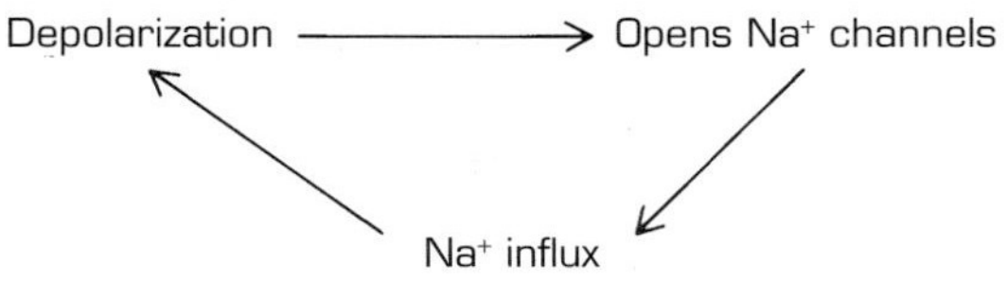

Figure 6-5 The explosive cycle leading to depolarizing phase of an action potential.

near E_{Na}. This explains the all-or-none behavior of the nerve action potential: once triggered, the process tends to run to completion.

Why should there be a threshold level of depolarization? Under the scheme discussed above, it might seem that any small depolarization would set the action potential off. However, in considering the effect of a depolarization, we must take into account the total current that flows across the membrane in response to the depolarization, not just the current carried by sodium ions. Recall that, at the resting E_m, p_K is very much greater than p_{Na}; therefore, flow of K^+ out of the cell can counteract the influx of Na^+ even if p_{Na} is moderately increased by a depolarization. Thus, for a moderate depolarization, the efflux of potassium might be larger than the influx of sodium, resulting in a net outward membrane current that keeps the membrane potential from depolarizing further and prevents the explosive cycle underlying the action potential. In order for the explosive process to be set in motion and an action potential to be generated, a depolarization must produce a net inward membrane current, which will in turn produce a further depolarization. A depolarization that produces an action potential must be sufficiently large to open quite a few sodium channels in order to overcome the efflux of potassium ions resulting from the depolarization. The **threshold potential** will be reached at that value of E_m where the influx of Na^+ exactly balances the efflux of K^+; any further depolarization will allow Na^+ influx to dominate, resulting in an explosive action potential.

Factors that influence the actual value of the threshold potential for a particular neuron include the density of voltage-sensitive sodium channels in the plasma membrane and the strength of the connection between depolarization and opening of those channels. Thus, if voltage-sensitive sodium channels are densely packed in the membrane, opening only a small fraction of them will produce a sizable inward sodium current, and we would expect that the threshold depolarization would be smaller than if the channels were sparse. Often, the density of voltage-sensitive sodium channels is highest just at the point (called the initial segment) where a neuron's axon leaves the cell body; this results in that portion of the cell having the lowest threshold for action potential generation. Another important factor in determining the threshold is the steepness of the relation between depolarization and sodium channel opening. In some cases the sodium channels have "hair triggers," and only a small depolarization from the resting E_m is required to open large numbers of channels. In such cases we would expect the threshold to be close to the resting membrane potential. In other neurons, larger depolarizations are necessary to open appreciable numbers of sodium channels, and the threshold is further from resting E_m.

Repolarization

What causes E_m to return to rest again following the regenerative depolarization during an action potential? There are two important factors: (1) the depolarization-induced increase in p_{Na} is transient; and (2) there is a delayed, voltage-dependent increase in p_K. These will be discussed in turn below.

The effect of depolarization on the voltage-dependent sodium channels is twofold. These effects can be summarized by the diagram in Figure 6-6, which illustrates the behavior of a single voltage-sensitive sodium channel in response to a depolarization. The channel acts as though the flow of Na^+ is controlled by two independent gates. One gate, called the *m* gate, is closed when E_m is equal to or more negative than the usual resting potential. This gate thus prevents Na^+ from entering the channel at the resting potential. The other gate, called the *h* gate, is open at the usual resting E_m. Both gates respond to depolarization, but with different speeds and in opposite directions. The *m* gate opens rapidly in response to depolarization; the *h* gate closes in response to depolarization, but does so slowly. Thus, immediately after a depolarization, the *m* gate is open, allowing Na^+ to enter the cell, but the *h* gate has not had time to respond to the depolarization and is thus still open. A little while later (about a millisecond or two), the *m* gate is still open, but the *h* gate has responded by closing, and the channel is again closed. The result of this behavior is that p_{Na} first increases in response to a depolarization, then declines again even if the depolarization were maintained in some way. This delayed decline in sodium permeability upon depolarization is called **sodium channel inactivation**. As shown in Figure 6-4, this return of p_{Na} to its resting level would alone be sufficient to bring E_m back to rest.

In addition to the voltage-sensitive sodium channels, there are voltage-sensitive potassium channels in the membranes of excitable cells. These channels are also closed at the normal resting membrane potential. Like the sodium channel *m* gates, the gates on the potassium channels open upon depolarization, so that the channel begins to conduct K^+ when the membrane potential is reduced. However, the gates of these potassium channels, which are called *n* gates, respond slowly to depolarization, so that p_K increases with a delay following a depolarization. The characteristic behavior of a single voltage-sensitive potassium channel is shown in Figure 6-7. Unlike the sodium channel, there is no gate on the potassium channel that closes upon depolarization; the channel remains open as long as the depolarization is maintained and closes only when membrane potential returns to its normal resting value.

These voltage-sensitive potassium channels respond to the depolarizing phase of the action potential and open at about the time sodium permeability returns to its normal low value as *h* gates close. Therefore, the repolarizing phase of the action potential is produced by the simultaneous decline of p_{Na} to its resting level and increase of p_K to a higher than normal level. Note that during this time, p_{Na}/p_K is actually smaller than its usual resting value. This explains the undershoot of membrane potential below its resting value at the

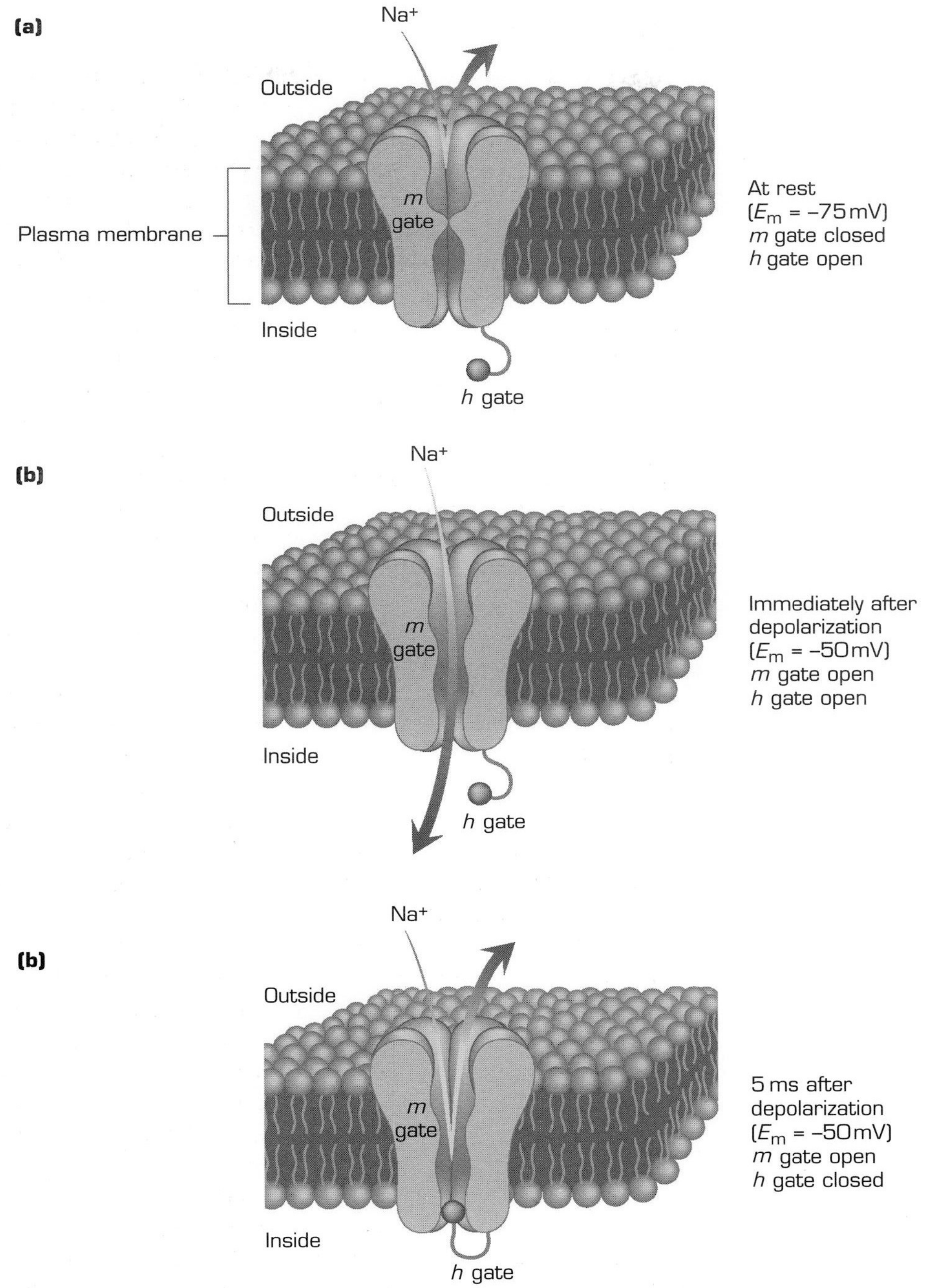

Figure 6-6 A schematic representation of the behavior of a single voltage-sensitive sodium channel in the plasma membrane of a neuron. (a) The state of the channel at the normal resting membrane potential. (b) Upon depolarization, the *m* gate opens rapidly and sodium ions are free to move through the channel. (c) After a brief delay, the *h* gate closes, returning the channel to a nonconducting state.

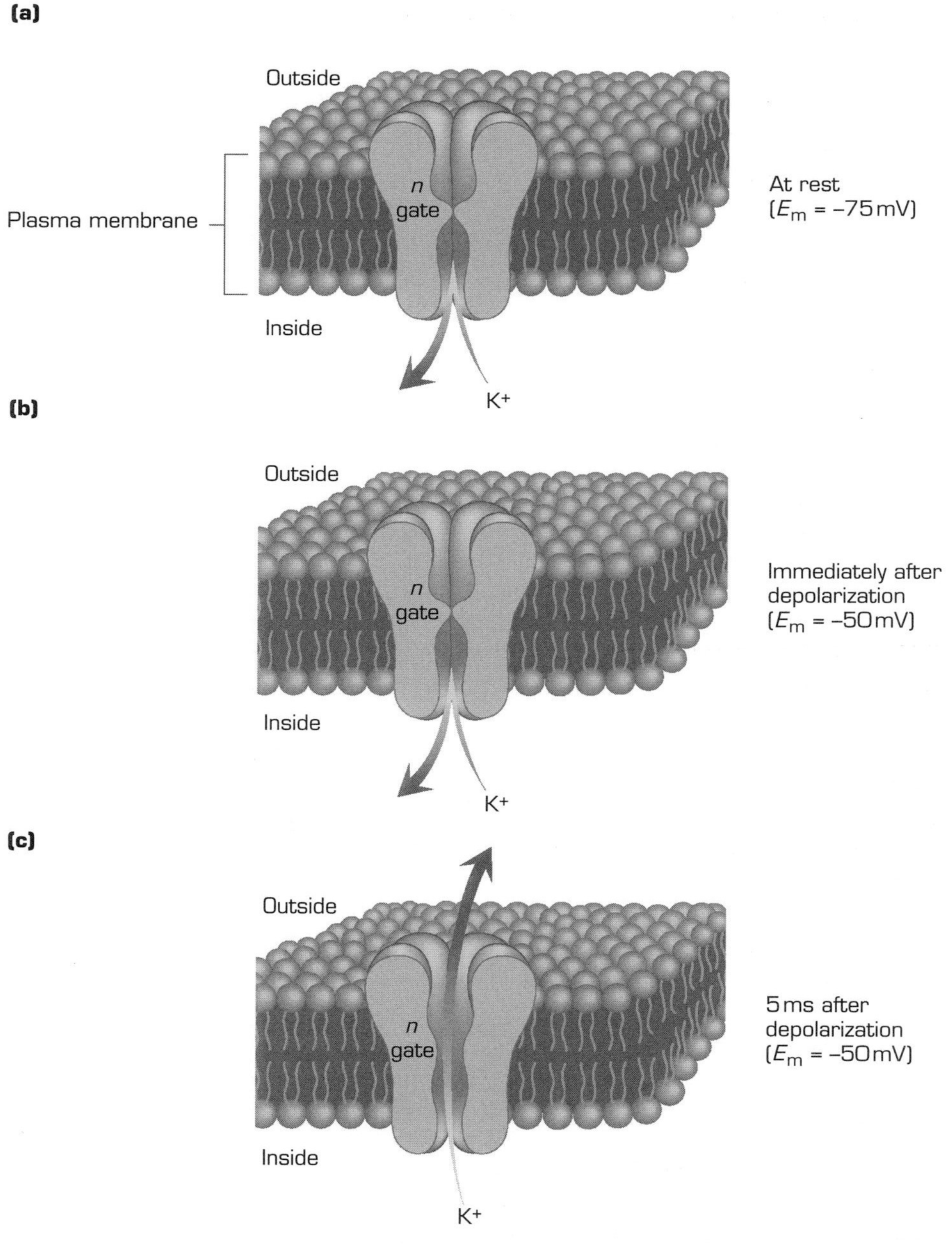

Figure 6-7 The behavior of a single voltage-sensitive potassium channel in the plasma membrane of a neuron. (a) At the normal resting membrane potential, the channel is closed. (b) Immediately after a depolarization, the channel remains closed. (c) After a delay, the *n* gate opens, allowing potassium ions to cross the membrane through the channel. The channel remains open as long as depolarization is maintained.

Table 6-1 Summary of responses of voltage-sensitive sodium and potassium channels to depolarization.

Type of channel	Gate	Response to depolarization	Speed of response
Sodium	*m* gate	Opens	Fast
Sodium	*h* gate	Closes	Slow
Potassium	*n* gate	Opens	Slow

end of an action potential: E_m approaches closer to E_K because p_K is still higher than usual while p_{Na} has returned to its resting state. Membrane potential returns to rest as the slow *n* gates have time to respond to the repolarization by closing and returning p_K to its normal value.

The sequence of changes during an action potential is summarized in Figure 6-8, and characteristics of the various gates are summarized in Table 6-1. An action potential would be generated in the sensory neuron of the patellar reflex in the following way. Stretch of the muscle induces depolarization of the specialized sensory endings of the sensory neuron (probably by increasing the relative sodium permeability). This depolarization causes the *m* gates of voltage-sensitive sodium channels in the neuron membrane to open, setting in motion a regenerative increase in p_{Na}, which drives E_m up near E_{Na}. With a delay, *h* gates respond to the depolarization by closing and potassium-channel *n* gates respond by opening. The combination of these delayed gating events drives E_m back down near E_K and actually below the usual resting E_m. Again with a delay, the repolarization causes the *h* gates to open and the *n* gates to close, and the membrane returns to its resting state, ready to respond to any new depolarizing stimulus.

The scheme for the ionic changes underlying the nerve action potential was worked out in a series of elegant electrical experiments by A. L. Hodgkin and A. F. Huxley of Cambridge University. Chapter 7 describes those experiments and presents a quantitative version of the scheme shown in Figure 6-8.

The Refractory Period

The existence of a refractory period would be expected from the gating scheme summarized in Figure 6-8. When the *h* gates of the voltage-sensitive sodium channels are closed (states C and D in Figure 6-8), the channels cannot conduct Na^+ no matter what the state of the *m* gate might be. When the membrane is in this condition, no amount of depolarization can cause the cell to fire an action potential; the *h* gates would simply remain closed, preventing the influx of Na^+ necessary to trigger the regenerative explosion. Only when enough time has passed for a significant number of *h* gates to reopen will the neuron be capable of producing another action potential.

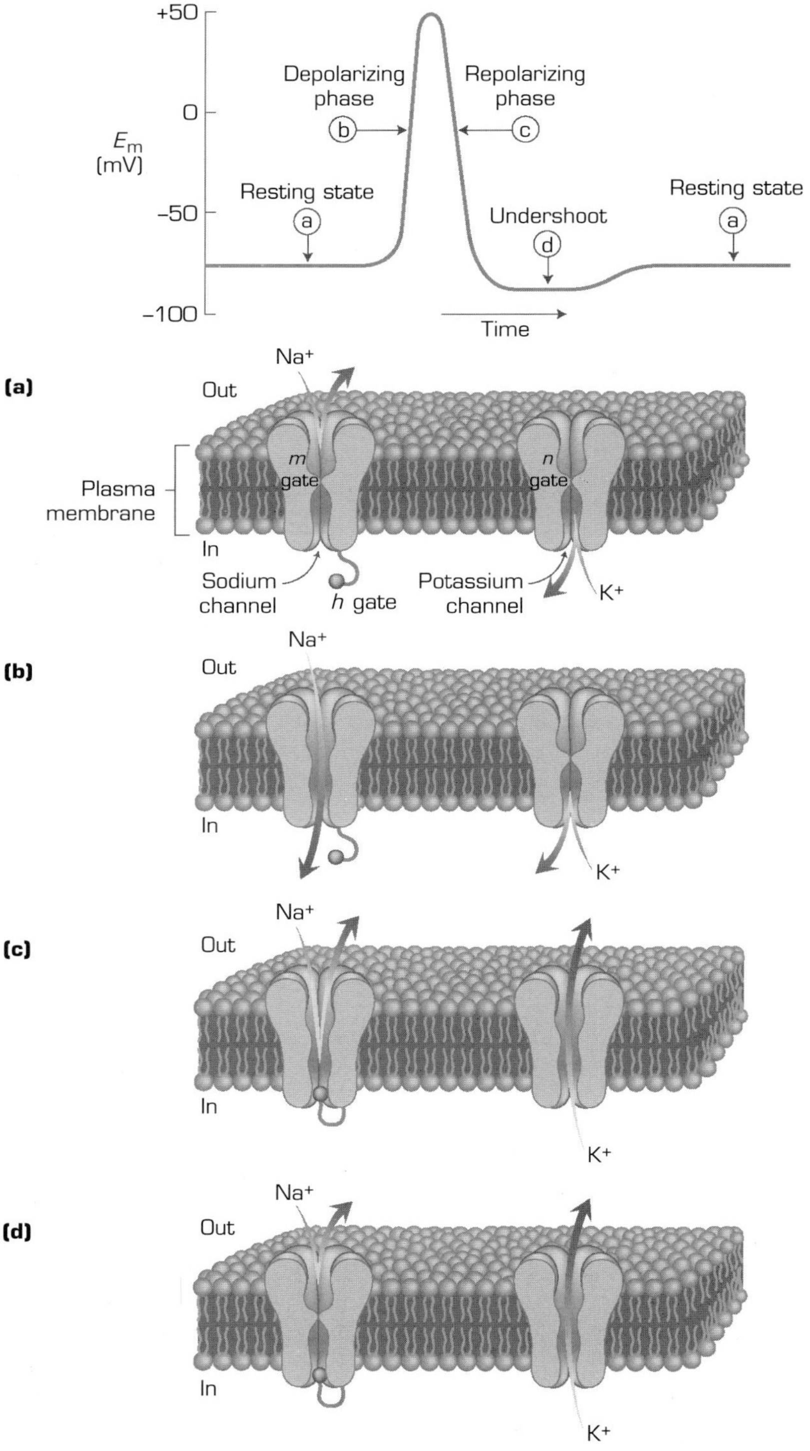

Figure 6-8 The states of voltage-sensitive sodium and potassium channels at various times during an action potential in a neuron. (a) At rest, neither channel is in a conducting state. (b) During the depolarizing phase of the action potential, the sodium channels open, but the potassium channels have not yet responded to the depolarization. (c) During the repolarizing phase, sodium permeability begins to return to its resting level as *h* gates respond to the preceding depolarizing phase. At the same time, potassium channels respond to the depolarization by opening. (d) During the undershoot, sodium permeability returns to its usual low level; potassium permeability, however, remains elevated because *n* gates respond slowly to the repolarization of the membrane. The resting state of the membrane is restored after *h* gates and *n* gates return to their resting configurations. (Animation available at www.blackwellscience.com)

Propagation of an Action Potential Along a Nerve Fiber

We can now see how an action potential arises as a result of a depolarizing stimulus, such as the muscle stretch in the case of the sensory neuron in the patellar reflex. How does that action potential travel from the ending in the muscle along the long, thin sensory fiber to the spinal cord? The answer to this question is inherent in the scheme for generation of the action potential just presented. As we've seen, the stimulus for an action potential is a depolarization of greater than about 10–20 mV from the normal resting level of membrane potential. The action potential itself is a depolarization much in excess of this threshold level. Thus, once an action potential occurs at one end of a neuron, the strong depolarization will bring the neighboring region of the cell above threshold, setting up a regenerative depolarization in that region. This will in turn bring the next region above threshold, and so on. The action potential can be thought of as a self-propagating wave of depolarization sweeping along the nerve fiber. When the sequence of permeability changes summarized in Figure 6-8 occurs in one region of a nerve membrane, it guarantees that the same gating events will be repeated in neighboring segments of membrane. In this manner, the cyclical changes in membrane permeability, and the resulting action potential, chews its way along the nerve fiber from one end to the other, as each segment of axon membrane responds in turn to the depolarization of the preceding segment. This behavior is analogous to that of a lighted fuse, in which the heat generated in one segment of the fuse serves to ignite the neighboring segment.

A more formal description of propagation can be achieved by considering the electrical currents that flow along a nerve fiber during an action potential. Imagine that we freeze an action potential in time while it is traveling down an axon, as shown in Figure 6-9a. We have seen that at the peak of the action potential, there is an inward flow of current, carried by sodium ions. This is shown by the inward arrows at the point labeled 1 in Figure 6-9a. The region of axon occupied by the action potential will be depolarized with respect to more distant parts of the axon, like those labeled 2 and 3. This difference in electrical potential means that there will be a flow of depolarizing current leaving the depolarized region and flowing along the inside of the nerve fiber; that is, positive charges will move out from the region of depolarization. In the discussion of the response to injected current in an axon (Figures 6-2 and 6-3), we saw that a voltage change produced by injected current decayed with distance from the point of injection. Similarly, the depolarization produced by the influx of sodium ions during an action potential will decay with distance from the region of membrane undergoing the action potential. This decay of depolarization with distance reflects the progressive leakage of the depolarizing current across the membrane, which occurs because the membrane is a leaky insulator. Figure 6-9b illustrates the profile of membrane potential that might be

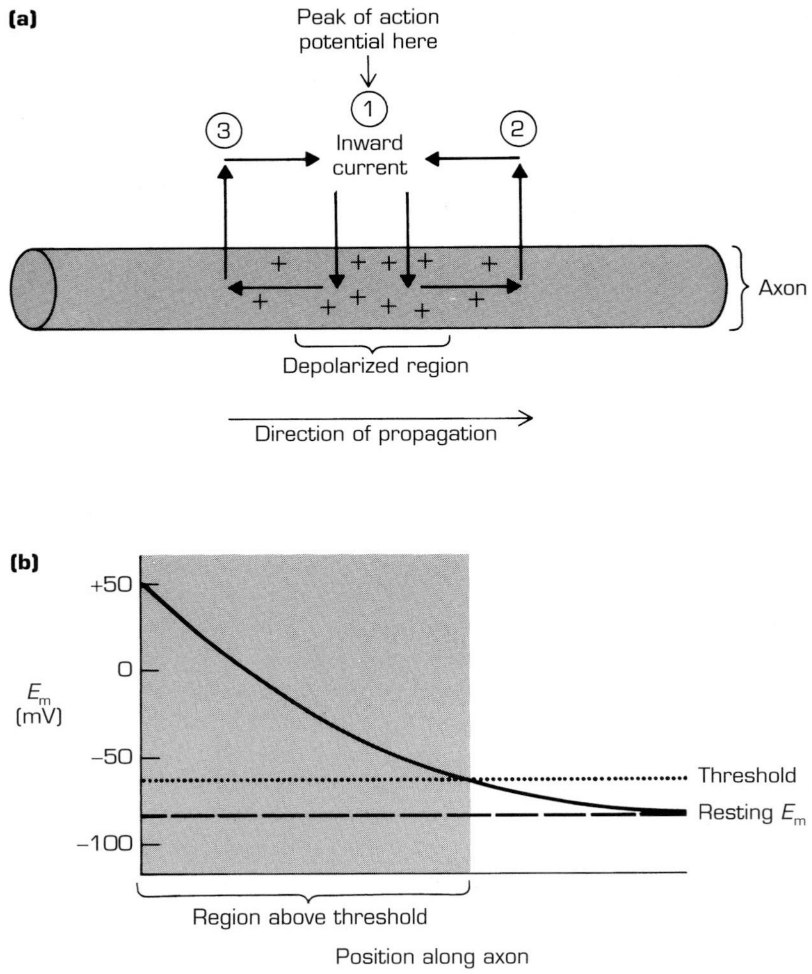

Figure 6-9 The decay of depolarization with distance from the peak of the action potential at a particular instant during the propagation of the action potential from left to right along the axon.

observed along the length of the axon at the instant the action potential at point 1 reaches its peak. Note that there is a region of axon over which the depolarization, although decaying, is still above the threshold for generating an action potential in that part of the membrane. Thus, if we "unfreeze" time and allow events to move along, the region that is above threshold will generate its own action potential. This process will continue as the action potential sweeps along the axon, bringing each successive segment of axon above threshold as it goes.

The flow of depolarizing current from the region undergoing an action potential is symmetrical in both directions along the axon, as shown in Figure 6-9a. Thus, current flows from point 1 to both point 2 and to point 3 in the figure. Nevertheless, the action potential in an axon typically moves in only one direction. That is because the region the action potential has just traversed, like point 3, is in the refractory period phase of the action potential cycle and is thus incapable of responding to the depolarization originating from the action potential at point 1. Of course, if a neurophysiologist comes along with an artificial situation, like that shown in Figure 6-2, and stimulates an action potential in the middle of a nerve fiber, that action potential will propagate in both directions along the fiber. The normal direction of propagation in an axon—the direction taken by normally occurring action potentials—is called the **orthodromic** direction; an abnormal action potential propagating in the opposite direction is called an antidromic action potential.

Factors Affecting the Speed of Action Potential Propagation

The speed with which an action potential moves down an axon varies considerably from one axon to another; the range is from about 0.1 m/sec to 100 m/sec. What characteristics of an axon are important in the determining the action potential propagation velocity? Examine Figure 6-9b again. Clearly, if the rate at which the depolarization falls off with distance is less, the region of axon brought above threshold by an action potential at point 1 will be larger. If the region above threshold is larger, then an action potential at a particular location will set up a new action potential at a greater distance down the axon and the rate at which the action potential moves down the fiber will be greater. The rate of voltage decrease with distance will in turn depend on the relative resistance to current flow of the plasma membrane and the intracellular path down the axon. Recall from the discussion of the response of an axon to injection of current (see Figure 6-3) that there are always two paths that current flowing down the inside of axon at a particular point can take: it can continue down the interior of the fiber or cross the membrane at that point. We said that the portion of the current taking each path depends on the relative resistances of the two paths. If the resistance of the membrane could be made higher or if the resistance of the path down the inside of the axon could be made lower, the path down the axon would be favored and a larger portion of the current would continue along the inside. In this situation, the depolarization resulting from an action potential would decay less rapidly along the axon; therefore, the rate of propagation would increase.

Thus, two strategies can be employed to increase the speed of action potential propagation: increase the electrical resistance of the plasma membrane to current flow, or decrease the resistance of the longitudinal path down the inside of the fiber. Both strategies have been adopted in nature. Among invertebrate animals, the strategy has been to decrease the longitudinal resistance of the axon interior. This can be accomplished by increasing the diameter of the axon. When a fiber is fatter, it offers a larger cross-sectional area to the internal flow of current; the effective resistance of this larger area is less because the current has many parallel paths to choose from if it is to continue down the interior of the axon. For the same reason, the electric power company uses large-diameter copper wire for the cables leaving a power-generating station; these cables must carry massive currents and thus must have low resistance to current flow to avoid burning up. Some invertebrate axons are the neuronal equivalent of these power cables: axons up to 1 mm in diameter are found in some invertebrates. As expected, these giant axons are the fastest-conducting nerve fibers of the invertebrate world.

Among vertebrate animals, there is also large variation in the size of axons, which range from less than 1 μm in diameter to as big as 30–50 μm in diameter. Thus, even the largest axons in a human nerve do not begin to rival the size of the giant axons of invertebrates. Nevertheless, the fastest-conducting vertebrate axons are actually faster than the giant invertebrate axons. Vertebrate animals have adopted the strategy of increasing the membrane resistance to current as well as increasing internal diameter. This has been accomplished by wrapping the axon with extra layers of insulating cell membrane: in order to reach the exterior, electrical current must flow not only through the resistance of the axon membrane, but also through the cascaded resistance of the tightly wrapped layers of extra membrane. Figure 6-10a shows a schematic cross-section of a vertebrate axon wrapped in this way. The cell that provides the spiral of insulating membrane surrounding the axon is a type of **glial cell**, a

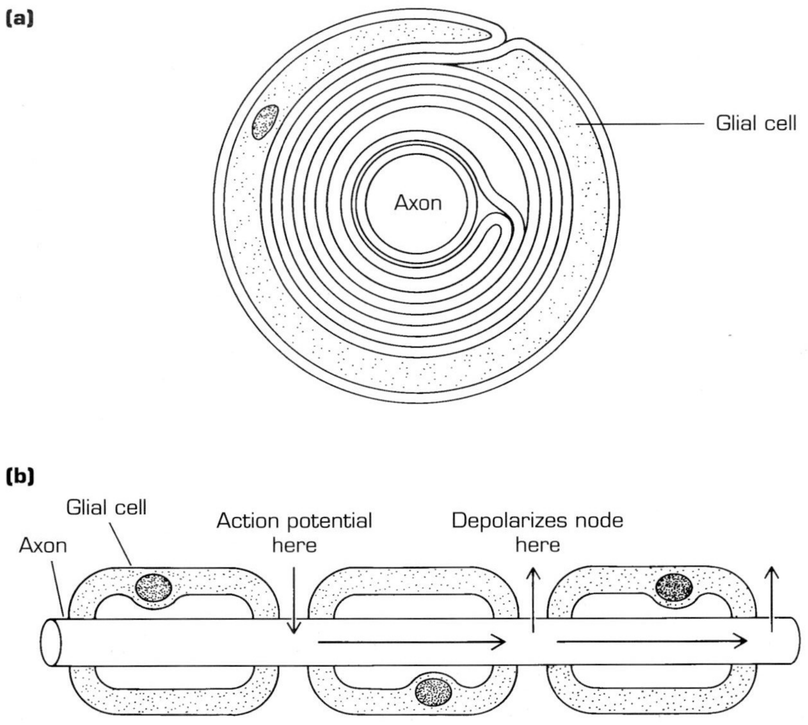

Figure 6-10 The propagation of an action potential along a myelinated nerve fiber. (a) Cross-section of a myelinated axon, showing the spiral wrapping of the glial cell membrane around the axon. (b) The depolarization from an action potential at one node spreads far along the interior of the fiber because the insulating myelin prevents the leakage of current across the plasma membrane. (Animation available at www.blackwellscience.com)

non-neuronal supporting cell of the nervous system that provides a sustaining mesh in which the neurons are embedded.

The insulating sheath around the axon is called **myelin**. By increasing the resistance of the path across the membrane, the myelin sheath forces a larger portion of the current flowing as the result of voltage change to move down the interior of the fiber. This increases the spatial spread of a depolarization along the axon and increases the rate at which an action potential propagates. In order to set up a new action potential at a distant point along the axon, however, the influx of sodium ions carrying the depolarizing current during the initiation of the action potential must have access to the axon membrane. To provide that access, there are periodic breaks in the myelin sheath, called **nodes of Ranvier**, at regular intervals along the length of the axon. This is diagrammed in Figure 6-10b. Thus, the depolarization resulting from an action potential at one node of Ranvier spreads along the interior of the fiber to the next node, where it sets up a new action potential. The action potential leaps along the axon, jumping from one node to the next. This form of action potential conduction is called **saltatory conduction**, and it produces a dramatic improvement in the speed with which a thin axon can conduct an action potential along its length.

The myelin sheath also has an effect on the behavior of the axon as an electrical capacitor. Recall from Chapter 3 that the cell membrane can be viewed as an insulating barrier separating two conducting compartments (the ICF and ECF). Thus, the cell membrane forms a capacitor. The capacitance, or charge-storing ability, of a capacitor is inversely related to the distance between the conducting plates: the smaller the distance, the greater the number of charges that can be stored on the capacitor in the presence of a particular voltage gradient. Thus, when the myelin sheath wrapped around an axon increases the distance between the conducting ECF and ICF, the effective capacitance of the membrane decreases. This means that a smaller number of charges needs to be added to the inside of the membrane in order to reach a particular level of depolarization. (If it is unclear why this is true, review the calculation in Chapter 3 of the number of charges on a membrane at a particular voltage.) An electrical current is defined as the rate of charge movement—that is, number of charges per second. In the presence of a particular depolarizing current, then, a given level of voltage will be reached faster on a small capacitor than on a large capacitor. Because the myelin makes the membrane capacitance smaller, a depolarization will spread faster, as well as farther, in the presence of myelin.

Molecular Properties of the Voltage-sensitive Sodium Channel

Ion channels are proteins, and like all proteins, the sequence of amino acids making up the protein of a particular ion channel is coded for by a particular gene. Thus, it is possible to study the properties of ion channels by applying techniques of molecular biology to isolate and analyze the corresponding gene. This has been done for an increasing variety of ion channels, including the voltage-sensitive sodium channel that underlies the action potential. The sodium channel is a large protein, containing some 2000 individual amino acids. A model of how the protein folds up into a three-dimensional structure has been developed, and this model is summarized schematically in Figure 6-11. According to the model, the protein consists of four distinct regions, called domains. Each domain consists of six separate segments that extend all the way across the plasma membrane (transmembrane segments), which are labeled S1 through S6. Within a domain, the protein threads its way through the membrane six times (Figure 6-11). The amino-acid sequences of each of the six transmembrane segments within a particular domain are similar to the corresponding segments in the other domains. Thus, the overall structure of the channel can be thought of as a series of six transmembrane segments, repeated four times.

It is thought that the four domains aggregate in a circular pattern as shown in Figure 6-11b to form the pore of the channel. The lining of the pore determines the permeation properties of the channel and gives the channel its selectivity for sodium ions. Interestingly, it seems that the lining is actually made up of the external loop connecting segments S5 and S6 within each of the four domains. In order for this external loop to form the transmembrane pore through which sodium ions cross the membrane, it must fold down into the pore in the manner shown schematically in Figure 6-11c.

One important question about the channel is what part of the protein is responsible for detecting changes in the membrane potential and thus imparts voltage sensitivity to the channel. Here, attention has focused on the fourth transmembrane segment of each domain, segment S4, which is marked with a + in Figure 6-11a. Segment S4 has an unusual accumulation of positive charge (because of positively charged arginine and lysine residues in that part of the protein), which should give S4 high sensitivity to the electric field across the membrane. Also, the positive charges in S4 are located within the membrane, which is the correct position to be acted upon by the transmembrane voltage. To test the idea that the charges in S4 are the voltage sensors, W. Stühmer and co-workers have constructed artificial sodium channels by altering the DNA so that one or more of the arginines or lysines in S4 was replaced with a neutral or negatively charged amino acid. These artificial channels were less voltage dependent than the normal channels, suggesting that the charges in S4 are indeed the voltage sensors that detect depolarization of the membrane and activate the opening of the *m* gate.

Another important issue is to establish the identity of the sodium inactivation gate, the *h* gate. Here, Stühmer and co-workers found that the part of the protein connecting domains III and IV (marked with * in Figure 6-11) is important. If that region was deleted or altered, the inactivation process was greatly impaired, though activation seemed normal. Note that this part of the protein is on the intracellular side of the membrane, which is where we have drawn the

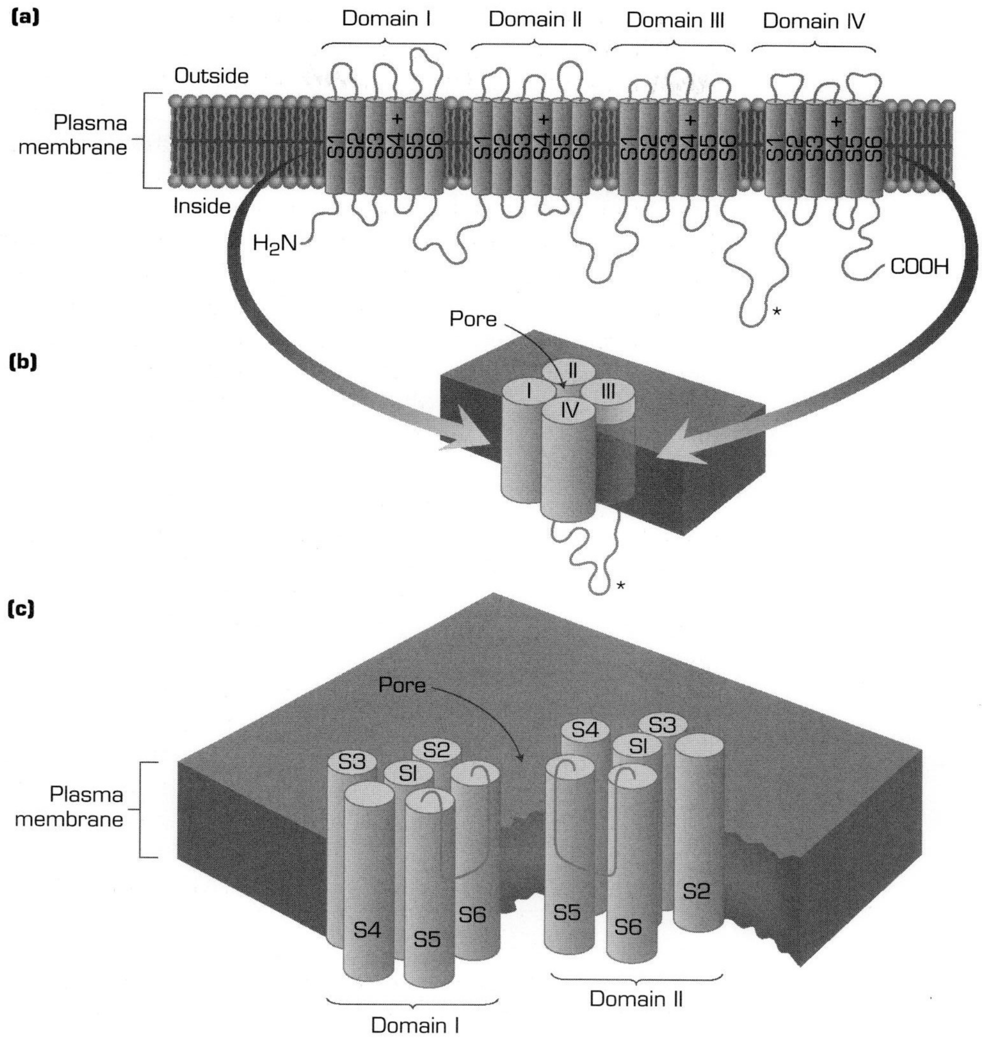

Figure 6-11 The molecular structure of the voltage-sensitive sodium channel. (a) The molecule consists of four domains of similar make-up, labeled with Roman numerals. Each domain has six transmembrane segments (S1–S6). The highly positively charged segment S4 is indicated in each domain by a plus sign (+). The linkage between domains III and IV, indicated by an asterisk (*), is involved in inactivation gating. (b) The domains are shown in a linear arrangement in (a), but in reality, the domains likely form a circular arrangement with the pore at the center. (c) The extracellular loop between S5 and S6 of each domain may fold in as indicated to line the entry to the pore. This region controls the ionic selectivity of the channel.

h gate in our cartoon diagrams of sodium channels in earlier figures in this chapter.

Molecular Properties of Voltage-dependent Potassium Channels

The DNA coding for various other voltage-activated channels, including voltage-activated potassium channels, has also been analyzed to reveal the sequence of amino acids making up those proteins. It is interesting that these voltage-activated channels all have similar (though, of course, not identical) amino-acid sequences, especially in segment S4, which seems to impart the voltage sensitivity. Thus, voltage-activated channels of various kinds represent a family of proteins coded by related genes that probably arose during the course of evolution from a single ancestral ion-channel gene that existed eons ago. Potassium channel genes, however, encode proteins that are much smaller than sodium channels. In fact, the protein encoded by potassium channel genes seems to correspond to a single one of the four domains present in the voltage-activated sodium channel (Figure 6-11). It is thought that functional potassium channels are formed by the aggregation of four of these individual protein subunits, so that the whole channel has an arrangement similar to that of the sodium channel shown in Figure 6-11b. In the sodium channel, however, the four domains are combined together into one large, continuous protein molecule, while in potassium channels each domain consists of a separate protein subunit.

Calcium-dependent Action Potentials

Action potentials are not unique to neurons. Action potentials are also found in non-neuronal excitable cells, such as muscle cells (as we will see in Part III of this book), and even in single-celled animals. Figure 6-12 shows that the protozoan, *Paramecium*, can produce action potentials similar to those of nerve cells, except that the action potential results from influx of calcium ions rather than sodium ions as in the typical nerve action potential. The depolarizing upstroke of the action potential is caused by influx of positively charged calcium ions, rather than influx of sodium ions. As with sodium ions, the equilibrium potential for calcium ions (with a valence of +2) is positive, so if the membrane potential is negative and a calcium channel opens, there will be an influx of calcium into the cell. In the case of the sodium-dependent action potential, sodium channels activated by depolarization provide the basis for the regenerative all-or-none depolarizing phase of the action potential. Similarly, in the case of calcium-dependent action potentials, calcium channels that open upon depolarization underlie the depolarizing phase of the action potential. Depolarization opens calcium channels, which allow influx of positively

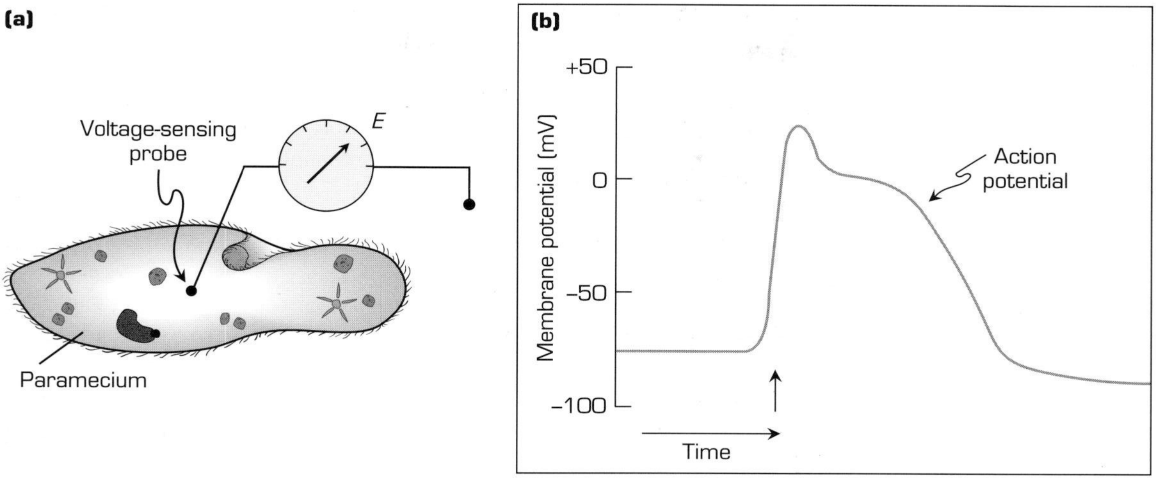

Figure 6-12 The single-celled protozoa, *Paramecium*, produces an action potential similar to a nerve action potential. (a) This diagram shows the recording configuration for intracellular recording. (b) The action potential elicited by an electrical stimulus (at the arrow). The action potential results from calcium influx through voltage-sensitive calcium channels.

charged calcium ions, which in turn produces more depolarization and opens more calcium channels (see Figure 6-5 for the analogous situation with depolarization-activated sodium channels). The calcium-dependent action potential in *Paramecium* is also similar to nerve action potentials in that it serves a coordinating function: it regulates the direction of ciliary beating and thus the movement of the cell. Mutant paramecia that lack the ion channels underlying the calcium action potential are unable to reverse the direction of ciliary beating and thus are unable to swim backwards when they encounter noxious environmental stimuli. Because these mutants can only swim forward, they are called "pawn" mutants, after the chess piece that can only move forward. Thus, some of the basic molecular machinery for electrical signaling, one of the hallmarks of nervous system function, predates by far the origin of the first neuron. This suggests that neural signaling arose by the evolutionary modification of pre-existing signaling mechanisms, found already in single-celled animals.

Voltage-dependent calcium channels are found in most neurons, and in some neurons, these voltage-activated calcium channels contribute significantly to the action potential. A comparison between the waveform of the sodium-dependent action potential and the waveform of an action potential with a component caused by calcium influx is shown in Figure 6-13. Often, the depolarization produced by calcium influx is slower and more sustained than the more spike-like action potential due to sodium and potassium channels alone. This is because the voltage-activated calcium channels commonly inactivate more slowly than voltage-activated sodium channels, so they produce a

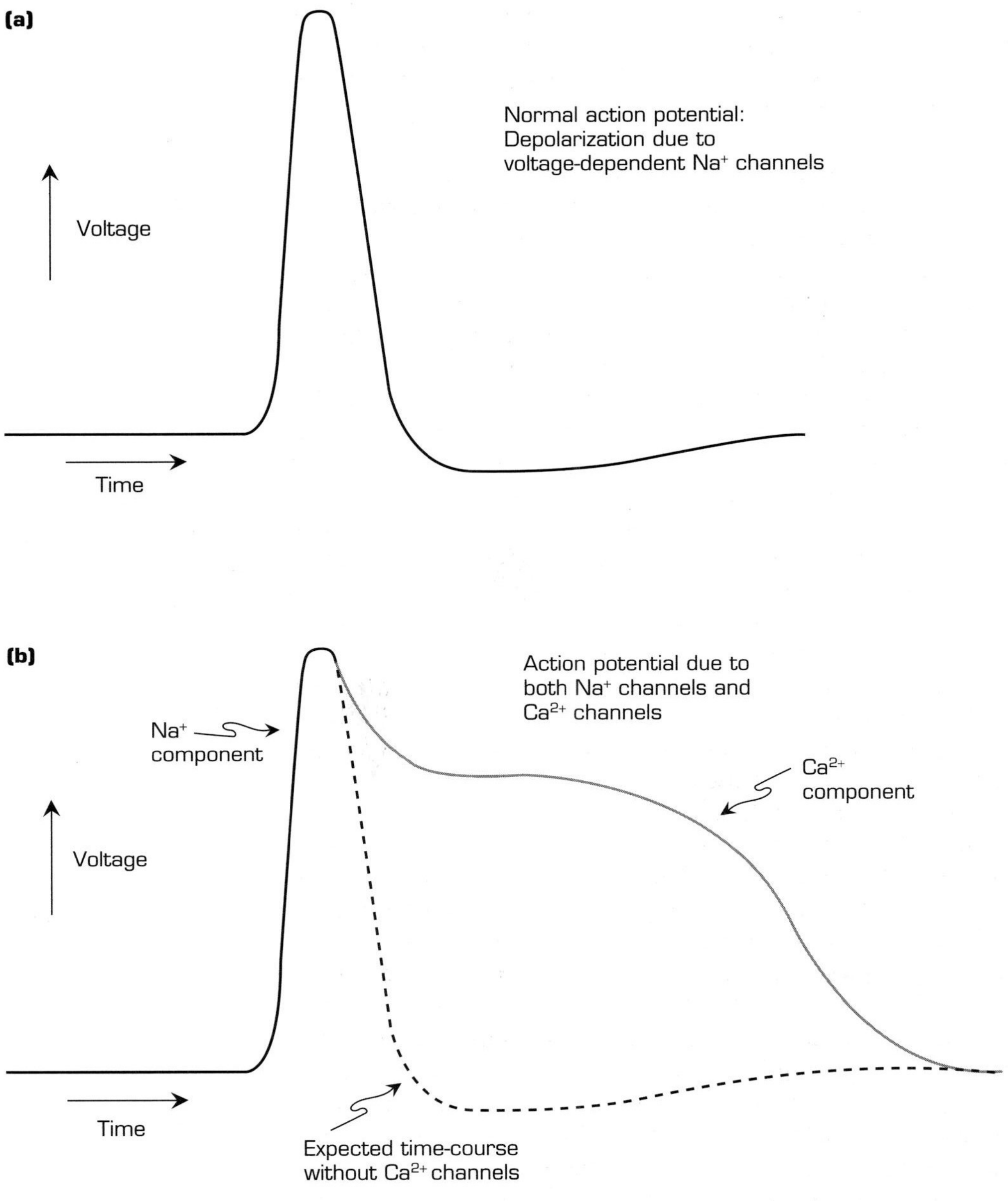

Figure 6-13 Comparison between action potentials in neurons without a contribution from voltage-dependent calcium channels (a) and with a calcium component (b). The rising phase of the action potential on the bottom is produced by depolarization-activated sodium channels, and the dashed black line shows the expected time-course of the action potential in the absence of calcium channels. The prolonged plateau depolarization is caused by the opening of voltage-sensitive calcium channels.

more sustained influx of positive charge, and thus a more prolonged depolarization. In neurons with a calcium-dependent component, then, the action potential has a rapid upstroke caused by the opening of sodium channels, followed by a longer duration plateau phase caused by the voltage-dependent calcium channels.

The influx of calcium ions through voltage-dependent calcium channels has functional consequences beyond contributing to the action potential. The increase in the intracellular concentration of calcium that results from the influx is an important cellular signal that allows depolarization of a cell to be coupled to the triggering of internal cellular events. For example, we will see in Chapter 8 that an increase in intracellular calcium is the trigger for release of neurotransmitter from the presynaptic terminal when an action potential arrives at the synaptic junction between two neurons. Another important effect of internal calcium is the activation of other kinds of ion channels. In addition to the potassium channels opened by depolarization, which we have discussed previously in this chapter, neurons frequently have potassium channels that are opened by an increase in internal calcium. Such calcium-activated potassium channels can contribute to action potential repolarization in neurons that have a calcium component in the action potential (e.g., Figure 6-13). As we have discussed earlier, an increase in potassium permeability accounts in part for the repolarizing phase of the action potential and produces the hyperpolarizing undershoot after repolarization. This increase in potassium permeability can be accomplished with voltage-activated potassium channels or with calcium-activated potassium channels. The activation scheme for calcium-activated potassium channels is summarized in Figure 6-14.

One important functional difference between voltage-activated and calcium-activated potassium channels is the amount of time the channels can remain open after the membrane potential has returned to its negative level at the end of the action potential. The action potential undershoot corresponds to the time after an action potential when the voltage-dependent potassium channels remain open, while sodium permeability has returned to rest; because the ratio p_{Na}/p_K is therefore *smaller* than the usual value, the membrane potential is driven even nearer to the potassium equilibrium potential than the normal resting potential. The period of hyperpolarization during the undershoot ends as the voltage-dependent potassium channels close in response to repolarization, which takes a few milliseconds or less. Calcium-activated potassium channels, however, remain open for as long as the intracellular calcium level remains elevated after the action potential. This can be hundreds of times longer than the undershoot produced by the voltage-dependent potassium channels, as shown in Figure 6-15. The longer-lasting hyperpolarization is called the **afterhyperpolarization** to distinguish it from the undershoot. The presence of an afterhyperpolarization requires both a significant calcium influx during the action potential (to produce an increase in internal calcium concentration) and significant numbers of calcium-activated potassium channels (to produce an increase in potassium permeability in response to the

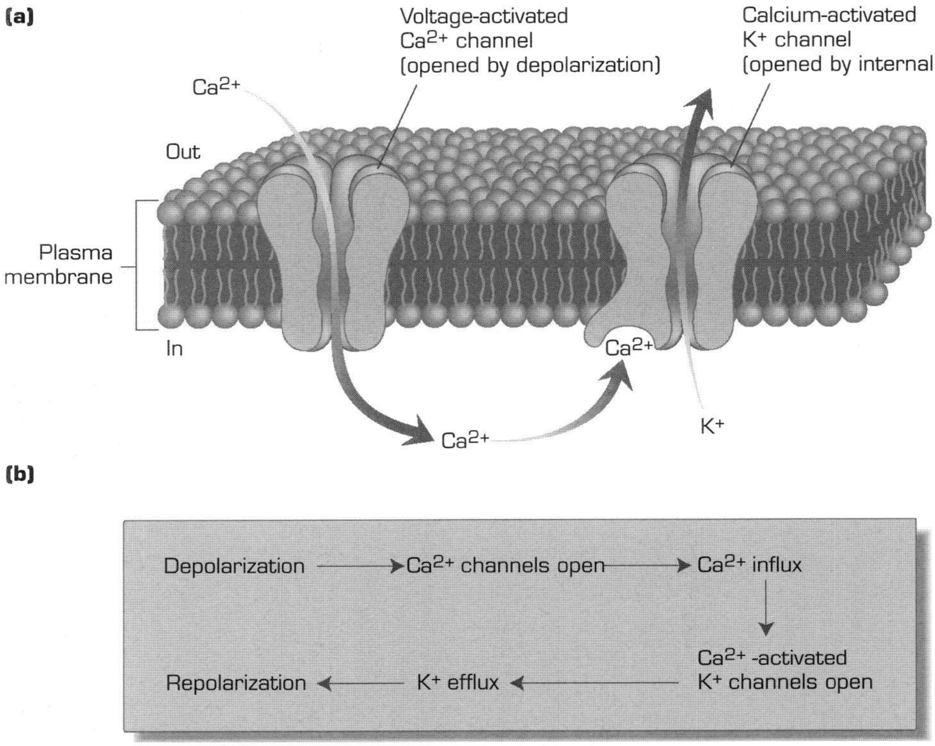

Figure 6-14 Activation of potassium channels by internal calcium ions. (a) Upon depolarization, voltage-dependent calcium channels open and calcium ions enter the cell from the extracellular fluid. The calcium ions then bind to and open calcium-activated potassium channels, which allow potassium ions to exit from the cell. (b) A summary of the sequence of events leading to the activation of calcium-activated potassium channels.

increase in internal calcium). Not all neurons possess these requirements and thus not all neurons show prolonged afterhyperpolarizations. In neurons that have only a small component of calcium influx during a single action potential, afterhyperpolarizations may still be observed if the cell fires a rapid burst of action potentials because the internal calcium contributed by each action potential may sum temporally to reach the calcium level necessary to activate calcium-activated potassium channels. The afterhyperpolarization is important in determining the temporal patterning of action potentials, because the long period of increased potassium permeability makes it more difficult for the neuron to fire action potentials in a rapid series. In neurons that require a burst of several action potentials to initiate the afterhyperpolarization, the calcium-activated potassium channels can be important in terminating the burst. This can be a mechanism for timed bursts of action potentials separated by silent periods in neurons that control rhythmic events.

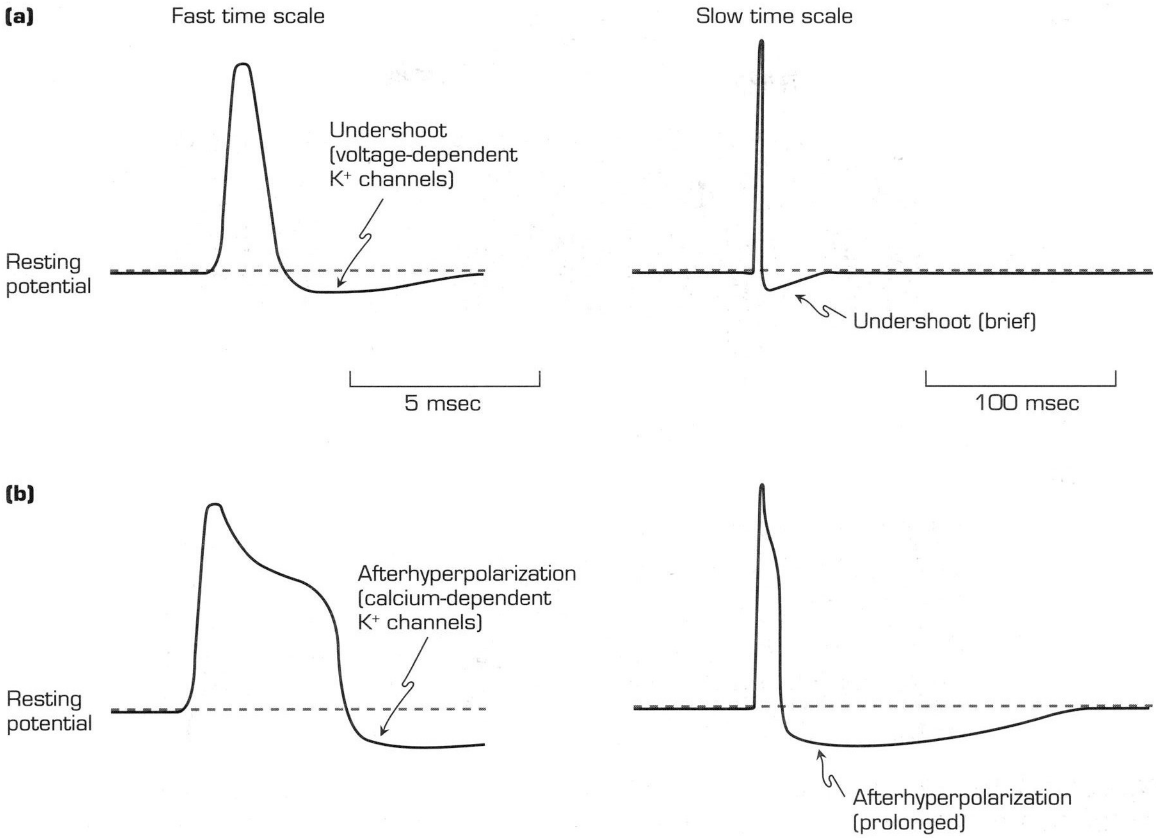

Figure 6-15 The time-course of the undershoot compared with the time-course of the afterhyperpolarization produced by calcium-activated potassium channels. (a) The action potential of a neuron with only voltage-dependent sodium and potassium channels. (b) The action potential of a neuron with voltage-dependent calcium channels and calcium-activated potassium channels in addition to the usual voltage-dependent sodium and potassium channels. The left traces in both (a) and (b) show the action potential on a fast time scale (milliseconds), while the right traces show the same action potentials on a slower time scale (hundreds of milliseconds).

Summary

The basic long-distance signal of the nervous system is a self-propagating depolarization called the action potential. The action potential arises because of a sequence of voltage-dependent changes in the ionic permeability of the neuron membrane. This voltage-dependent behavior of the membrane is due to gated sodium and potassium channels. The conducting state of the sodium channels is controlled by *m* gates, which are closed at the usual resting E_m and open rapidly upon depolarization, and by *h* gates, which are open at the usual resting E_m and close slowly upon depolarization. The voltage-sensitive potassium channels are controlled by a single type of gate, called the *n* gate, which is closed at the resting E_m and opens slowly upon depolarization. In response to depolarization, p_{Na} increases dramatically as *m* gates open, and E_m is driven up near E_{Na}. With a delay, *h* gates close, restoring p_{Na} to a low level, and *n* gates open, increasing p_K. As a result, p_{Na}/p_K falls below its normal resting value, and E_m is driven back to near E_K. The resulting repolarization restores the membrane to its resting state.

The behavior of the voltage-dependent sodium and potassium channels can explain (1) why depolarization is the stimulus for generation of an action potential; (2) why action potentials are all-or-none events; (3) how action potentials propagate along nerve fibers; (4) why the membrane potential becomes positive at the peak of the action potential; (5) why the membrane potential is transiently more negative than usual at the end of an action potential; and (6) the existence of a refractory period after a neuron fires an action potential.

Action potentials of some neurons have components contributed by voltage-dependent calcium channels, which open upon depolarization like voltage-dependent sodium channels but specifically allow influx of calcium ions. The influx of calcium ions through these channels can increase the intracellular concentration of calcium. Calcium-activated potassium channels open when internal calcium is elevated, contributing to the repolarization of the action potential and producing a prolonged period of elevated potassium permeability during which the membrane potential is more negative than the usual resting membrane potential.

The Action Potential: Voltage-clamp Experiments

7

In Chapter 6, we discussed the basic membrane mechanisms underlying the generation of the action potential in a neuron. We saw that all the properties of the action potential could be explained by the actions of voltage-sensitive sodium and potassium channels in the plasma membrane, both of which behave as though there are voltage-activated gates that control permeation of ions through the channel. In this chapter, we will discuss the experimental evidence that gave rise to this scheme for explaining the action potential. The fundamental experiments were performed by Alan L. Hodgkin and Andrew F. Huxley in the period from 1949 to 1952, with the participation of Bernard Katz in some of the early work. The Hodgkin–Huxley model of the nerve action potential is based on electrical measurements of the flow of ions across the membrane of an axon, using a technique known as voltage clamp. We will start by describing how the voltage clamp works, and then we will discuss the observations Hodgkin and Huxley made and how they arrived at the gated ion channel model discussed in the last chapter.

The Voltage Clamp

We saw in Chapter 6 that the permeability of an excitable cell membrane to sodium and potassium depends on the voltage across the membrane. We also saw that the voltage-induced permeability changes occur at different speeds for the different ionic "gates" on the voltage-sensitive channels. This means that the membrane permeability to sodium, for example, is a function of two variables: voltage and time. Thus, in order to study the permeability in a quantitative way, it is necessary to gain experimental control of one of these two variables. We can then hold that one constant and see how permeability varies as a function of the other variable. The voltage clamp is a recording technique that allows us to accomplish this goal. It holds membrane voltage at a constant value; that is, the membrane potential is "clamped" at a particular

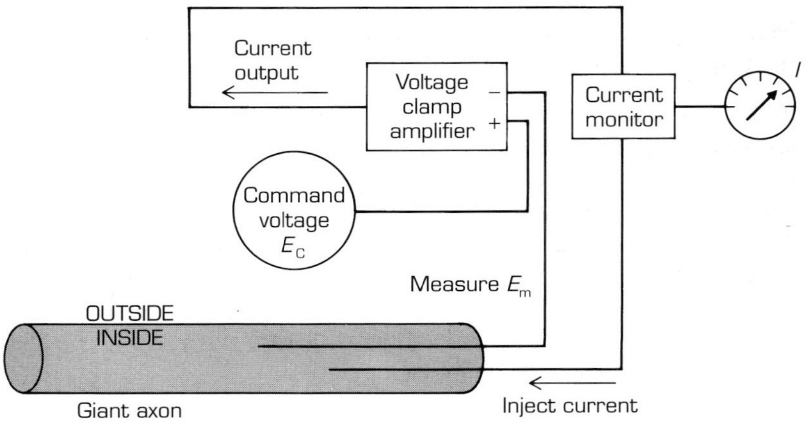

Figure 7-1 A schematic diagram of a voltage-clamp apparatus.

level. We can then measure the membrane current flowing at that constant membrane voltage and use the time-course of changes in membrane current as an index of the time-course of the underlying changes in membrane ionic conductance.

A diagram of the apparatus used to voltage clamp an axon is shown in Figure 7-1. Two long, thin wires are threaded longitudinally down the interior of an isolated segment of axon. One wire is used to measure the membrane potential, just as we have done in a number of previous examples using intracellular microelectrodes; this wire is connected to one of the inputs of the voltage-clamp amplifier. The other wire is used to pass current into the axon and is connected to the output of the voltage-clamp amplifier. The other input of the amplifier is connected to an external voltage source, the command voltage, that is under the experimenter's control. The command voltage is so named because its value determines the value of resting membrane potential that will be maintained by the voltage-clamp amplifier.

The amplifier in the voltage-clamp circuit is wired in such a way that it feeds a current into the axon that is proportional to the difference between the command voltage and the measured membrane potential, $E_C - E_m$. If that difference is zero (that is, if $E_m = E_C$), the amplifier puts out no current, and E_m will remain stable. If E_m does not equal E_C, the amplifier will pass a current into the axon to make the membrane potential move toward the command voltage. For example, if E_m is −70 mV and E_C is −60 mV, then $E_C - E_m$ is a positive number. Because the amplifier passes a current that is proportional to that difference, the current will also be positive. That is, the injected current will move positive charges into the axon and depolarize the membrane toward E_C. This would continue until the membrane potential equals the command potential of −60 mV. On the other hand, if E_C were more negative than E_m, $E_C - E_m$ would be a negative number, and the injected current would be negative. In this case, the current would hyperpolarize the axon until the membrane potential equaled the command voltage.

Measuring Changes in Membrane Ionic Conductance Using the Voltage Clamp

By inserting a current monitor into the output line of the amplifier, we can measure the amount of current that the amplifier is passing to keep the membrane voltage equal to the command voltage. How does this measured current give information about changes in ionic current and, therefore, changes in ionic conductance of the membrane? First of all, let's review what happens to membrane current and membrane potential without the voltage clamp, using the principles we discussed in Chapters 5 and 6. This is illustrated in Figure 7-2a, which shows the changes in transmembrane ionic current and membrane potential in response to a stepwise increase in p_{Na}, with p_K remaining constant. Under resting conditions, we have seen that the steady-state membrane potential will be between E_{Na} and E_K, at the membrane voltage at which the inward sodium current exactly balances the outward potassium current, so that the total membrane current is zero ($i_{Na} + i_K = 0$). When p_{Na} is suddenly increased, the steady state is perturbed, and there will be an increase in i_{Na}. This greater sodium

(a)

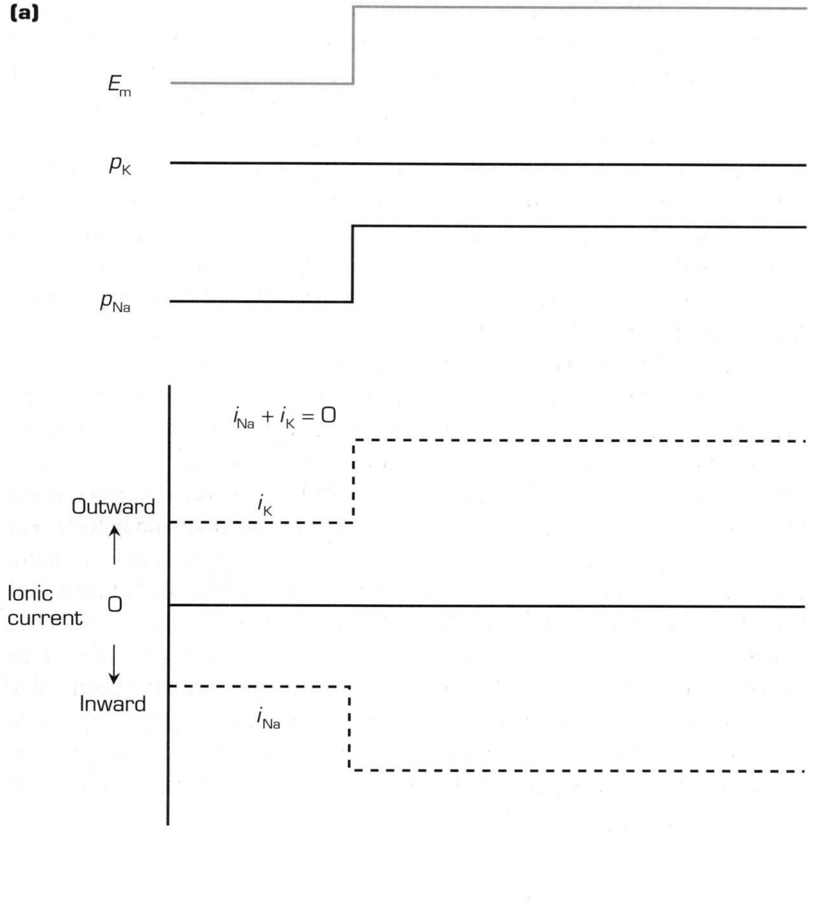

Figure 7-2 The ionic currents flowing in response to a stepwise change in p_{Na}, either without voltage clamp (a) or with voltage clamp (b). Without voltage clamp, both i_{Na} and i_K increase in response to the increase in p_{Na}, and a new steady-state membrane potential is reached at a more depolarized level. With voltage clamp, the membrane potential remains constant because the voltage-clamp apparatus injects current (i_{clamp}) that compensates for the increased sodium current. Potassium current remains constant because neither p_K nor E_m changes.

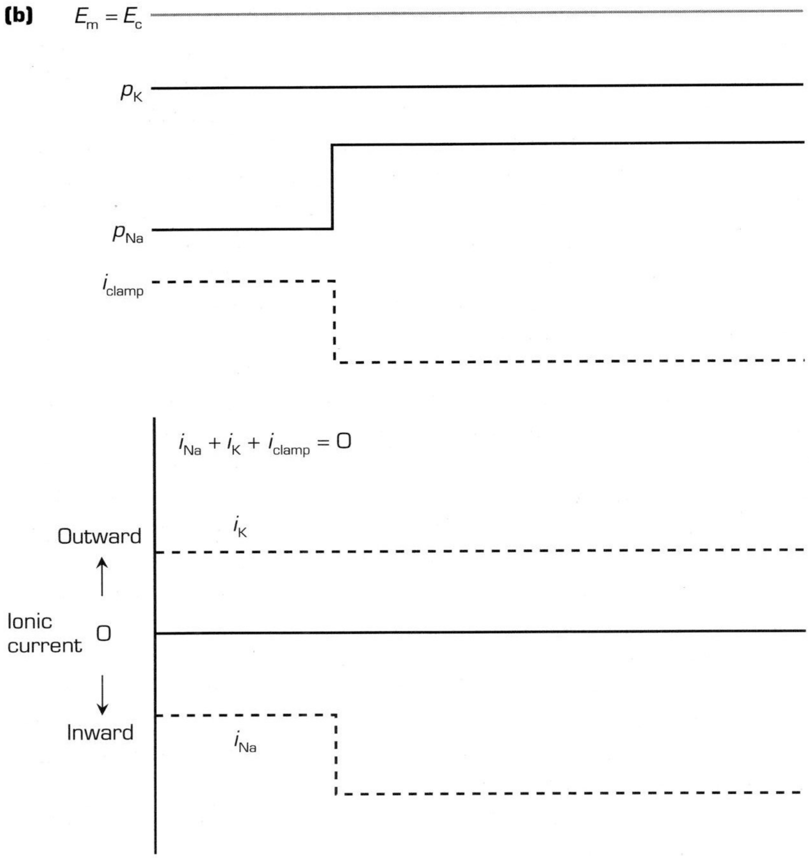

Figure 7-2 (cont'd)

influx causes E_m to move positive from its original resting value. With depolarization, however, potassium current increases because of the increasing difference between E_m and E_K. The membrane potential will reach a new steady state, governed by the new ratio of p_{Na}/p_K, at which both i_{Na} and i_K are larger than they were initially, but once again exactly balance each other. This is just a restatement of the basis of resting membrane potential discussed in detail in Chapter 5. Let's consider now what happens if the same change in p_{Na} occurs under voltage clamp, as shown in Figure 7-2b. Now we must consider an additional source of current: the current provided by the voltage-clamp apparatus (i_{clamp}). Suppose we set the command voltage, E_C, to be equal to the normal steady-state membrane potential of the cell and turn on the voltage-clamp apparatus. In this situation, E_m is already equal to E_C and the current injected by the voltage-clamp apparatus will be zero. Suppose that at some time after we turn on the apparatus, there is a sudden increase in the sodium permeability of the membrane. As we have just seen, this would normally cause the membrane potential to take up a new steady-state value closer to the sodium equilibrium potential; that is, the cell would depolarize because of the increase in inward sodium current across the membrane. However, now the voltage-clamp circuit will detect the depolarization as soon as it begins, and the voltage-clamp amplifier will inject negative current into the axon to counter the increased sodium current (see trace labeled i_{clamp} in Figure 7-2b). The voltage clamp will continue to inject this holding current to maintain E_m at its usual resting value for as long as the increased sodium permeability persists, so that E_m remains equal to E_C. Thus, the injected current will be equal in magnitude to the increase in sodium current resulting from the increase in sodium permeability. Notice that there is now no change in i_K, because there is now no change in E_m (as well as no change in p_K). If the potassium permeability, rather than the sodium permeability, were to undergo a stepwise increase from its normal resting value, then the voltage-clamp apparatus will respond as shown in Figure 7-3. In this case, the increased potassium permeability would normally drive E_m more *negative*, toward E_K, and the cell would *hyperpolarize*. However, the voltage-clamp amplifier will inject a depolarizing current of the right magnitude to counteract the hyperpolarizing potassium current leaving the cell. The point is that the current injected by the voltage clamp gives a direct measure of the change in ionic current resulting from a change in membrane permeability to an ion.

How do we relate the measured change in membrane current to the underlying change in membrane permeability? Recall from Chapter 5 that the ionic current carried by a particular ion is given by the product of the membrane conductance to that ion and the voltage driving force for that ion, which is the difference between the actual value of membrane potential and the equilibrium potential for the ion. For example, for sodium ions

$$i_{Na} = g_{Na}(E_m - E_{Na}) \tag{7-1}$$

Thus, we can calculate g_{Na} from the measured i_{Na} according to the relation

$$g_{Na} = i_{Na}/(E_m - E_{Na}) \tag{7-2}$$

In this calculation, E_m is equal to the value set by the voltage clamp, and E_{Na} can be computed from the Nernst equation or measured experimentally by setting E_C to different values and determining the setting that produces no change in ionic current upon a change in g_{Na} (that is, $E_m - E_{Na} = 0$). In this way, it is straightforward to obtain a measure of the time-course of a change in membrane ionic conductance from the time-course of the change in ionic current. As discussed in Chapter 5, conductance is not the same as permeability. However, for rapid changes in permeability like those underlying the action potential, we can treat the two as having the same time-course.

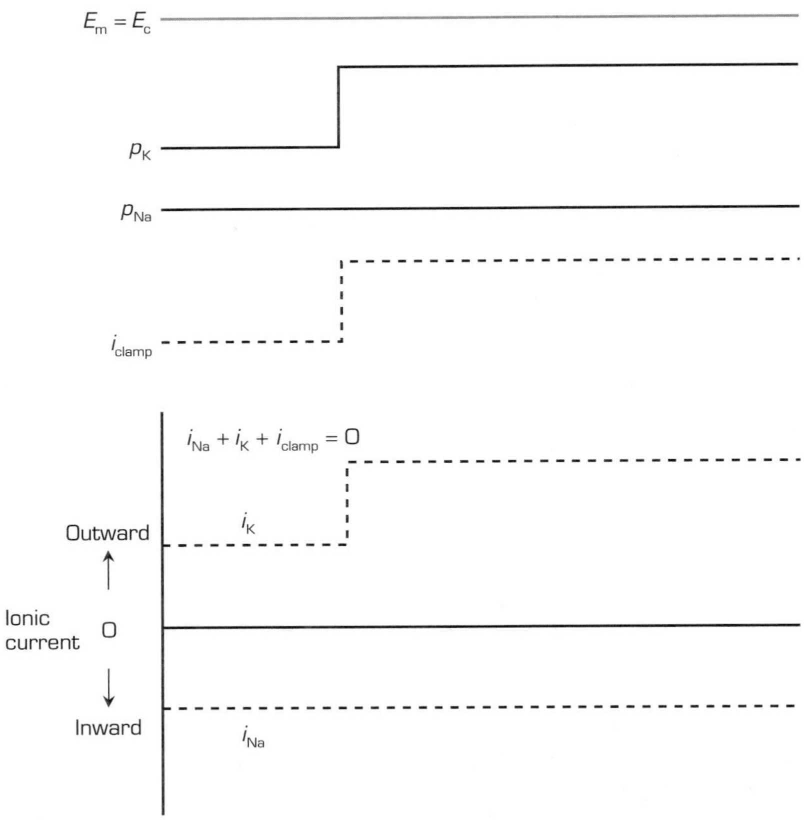

Figure 7-3 The changes in ionic current and injected current after a stepwise change in p_K under voltage clamp. Potassium current increases because of the increase in p_K, but the voltage-clamp amplifier injects compensating current to keep E_m constant.

The Squid Giant Axon

The experimental arrangement diagrammed in Figure 7-1 was technically feasible only because nature provided neurophysiology with an axon large enough to allow experimenters to thread a pair of wires down the inside. The axon used by Hodgkin and Huxley was the giant axon from the nerve cord of the squid. This axon can be up to 1 mm in diameter, large enough to be dissected free from the surrounding nerve fibers and subjected to the voltage-clamp procedure described above. The axon is so large that it is possible to squeeze the normal ICF out of the fiber—like toothpaste out of a tube—and replace it with artificial ICF of the experimenter's concoction. This allows the tremendous experimental advantage of being able to control the compositions of both the intracellular and the extracellular fluids.

Ionic Currents Across an Axon Membrane Under Voltage Clamp

The membrane currents flowing in a squid giant axon during a maintained depolarization can be studied in an experiment like that shown in Figure 7-4.

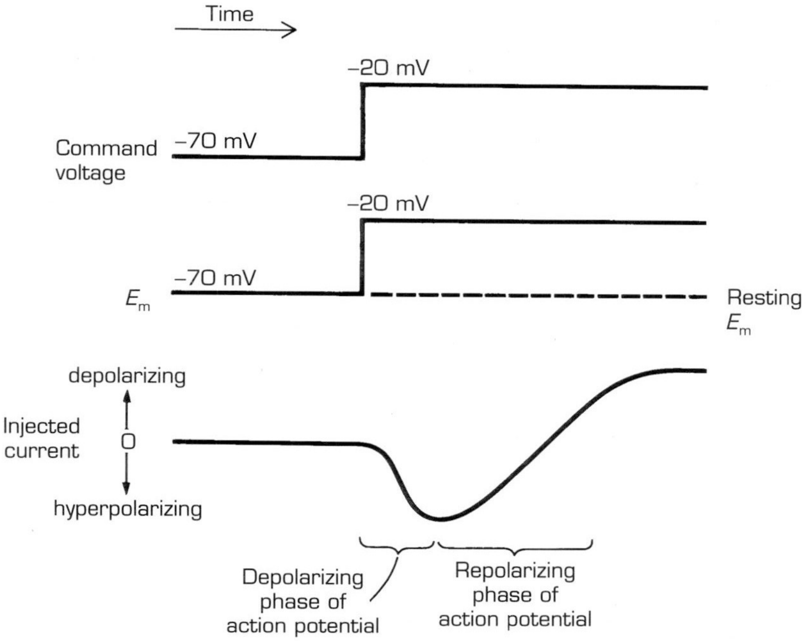

Figure 7-4 A diagram of the current injected by a voltage-clamp amplifier into an axon in response to a voltage step from −70 to −20 mV.

In this case, the command voltage to the voltage-clamp amplifier is first set to be equal to the normal resting potential of the axon, which is about −60 to −70 mV. The command voltage is then suddenly stepped to −20 mV, driving the membrane potential rapidly up to the same depolarized value. A depolarization of this magnitude is well above threshold for eliciting an action potential in the axon; however, the voltage-clamp circuit prevents the membrane potential from undergoing the usual sequence of changes that occur during an action potential. The membrane potential remains clamped at −20 mV.

What current must the voltage-clamp amplifier inject into the axon in order to keep E_m at −20 mV? The sodium permeability of the membrane will increase in response to the depolarization and an increased sodium current will enter the axon through the increased membrane conductance to sodium. In the absence of the voltage clamp, this would set up a regenerative depolarization that would drive E_m up near E_{Na}, to about +50 mV. In order to counter this further depolarization, the voltage-clamp amplifier must inject a hyperpolarizing current during the strong depolarizing phase of the action potential. With time, however, the sodium permeability of the membrane declines, and the potassium permeability increases in response to the depolarization of the membrane. Normally, this would drive E_m back down near E_K. To counter this tendency and maintain E_m at −20 mV, the voltage clamp then must pass a depolarizing current that is maintained as long as potassium permeability remains elevated. Thus, in response to a depolarizing step above threshold, the membrane of an excitable cell would be expected to show a transient inward current followed

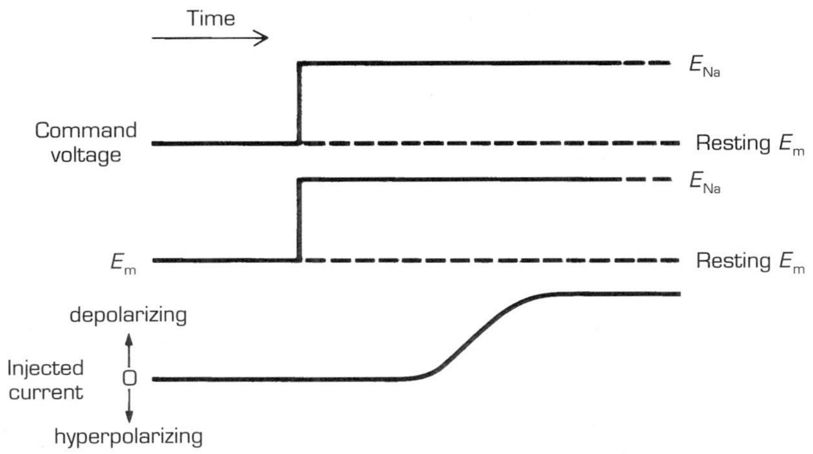

Figure 7-5 A diagram of the current injected by a voltage-clamp amplifier into an axon in response to a voltage step from the normal resting membrane potential to the sodium equilibrium potential. The initial sodium current is absent because there is no driving force for sodium current when E_m equals E_{Na}.

by a maintained outward current. The voltage-clamp records of membrane current illustrating this sequence of changes are shown in Figure 7-4.

What was the nature of the evidence that the initial inward current was carried by sodium ions? This was demonstrated by measuring the membrane current resulting from a series of voltage steps of different amplitudes. As we have seen previously, if the clamped value of membrane potential were equal to the sodium equilibrium potential, there would be no driving force for a net sodium current across the membrane. Therefore, if the initial current is carried by sodium ions, that component of the current should disappear when the command voltage is equal to E_{Na}. A sample of membrane current observed in response to a voltage step to E_{Na} is shown in Figure 7-5. The initial component of inward current disappears in this situation, leaving only the late outward current. Hodgkin and Huxley went one step further and systematically varied E_{Na} by altering the external sodium concentration; they found that the membrane potential at which the early current component disappeared was always E_{Na}. This is strong evidence that the inward component of current in response to a depolarization is carried by sodium ions. This notion also agrees with early observations that the membrane potential reached by the peak of the action potential was strongly influenced by the external sodium concentration.

The two components of membrane current can be separated by comparing the current observed following a voltage step to a particular voltage when that voltage is equal to E_{Na} and when E_{Na} has been moved to another value by altering the external sodium concentration. A specific example is shown in Figure 7-6. In this case, voltage-clamp steps are made to 0 mV in ECF containing normal sodium and in ECF with sodium reduced to be equal with internal sodium concentration. In the normal sodium ECF, E_{Na} will be positive to the command voltage; in the reduced sodium ECF, E_{Na} will equal the command potential and there will be no net sodium current across the membrane. When the observed current in reduced sodium ECF is subtracted from the current in

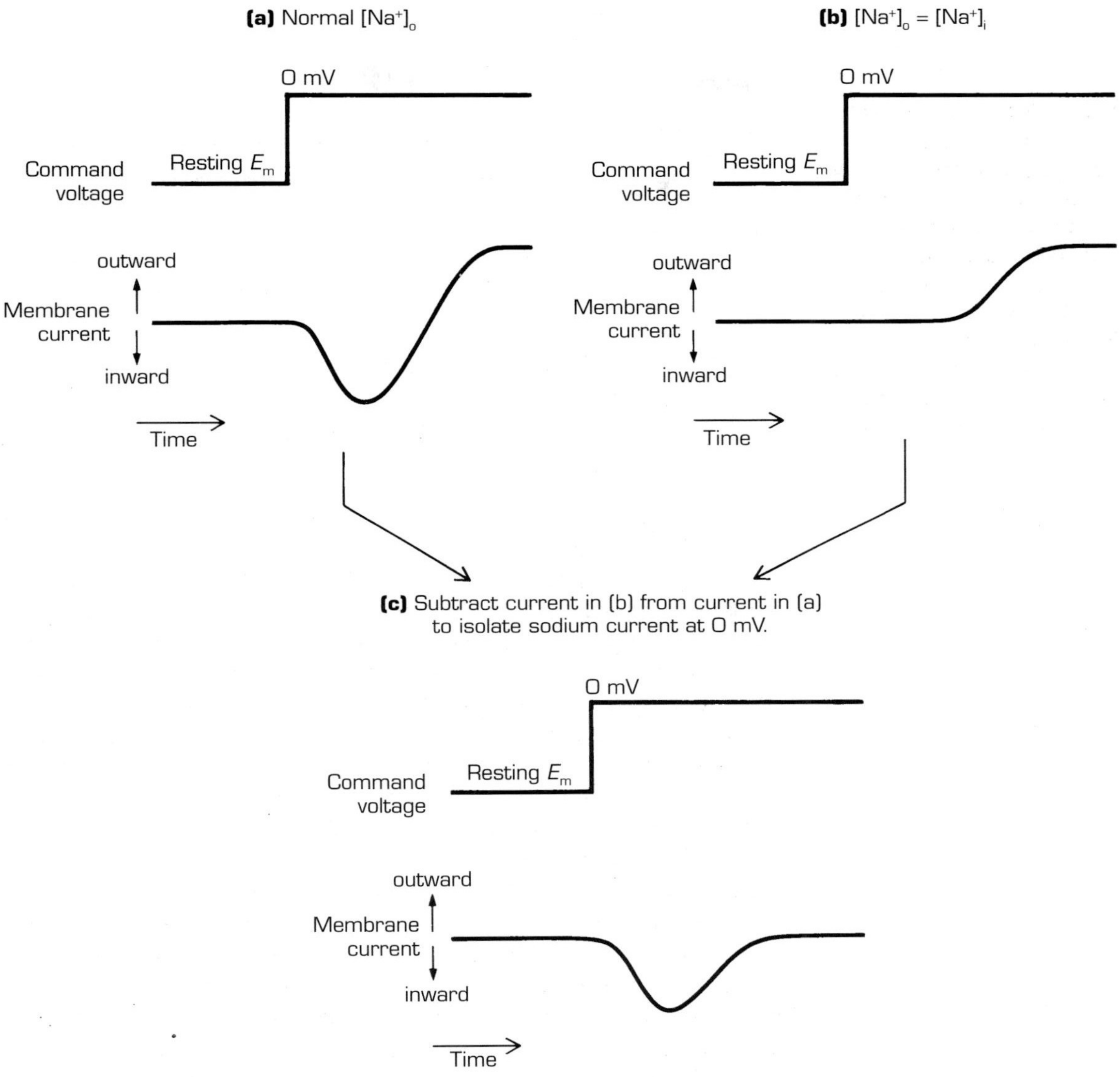

Figure 7-6 The procedure for isolating the sodium component of membrane current by varying external sodium concentration to alter the sodium equilibrium potential.

normal ECF, the difference will be the sodium component of membrane current in response to a step depolarization to 0 mV. This isolated sodium current is shown in Figure 7-6c. The membrane currents of Figure 7-6c can be converted to membrane conductance according to Equation (7-2), and the result gives the time-course of the membrane sodium and potassium conductances in response to a voltage-clamp step to 0 mV. This procedure can be repeated for a series of different values of command potential and E_{Na}, generating a full characterization of the sodium and potassium conductance changes as a function of both time and membrane voltage. The increase in sodium conductance in response to depolarization is transient, even if the depolarization is maintained. The increasing phase is called **sodium activation**, and the delayed fall is called **sodium inactivation**. We will discuss activation first and return later to the mechanism of inactivation. The onset of the increase in potassium conductance is slower than sodium activation and does not inactivate with maintained depolarization. Thus, at least on the brief time-scale relevant to the action potential, potassium conductance remains high for the duration of the depolarizing voltage step.

This rather involved procedure has been simplified considerably by the discovery of specific drugs that block the voltage-sensitive sodium channels and other drugs that block the voltage-sensitive potassium channels. The sodium channel blockers most commonly used are the biological toxins tetrodotoxin and saxitoxin. Both seem to interact with specific sites within the aqueous pore of the channel and physically plug the channel to prevent sodium movement. Potassium channel blockers include tetraethylammonium (TEA) and 4-aminopyridine (4-AP). Thus, the isolated behavior of the sodium current could be studied by treating an axon with TEA, while the isolated potassium current could be studied in the presence of tetrodotoxin.

The Gated Ion Channel Model

Membrane Potential and Peak Ionic Conductance

Hodgkin and Huxley discovered that the peak magnitude of the conductance change produced by a depolarizing voltage-clamp step depended on the size of the step. This established the voltage dependence of the sodium and potassium conductances of the axon membrane. The form of this dependence is shown in Figure 7-7 for both the sodium and potassium conductances. Note the steepness of the curves in both cases. For example, a voltage step to −50 mV barely increases g_{Na}, but a step to −30 mV produces a large increase in g_{Na}. Hodgkin and Huxley suggested a simple model that could account for voltage sensitivity of the sodium and potassium conductances. Their model assumes that many individual ion channels, each with a small ionic conductance, determine the behavior of the whole membrane as measured with the voltage-clamp procedure, and that each channel has two conducting states: an open state in which ions are free to cross through the pore, and a closed state in which the pore is blocked. That is, the channels behave as though access to the pore were controlled by a gate. The effect of membrane potential changes in this scheme is to alter the probability that a channel will be in the open, conducting state. With depolarization, the probability that a channel is open increases, so that a larger

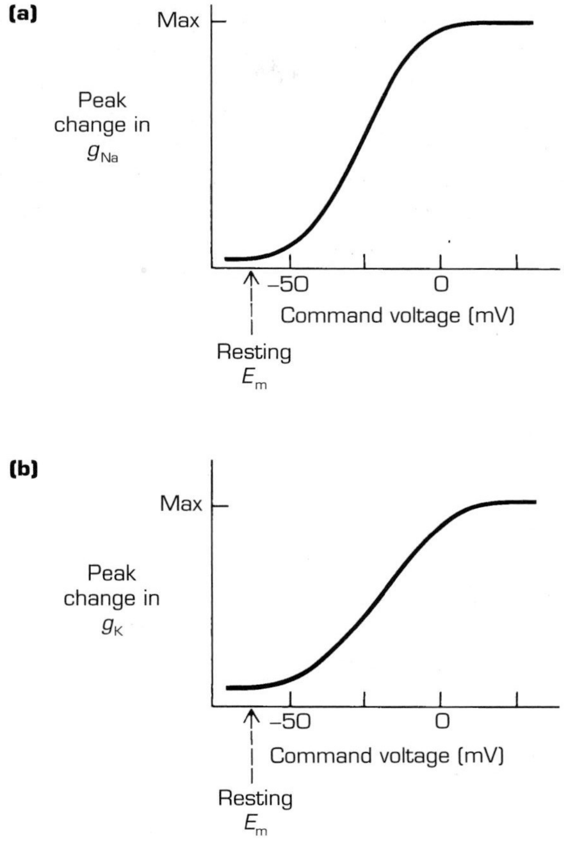

Figure 7-7 Voltage-dependence of peak sodium conductance (a) and potassium conductance (b) as a function of the amplitude of a maintained voltage step.

fraction of the total population of channels is open, and the total membrane conductance to that ion increases. The maximum conductance is reached when all the channels are open, so that further depolarization can have no greater effect.

In order for the conducting state of the channel to depend on transmembrane voltage, some charged entity that is either part of the channel protein or associated with it must control the access of ions to the channel. When the membrane potential is near the resting value, these charged particles are in one state that favors closed channels; when the membrane is depolarized, these charged particles take up a new state that favors opening of the channel. One scheme like this is shown in Figure 7-8. The charged particles are assumed to have a positive charge in Figure 7-8; thus, in the presence of a large, inside-negative electric field across the membrane, most of the particles would likely be near the inner face of the membrane. Upon depolarization, however, the distribution of charged particles within the membrane would become more even, and the fraction of particles on the outside would increase. The channel protein in Figure 7-8 is assumed to have a binding site on the outer edge of the membrane

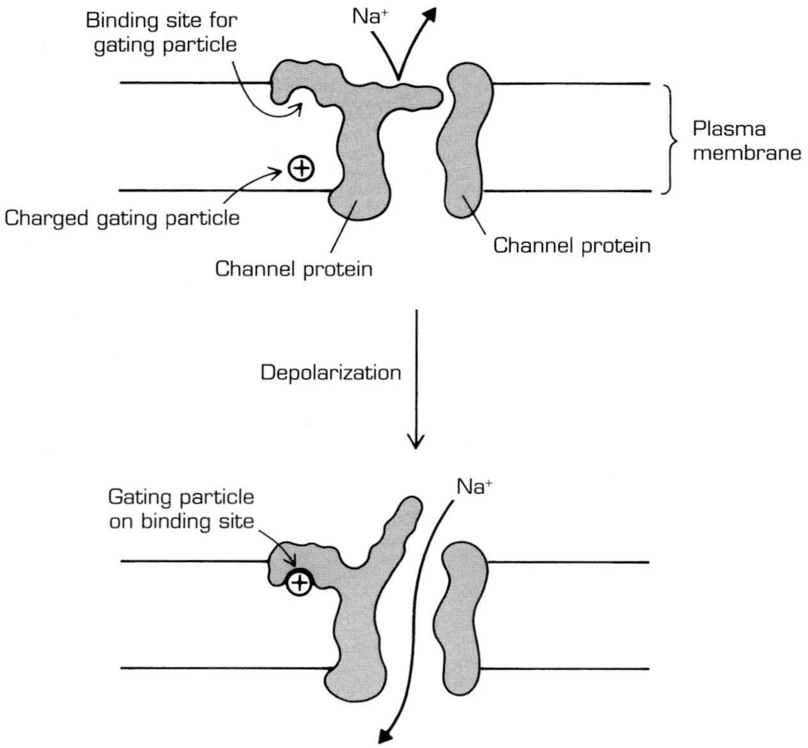

Figure 7-8 A schematic representation of the voltage-sensitive gating of a membrane ion channel. The conducting state of the channel is assumed in this model to depend on the binding of a charged particle to a site on the outer face of the membrane.

that controls the conformation of the "gating" portion of the channel. When the binding site is unoccupied, the channel is closed; when the site binds one of the positively charged particles (called **gating particles**), the channel opens. Thus, upon depolarization, the fraction of channels with a gating particle on the binding site will increase, as will the total ionic conductance of the membrane. It is important to emphasize that the drawings in Figure 7-8 are illustrative only; it is not clear, for example, that the gating particles are positively charged, although evidence from molecular studies suggests so. Negatively charged particles moving in the opposite direction or a dipole rotating in the membrane could accomplish the same voltage-dependent gating function. The molecular mechanism underlying the change in conducting state of the channel protein is unknown at present. It seems likely, however, that a conformation change related to charge distribution within the membrane is involved.

The S-shaped relationship between ionic conductance and membrane potential shown in Figure 7-7 is as expected from basic physical principles for the movement of charged particles under the influence of an electric field, as diagrammed schematically in Figure 7-8. The distribution of charged particles within the membrane will be related to the transmembrane electric field (i.e., the membrane potential) according to the Boltzmann relation:

$$P_o = \frac{1}{1 + e^{\left(\frac{W - z\varepsilon E_m}{kT}\right)}} \tag{7-3}$$

where P_o is the proportion of positive gating particles on the outside of the membrane, z is the valence of the gating charge, ε is the electronic charge, E_m is membrane potential, k is Boltzmann's constant, T is the absolute temperature, and W is a voltage-independent term giving the offset of the relation along the voltage axis. The steepness of the rise in P_o with depolarization depends on the valence, z, of the gating charge: the larger z becomes, the steeper is the rise of P_o (and thus of conductance) with depolarization. As we have noted earlier, the sodium and potassium conductances are steeply dependent on membrane potential, implying that the gating charge that moves in order to open a channel has a large valence. For example, in order to produce a rise in sodium conductance like that observed experimentally, the effective valence of the gating particle must be ~6 [i.e., $z \approx 6$ in the Boltzmann relation of Equation (7-3)].

Kinetics of the Change in Ionic Conductance Following a Step Depolarization

We saw in Chapter 6 that differences in the speed with which the three types of voltage-sensitive gates respond to voltage changes are important in determining the form of the action potential. For instance, the opening of the potassium channels must be delayed with respect to the opening of the sodium channels to avoid wasteful competition between sodium influx and potassium efflux during the depolarizing phase of the action potential. We will now consider how the time-course, or kinetics, of the conductance changes fit into the charged gating particle scheme just presented.

Hodgkin and Huxley assumed that the rate of change in the membrane conductance to an ion following a step depolarization was governed by the rate of redistribution of the gating particles within the membrane. That is, they assumed that the interaction between gating particle and binding site introduced negligible delay into the temporal behavior of the channel. As an example, we will consider the kinetics of opening of the sodium channel following a step depolarization. In formal terms, the movement of gating particles within the membrane can be described by the following first-order kinetic model:

$$m \underset{b_m}{\overset{a_m}{\rightleftarrows}} (1 - m) \tag{7-4}$$

Here, m is the proportion of particles on the outside of the membrane, where they can interact with the binding sites, and $1 - m$ is the proportion of particles on the inside of the membrane. The rate constant, a_m, represents the rate at which particles move from the inner to the outer face of the membrane, and b_m is the rate of reverse movement. Because of the charge on the particles, a step change in the membrane voltage will cause an instantaneous change in the rate constants a_m and b_m. For instance, a step depolarization would increase a_m and decrease b_m, leading to a net increase in m and therefore a decrease in $1 - m$.

The equation governing the rate at which the charges redistribute following a change in membrane potential will be

$$dm/dt = a_m(1 - m) - b_m m \tag{7-5}$$

In Equation (7-5), dm/dt is the net rate of change of the proportion of particles on the outside face of the membrane. In words, $a_m(1 - m)$ is the rate at which particles are leaving the inside of the membrane, and $b_m m$ is the rate at which particles are leaving the outside surface; the difference between those two rates is the net rate of change in m. If the distribution of particles is stable—as it would be if E_m had been constant for a long time—the rate at which particles move from inside to outside would equal the rate of movement in the opposite direction, and dm/dt would be zero. If the system is suddenly perturbed by a depolarization, a and b would change and the balance on the right side of Equation (7-5) would be destroyed. If the depolarization is maintained, the rate at which the system will approach a new steady distribution of particles will be governed by Equation (7-5).

The solution of a first-order kinetic expression like Equation (7-5) is an exponential function; that is, following a step change in membrane voltage m will approach a new steady value exponentially. The exponential solution can be written

$$m(t) = m_\infty - (m_\infty - m_0)\, e^{-(a_m + b_m)t} \tag{7-6}$$

This equation states that following a change in membrane potential, m will change exponentially from its initial value (m_0) to its final value (m_∞) at a rate governed by the rate constants (a_m and b_m) for movement of the gating particles at that new value of membrane potential. The behavior of m with time after a depolarization, as expected from Equation (7-6), is summarized in Figure 7-9. The number of binding sites occupied by gating particles would be expected to be proportional to m, the fraction of available particles on the outer face of the membrane. Thus, if the occupation of a single binding site causes the channel to open and if the coupling between binding of the gating particle and opening of the channel involves no significant delays, the number of open channels would be expected to follow the same exponential time-course as m after a step depolarization.

Because the total membrane sodium conductance is determined by the number of open sodium channels, sodium conductance measured with a voltage clamp would be expected to be exponential as well, given the assumption of a single gating particle leading to opening of the channel. This prediction, along with the actually observed kinetic behavior of g_{Na}, is diagrammed in Figure 7-10. Unlike the predicted exponential behavior, the rise in g_{Na} actually

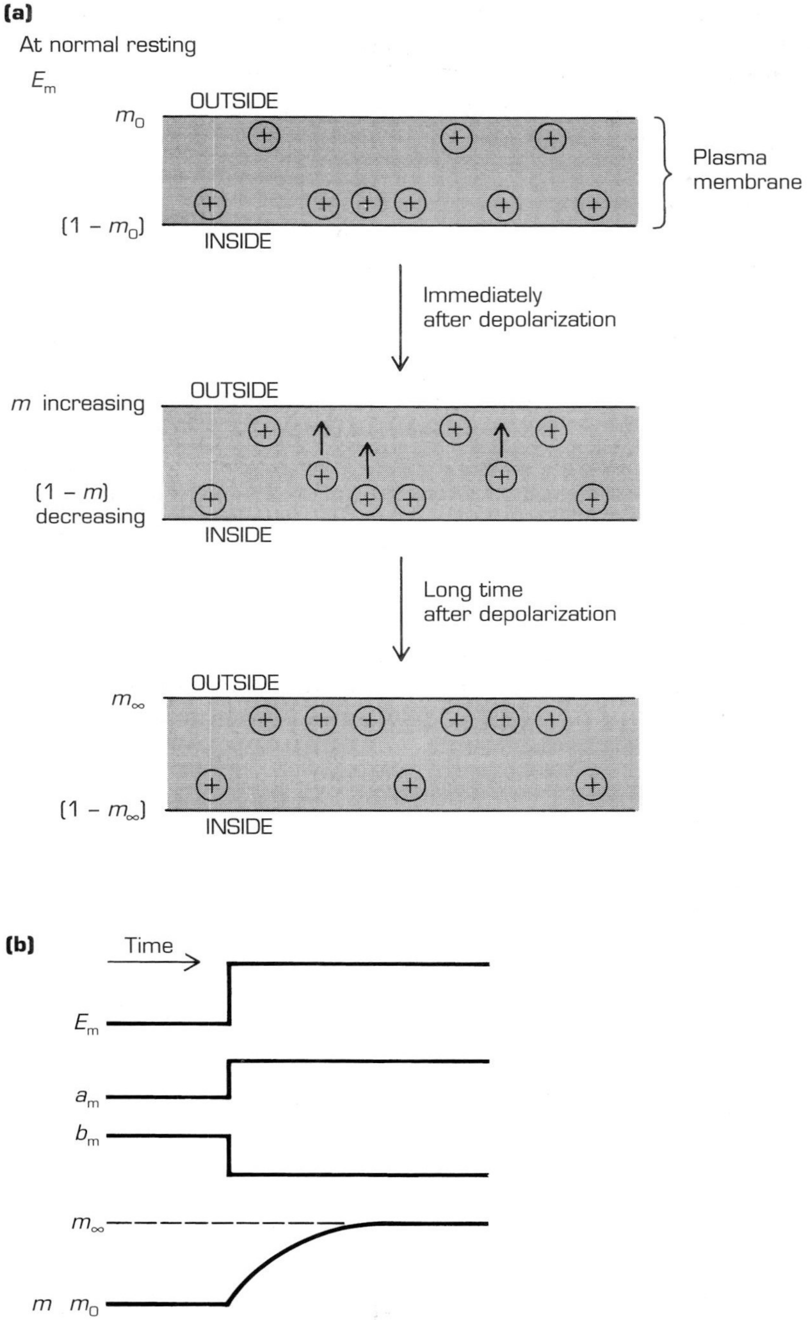

Figure 7-9 Change in the distribution of sodium channel gating particles after a depolarization of the membrane. (a) A schematic diagram of the distribution of charged gating particles at the normal resting potential and at different times after depolarization of the membrane. (b) The effect of a step change in membrane potential (top trace) on the rate constants for movement of the gating particles (middle traces) and on the proportion of particles on the outer side of the membrane (bottom trace).

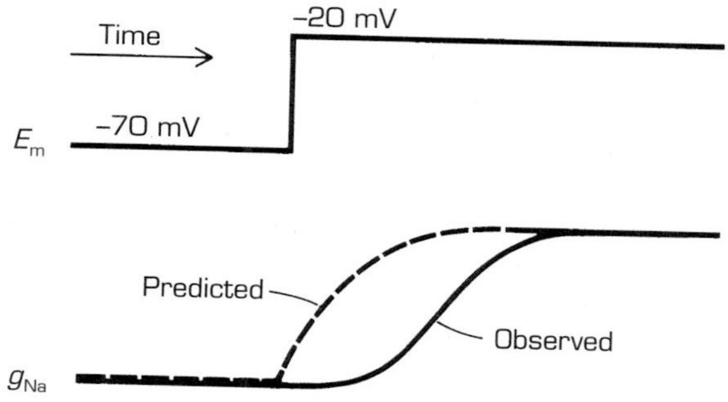

Figure 7-10 The predicted time-course of the change in sodium conductance following a depolarizing step (dashed line), assuming that the proportion of open channels—and hence the total sodium conductance—is directly related to the fraction of gating particles on the outer face of the membrane. The solid line shows the observed change in sodium conductance following a step depolarization.

exhibited a pronounced delay following the voltage step. The S-shaped increase in g_{Na} would be explained if more than one binding site must be occupied by gating particles before the channel will open. If the binding to each of several sites is independent, the probability that any one site is occupied will be proportional to m and will thus rise exponentially with time after a step voltage change, as discussed above. The probability that all of a number of sites will be occupied will be the product of the probabilities that each single site will bind a gating particle. That is, if there are two binding sites, the probability that both are occupied will be the product of the probability that site 1 binds a particle and the probability that site 2 binds a particle. Because each of these probabilities is proportional to m, the joint probability that both sites are occupied is proportional to m^2. Similarly, if there were x sites, the probability of channel opening would be proportional to m^x. The actual rise in sodium conductance following a depolarizing step suggested that $x = 3$ for the sodium channel: three binding sites must be occupied by gating particles before the channel will conduct. Thus, the turn-on of g_{Na} following a voltage-clamp step to a particular level of depolarization was proportional to m^3, and the temporal behavior of m was given by Equation (7-6).

A similar analysis was carried out for the change in potassium conductance following a step depolarization. The results suggested that $x = 4$ for the voltage-sensitive potassium channel of squid axon membrane. Thus, the gating charges for the potassium channel redistributed after a change in membrane potential according to a relation equivalent to Equation (7-5):

$$dn/dt = a_n(1 - n) - b_n n \tag{7-7}$$

By analogy with the sodium system, n is the proportion of potassium gating particles on the outside of the membrane, $1 - n$ is the proportion on the inner face of the membrane, and a_n and b_n are the rate constants for particle transition from one face to the other. Equation (7-7) has a solution equivalent to Equation (7-6):

$$n(t) = n_\infty - (n_\infty - n_0)\, e^{-(a_n + b_n)t} \qquad (7\text{-}8)$$

Here, n_0 and n_∞ are the initial and final values of n. The rise in potassium conductance following a step depolarization was found to be proportional to n^4; therefore, the potassium channel behaves as though four binding sites must be occupied by gating particles in order for the gate to open. A major difference between the potassium and the sodium channels is that the rate constants, a_n and b_n, are smaller for potassium channels. That is, the sodium channel gating particles appear to be more mobile than their potassium channel counterparts; this accounts for the greater speed of the sodium channel in opening after a depolarization, which we have seen is a crucial part of the action potential mechanism.

Sodium Inactivation

Recall that the change in sodium conductance following a maintained depolarizing step is transient. We have so far considered only the first part of that change: the increase in sodium conductance called sodium activation. We will now turn to the delayed decline in sodium conductance following depolarization. This delayed decline in conductance is called sodium inactivation. Following along in the vein used in the analysis of sodium and potassium channel opening, Hodgkin and Huxley assumed that sodium inactivation was caused by a voltage-sensitive gating mechanism. They supposed that the conducting state of the sodium channel was controlled by two gates: the activation gate whose opening we discussed above, and the inactivation gate. A diagram of this arrangement is shown in Figure 7-11. Like the activation gate, the inactivation gate is controlled by a charged gating particle; when the binding site on the gate is occupied, the inactivation gate is open. Unlike the activation gate, however, the inactivation gate is normally open and closes upon depolarization. If we keep the convention of the gating particle being positively charged, this behavior can be modeled by an arrangement with the inactivation gate and its binding site on the inner face of the membrane. Upon depolarization, the probability that a gating particle is on the inner face decreases, and so the probability that the gate closes will increase.

To study the voltage dependence of the sodium-inactivation process, Hodgkin and Huxley performed the type of experiment illustrated in Figure 7-12. They used a fixed depolarizing test step of a particular amplitude and measured the peak amplitude of the increase in sodium conductance that resulted from the test step. The test depolarization was preceded by a long-duration prepulse whose amplitude could be varied. As shown in Figure 7-12, they found that a depolarizing prepulse reduced the amplitude of the response to the test depolarization, while a hyperpolarizing prepulse increased the size of the test response. This implied that the depolarizing prepulses closed the inactivation gates of some portion of the sodium channels, so that those channels did not conduct even when the activation gates were opened by the

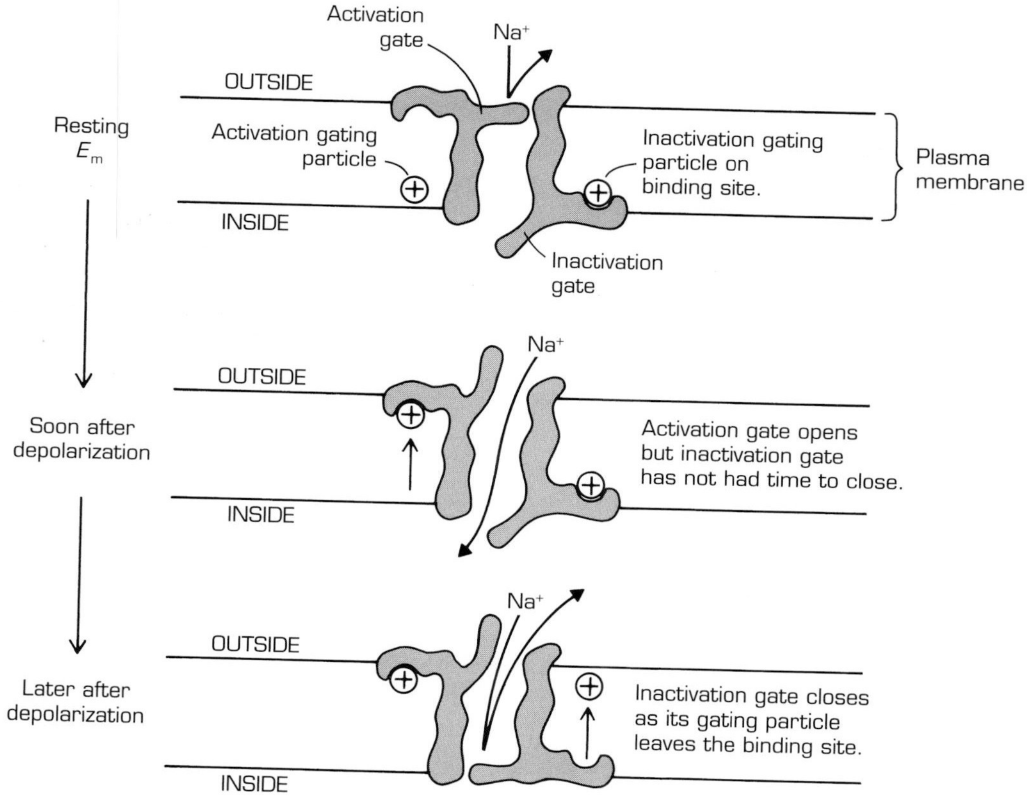

Figure 7-11 A diagram of the sodium channel protein, showing the gating particles for both the activation and the inactivation gates.

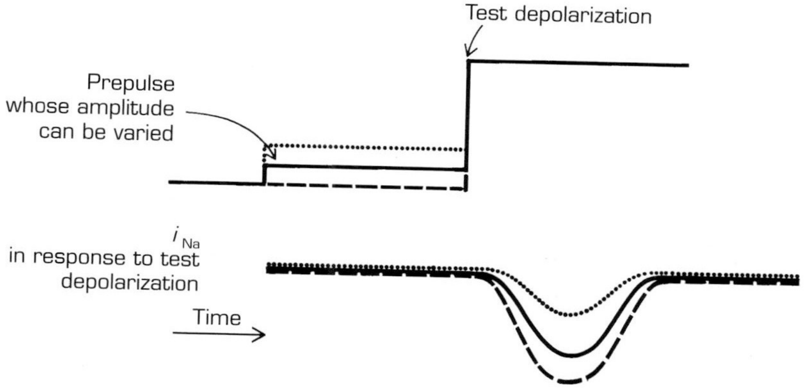

Figure 7-12 The procedure for measuring the voltage dependence of sodium channel inactivation.

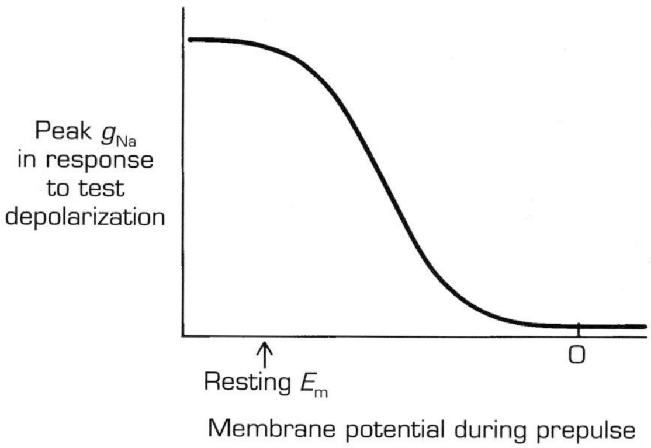

Figure 7-13 The relation between amplitude of an inactivating prepulse and the peak sodium conductance in response to a subsequent test depolarization.

subsequent depolarization; therefore, there was a smaller increase in sodium conductance during the test step. The finding that hyperpolarizing prepulses increased the test response suggests that the inactivation gates of some portion of the sodium channels are already closed at the normal resting potential; increasing E_m causes those gates to open, and the channels are then able to conduct in response to the test depolarization. By varying the amplitude of the prepulse, Hodgkin and Huxley were able to establish the dependence of the inactivation gate on membrane potential. The relation between E_m during the prepulse and the peak sodium conductance during the test depolarization is shown in Figure 7-13. Note that all the inactivation gates close when the membrane potential reaches about 0 mV, and that even a small depolarization can cause a significant reduction in the peak change in sodium conductance.

The time-course of sodium inactivation was studied by varying the duration of the prepulse, rather than its amplitude. With short prepulses, there was not much time for the inactivation gates to close, and the response to the test depolarization was only slightly reduced. With longer prepulses, there was a progressively larger effect. This relation between prepulse duration and peak sodium conductance during the test step is shown in Figure 7-14. It was found that the data were described by a single exponential equation, rather than the powers of exponentials that were necessary to describe the kinetics of sodium and potassium activation. Recall from the discussion of the voltage-dependent opening of the sodium channel that a single exponential is what would be expected if the state of the gate is controlled by a single gating particle. Thus, the closing of the inactivation gate seems to occur when a single particle comes off a single binding site on the gating mechanism. An equation analogous to Equations (7-6) and (7-8) can be written to describe the temporal behavior of the inactivation gate:

$$h(t) = h_\infty - (h_\infty - h_0)\, e^{-(a_h + b_h)t} \qquad (7\text{-}9)$$

In this case, however, the parameter h decreases with depolarization; that is, upon depolarization, h declines exponentially from its original value (h_0) to its

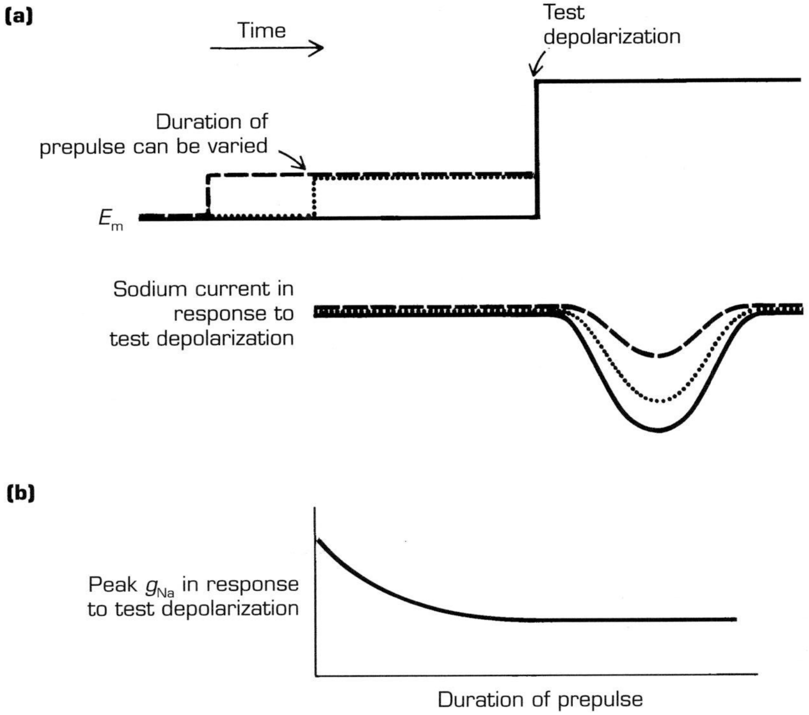

Figure 7-14 (a) The procedure for measuring the time-course of sodium channel inactivation by varying the duration of depolarizing prepulses. (b) The resulting exponential time-course of the closing of the inactivation gate of the sodium channel.

final value (h_∞). The rate of that decline is governed by the rate constants, a_h and b_h, for movement of the inactivation gating particle through the membrane. As expected from the discussion in Chapter 6, the closing of the inactivation gate is slower than the opening of the activation gate, implying that the inactivation gating particle is less mobile (i.e., the rate constants are smaller).

Is there any reason to suppose that the activation and inactivation gates are separate entities, as drawn in Figure 7-11 and throughout Chapter 6? After all, we could get the same behavior of the channel with a single gate that first opens, then closes upon depolarization. There is evidence, however, that the processes of activation and inactivation of the sodium channel are controlled by distinct and separable parts of the channel protein molecule. If, for example, we apply a proteolytic enzyme, such as trypsin or pronase, to the intracellular membrane face, we can selectively eliminate sodium channel inactivation while leaving activation intact. The sodium current observed in such an experiment is shown in Figure 7-15. As we have seen previously, in the normal situation the sodium current first increases, then decreases after a step depolarization as the channels open and then close with a delay (Figure 7-15a). After applying a protease to the internal face of the membrane (Figure 7-15b), the sodium current increases upon depolarization, as before, but now the current remains on for the duration of the depolarization: the inactivation gate has been

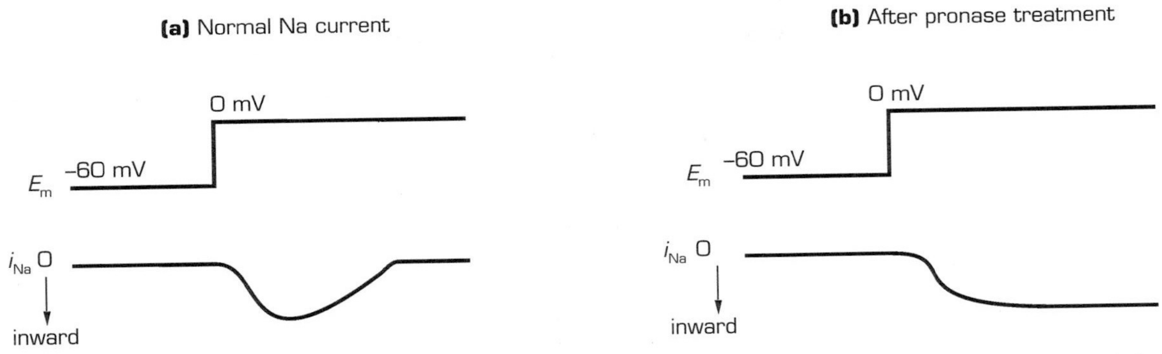

Figure 7-15 Removal of the inactivation gate by treating the inside of the membrane with a proteolytic enzyme, pronase. (a) Normal sodium current. The current rises (activates), then declines (inactivates) during a maintained depolarization. (b) Sodium current after pronase treatment. The current activates normally, but fails to inactivate during a maintained depolarization.

destroyed but activation is normal. This supports the idea that there are two separate gates controlling access to the sodium channel pore. It also suggests that the inactivation gate is on the intracellular part of the channel protein molecule, because the proteolytic enzyme is ineffective on the outside of the membrane.

The Temporal Behavior of Sodium and Potassium Conductance

The gating parameters m, n, and h specify the change in sodium and potassium conductance following a depolarizing voltage-clamp step. The sodium and potassium conductance is given by

$$g_{Na} = \bar{g}_{Na} m^3 h \qquad (7\text{-}10)$$

The potassium conductance is given by

$$g_K = \bar{g}_K n^4 \qquad (7\text{-}11)$$

where $\bar{g}_{Na}$ and $\bar{g}_K$ are the maximal sodium and potassium conductances, and m, n, and h are given by Equations (7-6), (7-8), and (7-9), respectively. Thus, following a depolarization, the sodium conductance rises in proportion to the third power of the activation parameter m and falls in direct proportion to the decline in the inactivation parameter, h. Figure 7-16a summarizes the responses of each gating parameter separately and also shows the product m^3h, which governs the time-course of the sodium conductance after depolarization. The potassium conductance rises as the fourth power of its activation parameter, n, and does not inactivate, as shown in Figure 7-16b. The names

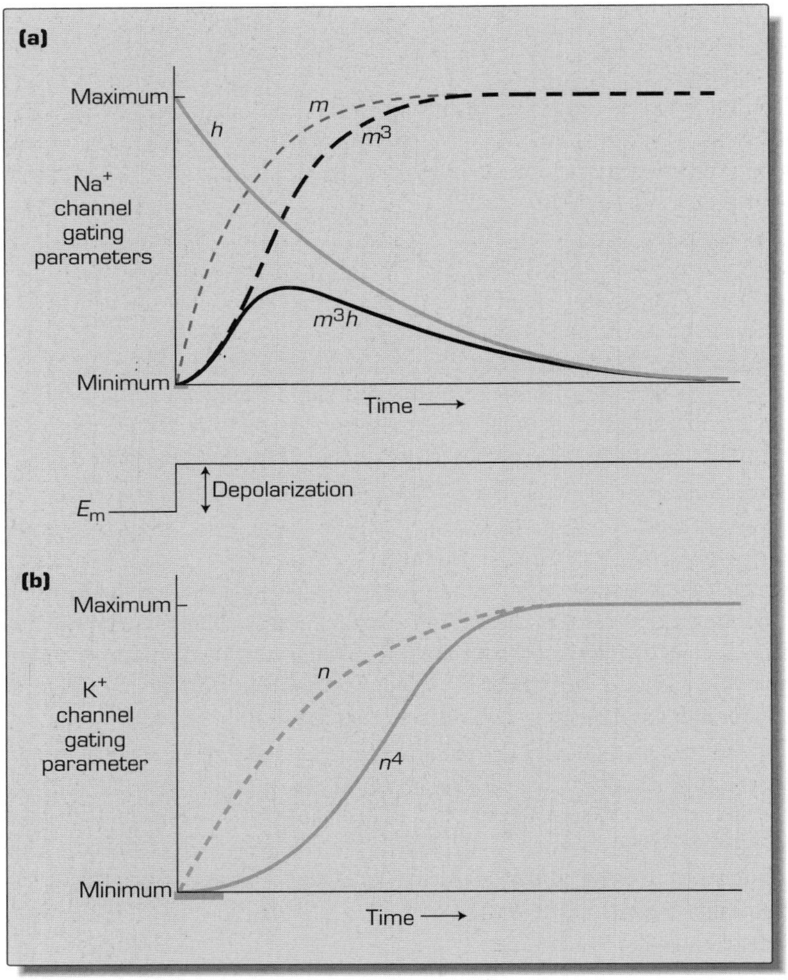

Figure 7-16 The time-courses of sodium conductance and potassium conductance following a step depolarization.
(a) Sodium conductance reflects the time-course of both inactivation (h) and activation (m). In the case of activation, channel opening is proportional to the third power of m. The rise and fall of sodium conductance is proportional to m^3h.
(b) The rise of potassium conductance is proportional to the fourth power of the activation parameter, n.

used in Chapter 6 for the various voltage-sensitive gates of the potassium and sodium channels derive from the variables chosen by Hodgkin and Huxley to represent these activation and inactivation parameters. The sodium activation gate is called the *m* gate, the sodium inactivation gate the *h* gate, and the potassium gate the *n* gate to reflect the roles of those parameters in Equations (7-10) and (7-11).

The surest test of a theory like the Hodgkin and Huxley theory of the action potential is to see if it can quantitatively describe the event it is supposed to explain. Hodgkin and Huxley tested their theory in this way by determining if they could quantitatively reconstruct the action potential of a squid giant axon using the system of equations they derived from their analysis of voltage-clamp data. Because the action potential does not occur under voltage-clamp conditions, this required knowing both the voltage dependence and the time dependence of a large number of parameters. This included knowing how the rate constants for all three gating particles and how the maximum values of h, m, and n depend on the membrane voltage. All of these parameters could be determined experimentally from a complete set of voltage-clamp experiments, allowing Hodgkin and Huxley to calculate the action potential that would occur if their axon were not voltage clamped. They then compared their calculated action potential with the action potential recorded from the same axon when the voltage-clamp apparatus was switched off. They found that the calculated action potential reproduced all the features of the real one in exquisite detail, confirming that they had covered all the relevant features of the nerve membrane involved in the generation of the action potential.

Gating Currents

Hodgkin and Huxley realized that their scheme for the gating of the sodium and potassium channel predicted that there should be an electrical current flow within the membrane associated with the movement of the charged gating particles. When a step change in membrane potential is made, the charged gating particles redistribute within the membrane; because the movement of charge through space is an electrical current (by definition), this redistribution of charges from one face of the membrane to the other should be measurable as a rapid component of membrane current in response to the voltage change. A current of this type flowing within a material is called a **displacement current**. The equipment available to Hodgkin and Huxley was inadequate to detect this small current, however. Almost 20 years later, Armstrong and Bezanilla managed to measure the displacement current associated with the movement of the gating particles.

The procedure for measuring the displacement currents, which have come to be called **gating currents** because of their presumed function in the membrane, is illustrated in Figure 7-17. The basic idea is to start by holding the membrane potential at a hyperpolarized level; this insures that all the gating particles are on the inner face of the membrane (assuming, once again, that the gating particles are positively charged). In addition, all the sodium and potassium currents through the channels are blocked by drugs, like tetrodotoxin and tetraethylammonium. A step is then made to a more hyperpolarized level, say 30 mV more negative. Because all the gating charges are already on the inner face of the membrane, no displacement current will flow as the result of this hyperpolarizing step. The only current flowing in this situation will be the rapid influx of negative charge necessary to step the voltage down. The voltage is then returned to the original hyperpolarized holding level, and a 30 mV depolarizing step is made. The influx of positive charge necessary to depolarize by 30 mV will be equal in magnitude, but opposite in sign, to the influx of negative charge necessary to make the previous 30 mV hyperpolarizing step. However, the depolarizing step will in addition cause some gating charges to move from the inner to the outer face of the membrane. Thus, there will be an extra

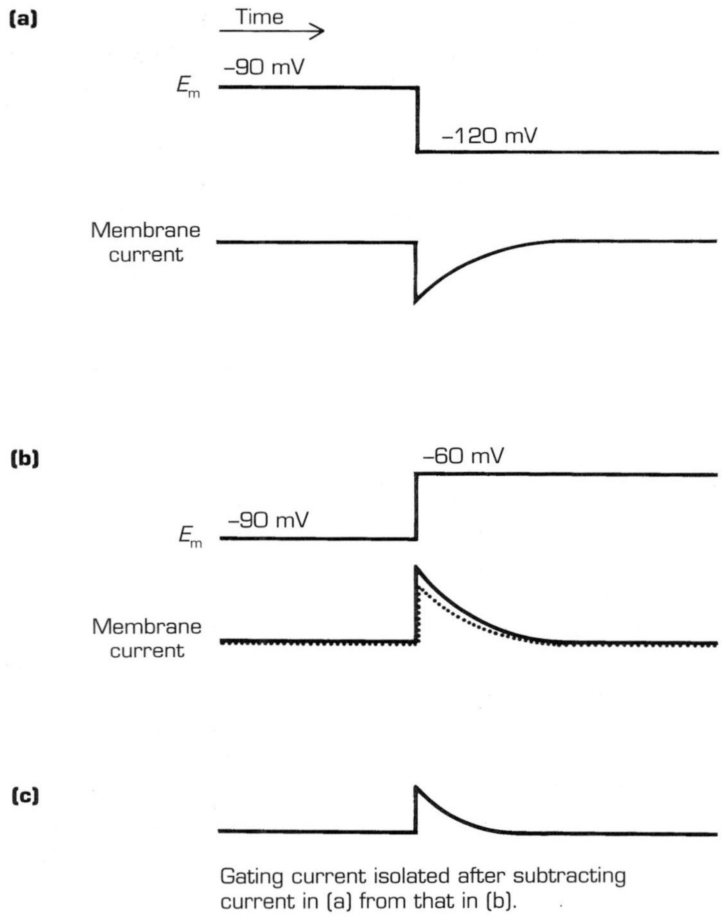

Figure 7-17 The procedure for isolating the gating current associated with the opening of voltage-sensitive sodium channels of an axon membrane. (a) Membrane voltage is stepped negative from a hyperpolarized level. With all ion channels blocked, the only current flowing is that required to move the membrane voltage more negative. (b) Membrane voltage is stepped positive from a hyperpolarized level. The current necessary to move the potential in the positive direction (dotted trace) will be the same amplitude, but opposite sign, as in (a). In addition, there will be an extra component of current in (b) caused by the movement of the charged gating particles in response to the depolarization. This component is seen in (c) on an expanded vertical scale.

component of current, due to the movement of gating charges, in response to the depolarizing step. By subtracting the current in response to the hyperpolarizing step from the depolarizing current, this extra gating current can be isolated. Experiments on this gating current suggest that it has the right voltage dependence and other properties to indeed represent the charge displacement underlying the gating scheme suggested by Hodgkin and Huxley. This is an important piece of evidence validating a basic feature of Hodgkin and Huxley's model of the membrane of excitable cells.

Summary

Hodgkin and Huxley made the fundamental observations on which our current understanding of the ionic basis of the action potential is based. In their experiments, they measured the ionic currents flowing across the membrane of a squid giant axon in response to changes in membrane voltage. This was done using the voltage-clamp apparatus, which provides a means of holding membrane potential constant in the face of changes in the ionic conductance of the axon membrane. By analyzing these ionic currents, Hodgkin and Huxley derived equations specifying both the voltage dependence and the time-course of changes in sodium and potassium conductance of the membrane. During a maintained depolarization, the sodium conductance increased rapidly, then declined, while potassium conductance showed a delayed but maintained increase. Analysis of the change in sodium conductance suggested that the conducting state of the sodium channel was controlled by a rapidly opening activation gate, called the *m* gate, and a slowly closing inactivation gate, called the *h* gate. The gates behave as though they are controlled by charged gating particles that move within the plasma membrane; when the gating particles occupy binding sites associated with the channel gating mechanism, the gates open. The kinetics of the observed gating behavior would be explained by the kinetics of the redistribution of the charged gating particles within the membrane following a step change in the transmembrane potential. The sodium activation gate appears to open when three independent binding sites are occupied by gating particles, while the inactivation gate closes when a single particle leaves a single binding site. The potassium channel is controlled by a single gate, the *n* gate, which opens when four binding sites are occupied. The rate at which the gating particles redistribute following a depolarization is different for the three types of gate, with sodium-activation gating being faster than sodium-inactivation or potassium-activation gating. Tiny membrane currents associated with the movement of the charged gating particles within the membrane have been detected. Experiments combining molecular biology with electrical measurements promise to establish the correspondence between Hodgkin and Huxley's gating mechanisms and actual parts of the ion-channel protein molecule.

8 Synaptic Transmission at the Neuromuscular Junction

Chapter 6 was concerned with the ionic basis of the action potential, the electrical signal that carries messages long distances along nerve fibers. Using the patellar reflex as an example, we discussed the mechanism that allows the message that the muscle was stretched to travel along the membrane of the sensory neuron from the sensory endings in the muscle to the termination of the sensory fiber in the spinal cord. After the message is passed to the motor neuron within the spinal cord, action potentials also carry the electrical signal back down the nerve to the muscle, to activate the reflexive contraction of the muscle. This chapter will be concerned with the mechanism by which action potential activity in the motor neuron can be passed along to the cells of the muscle, causing the muscle cells to contract. In Chapter 9, we will consider how action potentials in the sensory neuron influence the activity of the motor neuron in the spinal cord.

Chemical and Electrical Synapses

The point where activity is transmitted from one nerve cell to another or from a motor neuron to a muscle cell is called a **synapse**. In the patellar reflex, there are two synapses: one between the sensory neuron and the motor neuron in the spinal cord, and another between the motor neuron and the cells of the quadriceps muscle. There are two general classes of synapse: electrical synapses and chemical synapses. In both types, special membrane structures exist at the point where the input cell (called the **presynaptic cell**) comes into contact with the output cell (called the **postsynaptic cell**).

At a chemical synapse, an action potential in the presynaptic cell causes it to release a chemical substance (called a **neurotransmitter**), which diffuses through the extracellular space and changes the membrane potential of the postsynaptic cell. At an electrical synapse, a change in membrane potential (such as the depolarization during an action potential) in the presynaptic cell spreads directly to the postsynaptic cell without the action of an intermediary chemical. Both synapses in the patellar reflex, are chemical synapses. At a chemical synapse, the membranes of the presynaptic and postsynaptic cells come close to each other but are still separated by a small gap of extracellular space. At an electrical synapse, the presynaptic and postsynaptic membranes touch and the cell interiors are directly interconnected by means of special ion channels called gap junctions that allow flow of electrical current from one cell to another. We will concentrate in this chapter on chemical synaptic transmission. Electrical synaptic transmission will be described in more detail in Chapter 12.

The Neuromuscular Junction as a Model Chemical Synapse

The best understood chemical synapse is that between a motor neuron and a muscle cell. This synapse is given the special name **neuromuscular junction** (also sometimes called the **myoneural junction**). Although the fine details may differ somewhat, the basic scheme that describes the neuromuscular junction applies to all chemical synapses. Therefore, this chapter will concentrate on the characteristics of this special synapse at the output end of the patellar reflex. In the next chapter, we will consider some of the differences between the synapse at the neuromuscular junction and synapses in the central nervous system, such as the synapse between the sensory neuron and motor neuron in the spinal cord in the patellar reflex.

Transmission at a Chemical Synapse

The sequence of events during neuromuscular synaptic transmission is summarized in Figure 8-1. When an action potential arrives at the end of the motor neuron nerve fiber, it invades a specialized structure called the **synaptic terminal**. Depolarization of the synaptic terminal induces release of a chemical messenger, which is stored inside the terminal. At the vertebrate neuromuscular junction, this chemical messenger is **acetylcholine**; the chemical structure of acetylcholine (abbreviated ACh) is shown in Figure 8-2. The ACh diffuses across the space separating the presynaptic motor neuron terminal from the postsynaptic muscle cell and alters the ionic permeability of the muscle cell. This change in ionic permeability then depolarizes the muscle cell membrane. The remainder of this chapter will be concerned with a detailed description of this basic sequence of events.

Presynaptic Action Potential and Acetylcholine Release

The trigger for ACh release is an action potential in the synaptic terminal. The key aspect of the action potential is that it depolarizes the synaptic terminal,

1. Presynaptic action potential

↓

2. Depolarization of synaptic terminal

↓

3. Release of chemical neurotransmitter molecules

↓

4. Neurotransmitter molecules bind to special receptors on postsynaptic cell

↓

5. Change in ionic permeability of postsynaptic cell

↓

6. Change in membrane potential of postsynaptic cell

Figure 8-1 The sequence of events during transmission at a chemical synapse.

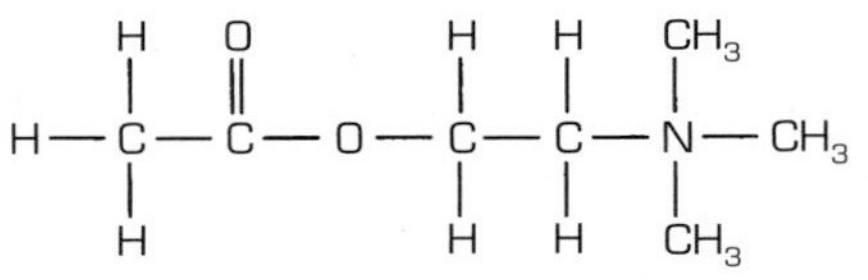

Figure 8-2 The chemical structure of acetylcholine (ACh), the chemical neurotransmitter at the neuromuscular junction.

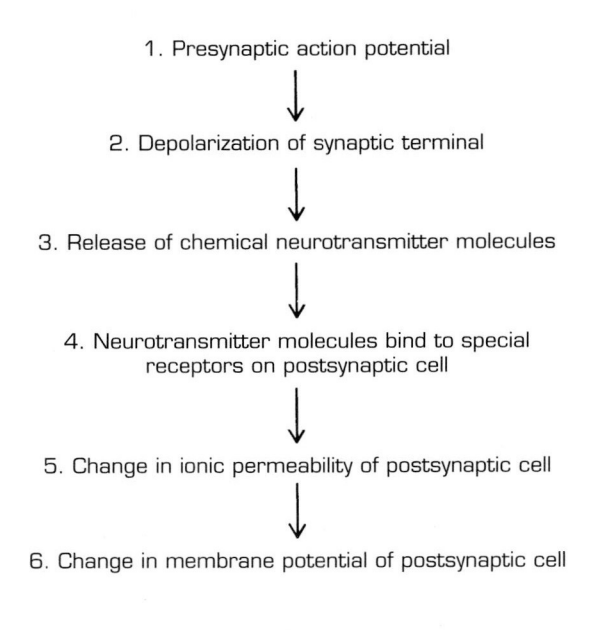

and any stimulus that depolarizes the synaptic terminal causes ACh to be released. The coupling between depolarization and release is not direct, however. The signal that mediates this coupling is the influx into the synaptic terminal of an ion in the ECF that we have largely ignored to this point—calcium ions.

Calcium is present at a low concentration in the ECF (1–2 m*M*) and is not important in resting membrane potentials or in most nerve action potentials, although some action potentials have a contribution from calcium influx (see Chapter 6). However, calcium ions must be present in the ECF in order for release of chemical neurotransmitter to occur. If calcium ions are removed from the ECF, depolarization of the synaptic terminal can no longer induce release of ACh. Depolarization causes external calcium ions to enter the synaptic terminal, and the calcium in turn causes ACh to be released from the terminal.

What mechanism provides the link between depolarization of the terminal and influx of calcium ions? As we've seen in earlier chapters, ions cross membranes through specialized transmembrane channels, and calcium ions are no different in this regard. The membrane of the synaptic terminal contains calcium channels that are closed as long as E_m is near its normal resting level. These channels are similar in behavior to the voltage-dependent potassium

1. Presynaptic action potential

↓

2. Depolarization of synaptic terminal

↓

3. Voltage-sensitive calcium channels open

↓

4. Calcium enters synaptic terminal

↓

5. Release of chemical neurotransmitter

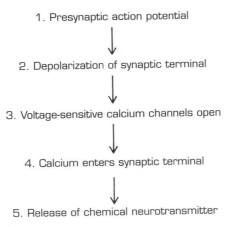

Figure 8-3 The sequence of events between the arrival of an action potential at a synaptic terminal and the release of chemical transmitter.

channels of nerve membrane; they open upon depolarization and close again when the membrane potential repolarizes. Thus, when an action potential invades the synaptic terminal, the calcium permeability of the membrane increases during the depolarizing portion of the action potential and declines again as membrane potential returns to normal.

Although the external calcium concentration is low (1–2 mM), the internal concentration of calcium ions in the ICF is much lower ($<10^{-6}$ M). From the Nernst equation, then, the equilibrium potential for calcium would be expected to be positive. Therefore, both the concentration and electrical gradients drive calcium into the terminal, and when calcium permeability increases, there will be an influx of calcium. During a presynaptic action potential, there is a spike of calcium entry into the terminal, resulting in release of neurotransmitter into the extracellular space. This sequence is summarized in Figure 8-3.

Effect of Acetylcholine on the Muscle Cell

The goal of synaptic transmission at the neuromuscular junction is to cause the muscle cell to contract. Acetylcholine released from the synaptic terminal accomplishes this goal by depolarizing the muscle cell. Because muscle cells are excitable cells like neurons, this depolarization will set in motion an all-or-none, propagating action potential if the depolarization exceeds threshold. The coupling between the muscle action potential and contraction will be the subject of Chapter 10. This section will discuss the effect of ACh on the muscle cell membrane, leading to depolarization.

The region of muscle membrane where synaptic contact is made is called the **end-plate** region, and it possesses special characteristics. In particular, the end-plate membrane is rich in a transmembrane protein that acts as an ion channel. Unlike the voltage-dependent channels discussed in Chapter 6, however, this channel is little affected by membrane potential. Instead, this channel is sensitive to ACh: it opens when it binds ACh. Thus, ACh released from the

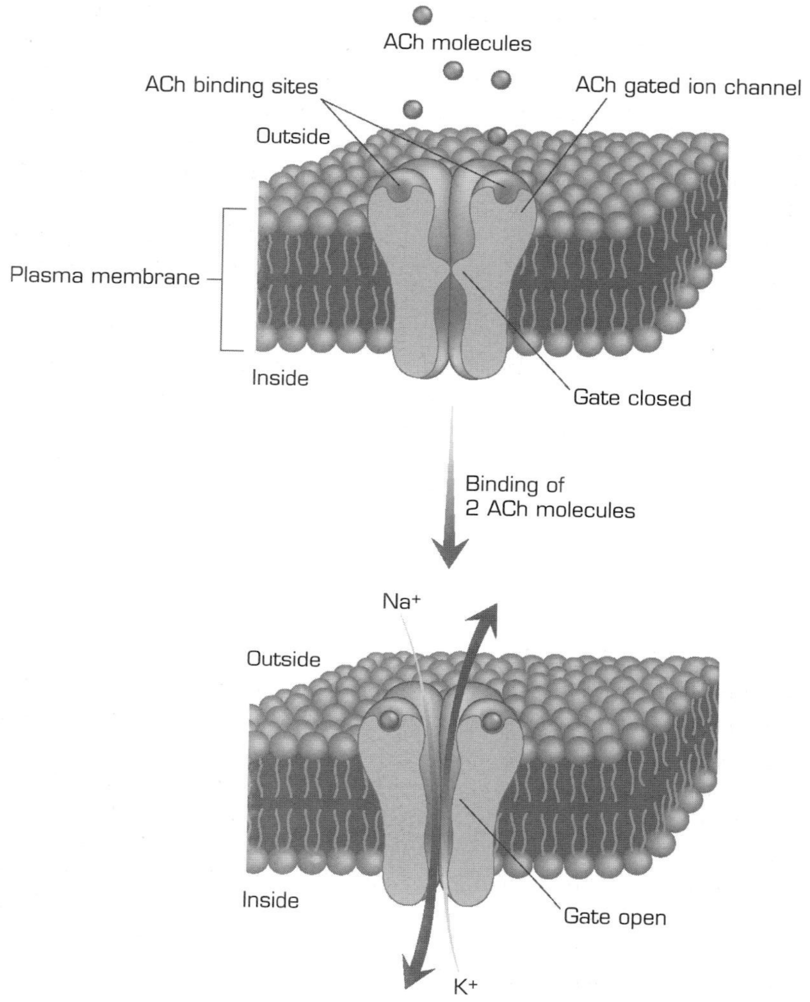

Figure 8-4 Schematic representation of the behavior of the ACh-sensitive channel in the end-plate membrane. The binding of two molecules of ACh to sites on the channel opens the gate, allowing sodium and potassium ions to flow through the channel. (Animation available at www.blackwellscience.com)

synaptic terminal diffuses across the synaptic cleft to the muscle membrane, where it combines with specific receptor sites associated with the ion channel. As shown schematically in Figure 8-4, the gate on the channel is closed in the absence of ACh. When the receptor sites are occupied, however, the gate opens, and the channel allows ions across the membrane. Two ACh molecules must bind to the channel in order for the gate to open (Figure 8-4). The ACh-binding site is highly specific; only ACh or a small number of structurally related compounds can bind to the site and cause the channel to open.

The ACh-activated channel of the muscle end-plate allows both sodium and potassium to cross the membrane about equally well. Thus, when ACh is present, the membrane permeability to both sodium and potassium increases. How can such a permeability increase produce a depolarization of the muscle cell? To see this, consider the situation diagrammed in Figure 8-5. Recall from Chapter 5 that membrane potential depends on the relative sodium and potassium permeabilities of the membrane (the Goldman equation). For the cell of Figure 8-5, p_{Na}/p_K is 0.02 at rest and E_m would be about −74 mV, assuming typical ECF and ICF (Table 2-1). In the presence of ACh, however, p_{Na} and p_K increase by equal amounts; p_{Na}/p_K increases to 0.51 and E_m depolarizes to about −17 mV.

The ACh-activated channels are packed densely in the end-plate region of the muscle, as illustrated in Figure 8-6a. The membrane is studded with ring-shaped particles that are found only at the region of synaptic contact. These particles have been biochemically isolated from the postsynaptic membrane and identified as the ACh-binding receptor molecule and its associated channel. The isolated receptor/channel complex can be inserted into artificial membranes, where they retain their function and their appearance through the electron microscope (Figure 8-6b). The hole in the middle of each particle is probably the aqueous pore through which the sodium and potassium ions cross the membrane.

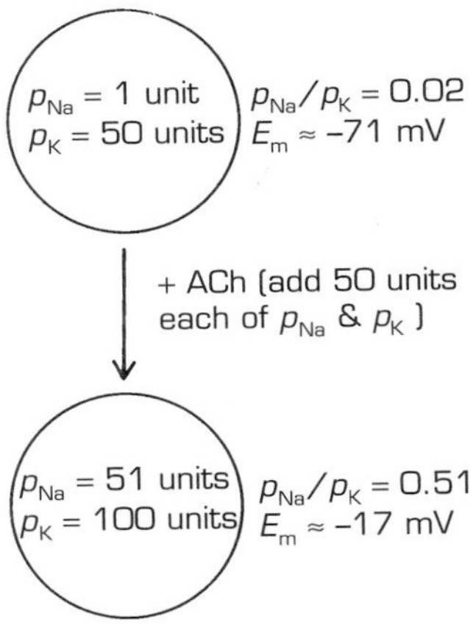

Figure 8-5 Opening a channel that allows both potassium and sodium to cross the membrane results in a higher value for p_{Na}/p_K and causes depolarization.

Neurotransmitter Release

We now return to the synaptic terminal for a more detailed examination of the mechanism of neurotransmitter release. Acetylcholine is released from the motor nerve terminal in quanta consisting of many molecules. Thus, the basic unit of release is not a single molecule of ACh, but the quantum. At the neuromuscular junction, it is estimated that a single quantum of ACh contains about 10,000 molecules. An individual quantum is either released all together or not released at all. The release of ACh during neuromuscular transmission can be thought of as the sudden appearance of a "puff" of ACh molecules in the extracellular space as the entire contents of a quantum is released. A single presynaptic action potential normally causes the release of more than a hundred quanta from the synaptic terminal.

The original suggestion that ACh is released in multimolecular quanta was made on the basis of a statistical analysis of the response of the postsynaptic muscle cell to action potentials in the presynaptic motor neuron. This analysis was first carried out by P. Fatt and B. Katz, and it initiated a series of studies by Katz and coworkers that gave rise to the basic scheme for chemical neurotransmission presented in this chapter. Experimentally, the analysis was accomplished by reducing the extracellular calcium concentration to the point where the influx of calcium ions into the synaptic terminal during an action potential was much less than usual. Under these conditions, a single presynaptic action potential released on average only one or two quanta of ACh instead of more than a hundred. Examples of end-plate potentials recorded in a muscle cell in response to a series of presynaptic action potentials are shown in Figure 8-7. Because only a small number of quanta are released per action potential, the

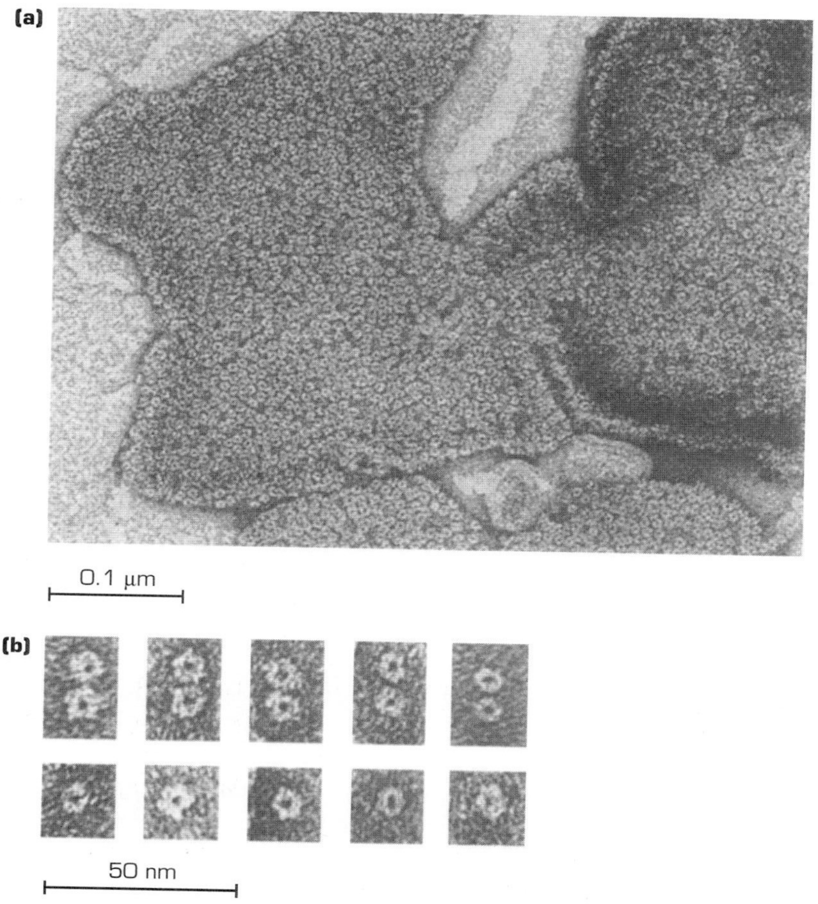

Figure 8-6 (a) A view through the electron microscope at the face of the postsynaptic membrane of the electric organ of the electrical skate, *Torpedo*. This organ, which is a rich source of ACh receptors for biochemical study, is a specialized type of muscle tissue. The membrane particles are the ACh-activated channels of the postsynaptic membrane. (b) Several views of individual ACh receptors that have been chemically isolated from preparations like that in (a), then placed in artificial membranes. (Courtesy of J. Cartaud of the Institut Jacques Monod, CNRS/ Universitè Paris 7, France.)

end-plate potentials in the reduced calcium ECF are much smaller than usual and do not reach threshold for generating an action potential in the muscle cell. Notice that the amplitude of the depolarization of the muscle cell fluctuates considerably over the series of presynaptic action potentials: sometimes there was a large response and other times there was no response at all. Fatt and Katz measured a large number of such responses and found that the amplitudes clustered around particular values that were integral multiples of the smallest observed response. For example, as shown in Figure 8-7b, there might be a cluster of responses that were 1 mV in amplitude, another cluster at 2 mV, and another at 3 mV. This indicates that the response was quantized in irreducible units of 1 mV, and that the presynaptic action potential released ACh in corresponding quantal units. Thus, a given presynaptic action potential might release three, two, one, or no quanta, but not 0.5 or 1.5 quanta.

Fatt and Katz also observed occasional, small depolarizations that occurred in the absence of any presynaptic action potentials. These spontaneous

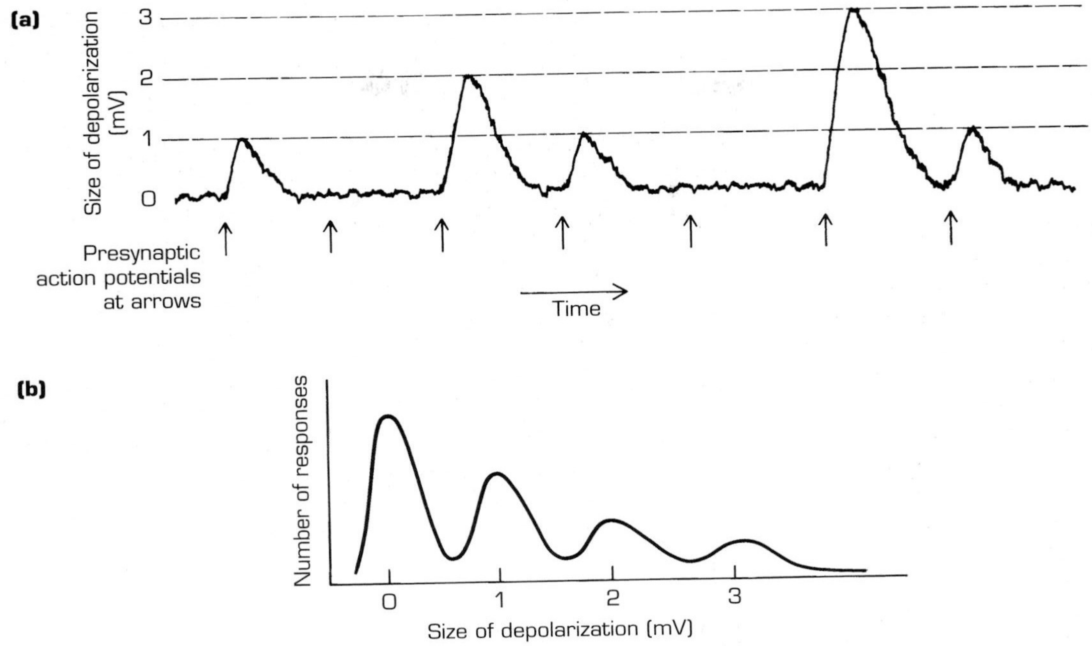

Figure 8-7 Quantized responses of muscle cell to action potentials in the presynaptic motor neuron. Arrows give timing of the presynaptic action potentials. (b) The graph shows the peak response amplitudes recorded in response to a series of several hundred presynaptic action potentials like those shown in (a).

depolarizations had approximately the same amplitude as the single quantum response produced by presynaptic action potentials in low-calcium ECF. That is, if the irreducible unit of evoked muscle depolarization was 1 mV, then the spontaneous events also were about 1 mV in amplitude. Figure 8-8 shows several of these spontaneous depolarizations recorded inside a muscle cell. These events are called miniature end-plate potentials, and they are assumed to result from spontaneous release of single quanta of ACh from the synaptic terminal. Under normal conditions, these spontaneous events occur at a low rate—about 1 or 2 per second; however, any manipulation that depolarizes the nerve terminal increases their rate of occurrence, confirming that their source is the process that couples depolarization to quantal ACh release during the normal functioning of the nerve terminal.

The Vesicle Hypothesis of Quantal Transmitter Release

To understand the basis of the packaging of ACh in quanta, it is necessary to look at the structure of the synaptic terminal, which is shown schematically in Figure 8-9. The terminal contains a large number of tiny, membrane-bound structures called **synaptic vesicles**. These vesicles contain ACh, and it is

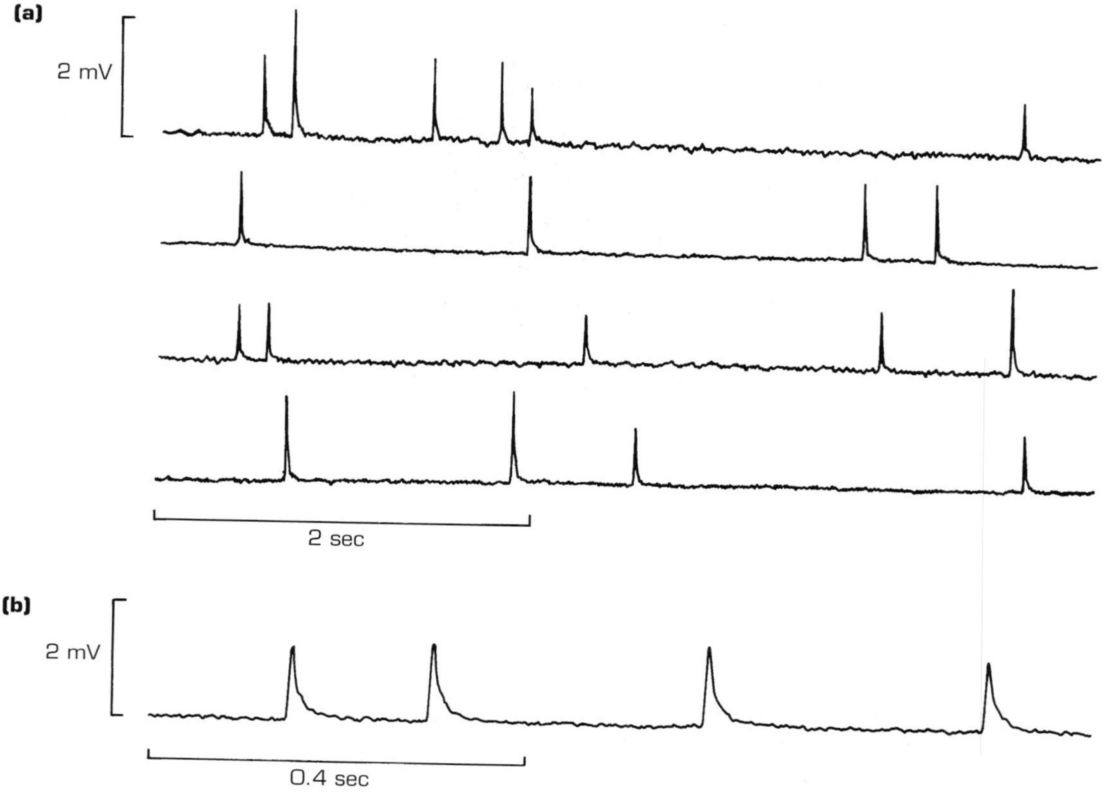

Figure 8-8 Spontaneous miniature end-plate potentials recorded from the end-plate region of a muscle cell. These randomly occurring small depolarizations of the muscle cell are caused by spontaneous release of single quanta of ACh from the synaptic terminal of the motor neuron. (a) Four 5-sec samples of muscle cell E_m, measured via an intracellular microelectrode. The spontaneous depolarizations occur at a rate of approximately one per second. (b) Spontaneous miniature end-plate potentials viewed on an expanded time-scale to show the shape of the events more clearly.

natural to assume that they represent the packets of ACh that are released in response to a presynaptic action potential. Indeed, these vesicles are depleted by any manipulation, such as prolonged depolarization or firing of large numbers of action potentials, which causes release of large amounts of ACh. It is now generally accepted that release of ACh is accomplished by the fusion of the vesicle membrane with the plasma membrane of the terminal, so that the contents of the vesicle are dumped into the extracellular space between the terminal and the muscle cell. The vesicles do not fuse with the plasma membrane just anywhere; rather, they apparently fuse only at specialized membrane regions, called **release sites** or **active zones**, that are found only on the membrane face opposite the postsynaptic muscle cell. Thus, quanta of ACh are released only into the narrow space, the **synaptic cleft**, separating the pre- and postsynaptic cells. With freeze-fracture electron microscopy, the active zone of

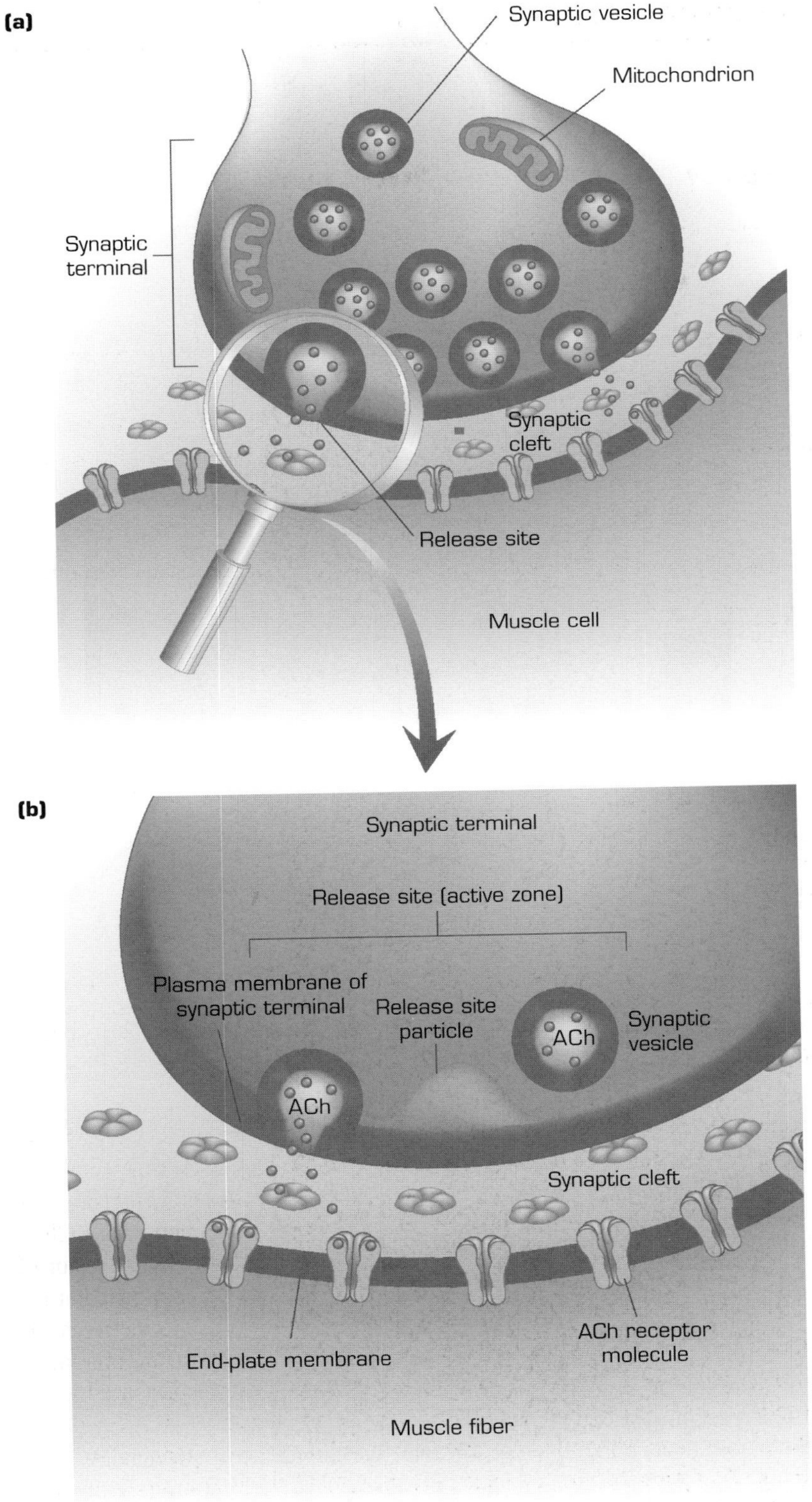

Figure 8-9 A schematic diagram of synaptic vesicles fusing with the plasma membrane to release ACh at the neuromuscular junction. Release occurs at specialized active zones in the presynaptic terminal. (Animation available at www.blackwellscience.com)

the presynaptic terminal appears as a double row of large membrane particles, which are probably membrane proteins involved in the fusion between the membrane of the synaptic vesicle and the presynaptic plasma membrane. Examples of these active zone particles can be seen in Figure 8-10.

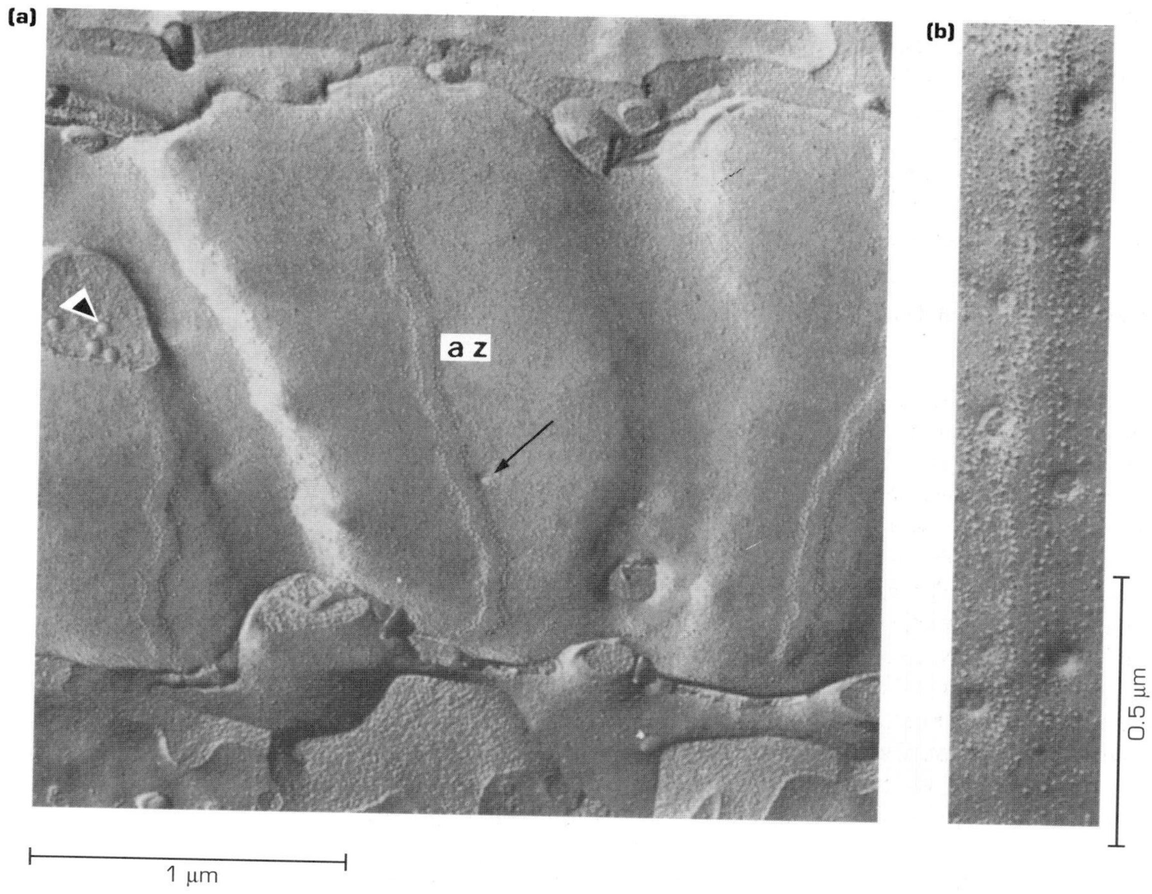

Figure 8-10 Electronmicrographs of the freeze-fractured face of a presynaptic terminal at the neuromuscular junction. (a) An unstimulated nerve terminal. Note the double row of particles defining a presynaptic release site or active zone (az). The arrow points to what appears to be a synaptic vesicle spontaneously fusing with the presynaptic membrane. Such spontaneous fusions presumably underlie the spontaneous miniature end-plate potentials shown in Figure 8-8. The arrowhead at the left points to a synaptic vesicle visible in a region where the membrane fractured all the way through to reveal a portion of the intracellular fluid. (b) A higher-power view of an active zone of a nerve terminal frozen during release of ACh stimulated by presynaptic action potentials. The ice-filled depressions arrayed along either side of the active zone correspond to regions where synaptic vesicles are in the process of fusing with the presynaptic membrane. (Reproduced from C.-P. Ko, Regeneration of the active zone at the frog neuromuscular junction. *Journal of Cell Biology* 1984;98:1685–1695; by copyright permission of the Rockefeller University Press.)

Anatomical evidence supporting vesicle exocytosis as the mechanism of ACh release was provided by freeze-fracture electron microscopy. In these experiments, a muscle and its attached nerve were placed in an apparatus that could very rapidly freeze the nerve and muscle. Then, the release process was literally frozen at the instant just after arrival of an action potential in the synaptic terminal, when ACh was being released. At this stage of transmission, synaptic vesicles can be seen in the process of fusing with the plasma membrane, as shown in Figure 8-10. The fusing vesicles appear as ice-filled pits or depressions in the presynaptic membrane, lined up along the presynaptic release sites. Fusing vesicles were observed only when ACh release should have been occurring, not before or after the action potential in the terminal. Further, the fusion occurred only when calcium was present in the ECF, which we have seen is prerequisite for release to occur.

Mechanism of Vesicle Fusion

The fusion of vesicle membrane with the plasma membrane is not a unique feature of synaptic transmission. Many other cellular processes require the fusion of intracellular vesicles with the plasma membrane. For instance, plasma membrane proteins are synthesized intracellularly within the Golgi apparatus and are then conveyed to their target sites by transport vesicles, which must then fuse with the plasma membrane to deliver their cargo. Also, secretion of substances to the extracellular space frequently occurs via exocytosis. The molecular mechanism of synaptic vesicle exocytosis shares common features with other forms of exocytosis. However, the requirement for rapid triggering of exocytosis in response to Ca^{2+} influx sets synaptic vesicle exocytosis apart from other forms of exocytosis. The delay time between a presynaptic action potential and the first appearance of the postsynaptic response is <0.5 msec. Therefore, there is little time for complex, multistage processes to prepare vesicles for membrane fusion. For this reason, vesicles must be placed very near the membrane at the active zone (Figure 8-10), ready for fusion when Ca^{2+} enters during an action potential.

Three membrane proteins that play a central role in synaptic vesicle fusion are **synaptobrevin**, which is associated with the vesicle membrane, and two plasma membrane proteins, **syntaxin** and **SNAP-25**. These proteins bind to each other to form the **core complex**, which brings the vesicle in close proximity to the plasma membrane, as shown in Figure 8-11. Formation of the core complex is required for neurotransmitter release. It is not yet clear, however, whether the core complex is directly involved in fusion or plays a vital role in preparing vesicles for fusion, a process called **priming**. Energy to prime vesicles for fusion is provided by hydrolysis of ATP, which is carried out by an ATPase called **NSF** that interacts with proteins of the core complex.

In other forms of exocytosis, fusion follows immediately after priming. Primed synaptic vesicles, however, must be prevented from fusing until influx of Ca^{2+} triggers the process. Therefore, the molecular machinery of fusion

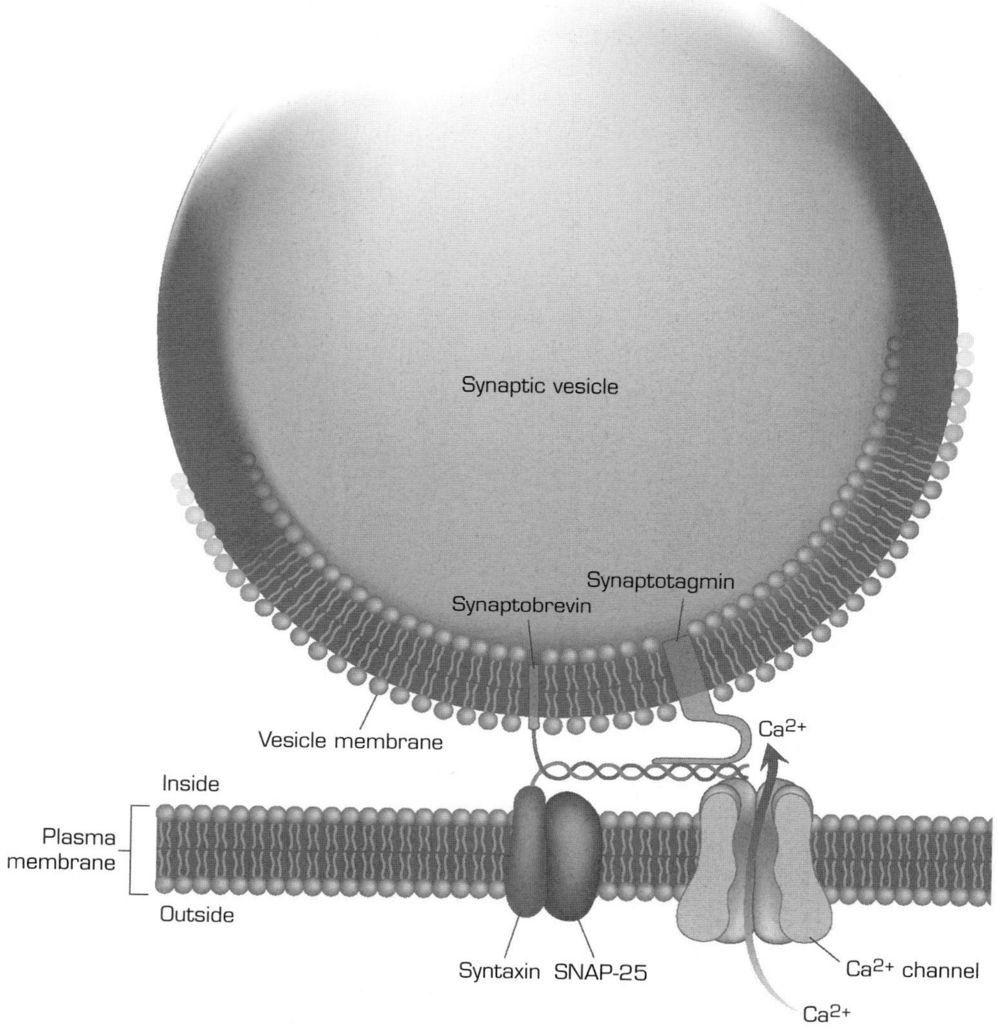

Figure 8-11 Proteins of the synaptic vesicle and the plasma membrane participate in synaptic vesicle exocytosis at the active zone in the presynaptic terminal.

requires a brake, which is removed when Ca^{2+} enters during an action potential. This role is carried out by **synaptotagmin**, a protein associated with the synaptic vesicle (Figure 8-11). Synaptotagmin includes two binding sites for Ca^{2+} and also interacts with the proteins of the core complex. This interaction prevents fusion from proceeding until calcium ions bind to synaptotagmin. If the gene for synaptotagmin is knocked out by genetic manipulation, rapid coupling between calcium influx and neurotransmitter release is lost.

The final component of the complex of proteins that regulate calcium-dependent fusion of synaptic vesicles is the calcium channel itself. Voltage-dependent Ca^{2+} channels of the synaptic terminal directly bind to syntaxin, which is part of the core complex. Thus, the source of the calcium ions that trigger neurotransmitter release is held in close proximity to the calcium sensor molecule (synaptotagmin) and the rest of the fusion machinery.

Recycling of Vesicle Membrane

If the membranes of synaptic vesicles fuse with the plasma membrane of the terminal during transmitter release, we might expect the area of the terminal membrane to increase with use. Indeed, over the life span of an animal, millions of synaptic vesicles might fuse with the terminal membrane, so that the terminal might become huge. However, this does not happen because the vesicle membrane does not remain part of the plasma membrane; instead, the fused vesicles are recycled. The scheme is summarized in Figure 8-12. After fusion, the vesicles pinch off again from the plasma membrane, are refilled with ACh, and are ready to be used again to transfer neurotransmitter into the synaptic cleft.

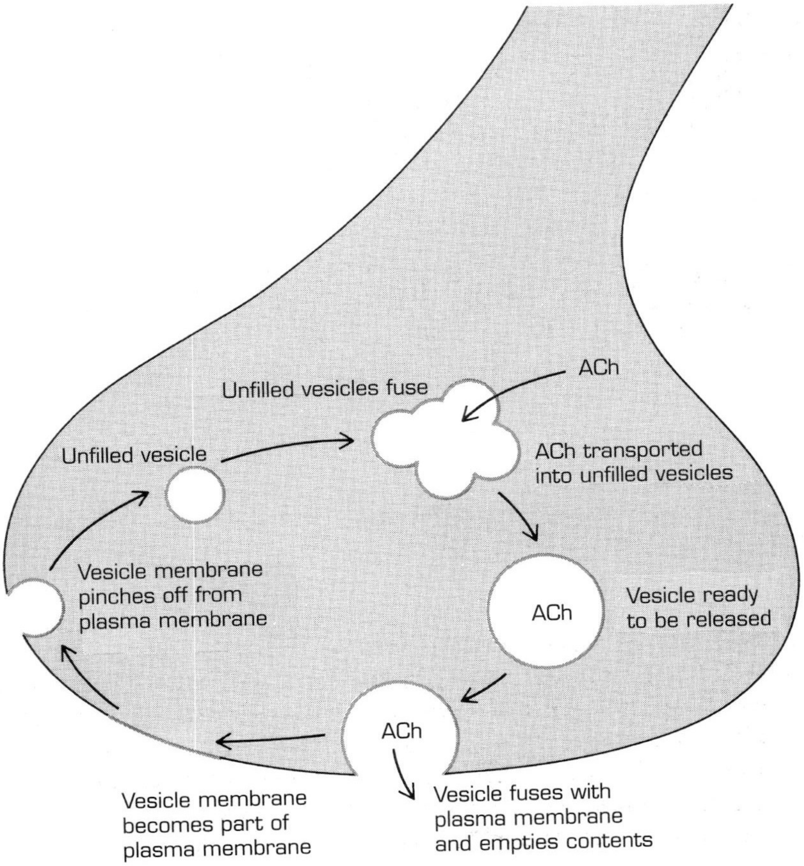

Figure 8-12 The recycling of vesicle membrane in the presynaptic terminal at the neuromuscular junction.

Inactivation of Released Acetylcholine

We have seen how ACh is released from the synaptic terminal and how it depolarizes the postsynaptic muscle cell. How is the action of ACh terminated so that the end-plate region returns to its resting state? The answer is that there is another specialized ACh-binding protein in the end-plate region. This protein is the enzyme acetylcholinesterase, which splits ACh into acetate and choline. Because neither acetate nor choline can bind to and activate the ACh-activated channel, the acetylcholinesterase effectively halts the action of any ACh it encounters.

When a puff of ACh is released in response to an action potential in the synaptic terminal, the concentration of ACh in the synaptic cleft abruptly rises. Some of the released ACh molecules will bind to ACh-activated channels, causing them to open and increasing the sodium and potassium permeability of the end-plate membrane; other ACh molecules will bind to acetylcholinesterase and be inactivated. Even though the binding of ACh to the postsynaptic channel is highly specific, it is readily reversible; the binding typically lasts for only about 1 msec. When an ACh molecule comes off a gate, the channel closes. The newly freed ACh molecule may then bind again to an ACh-activated channel, or it might bind to acetylcholinesterase and be inactivated. With time following release of the puff, the concentration of ACh in the cleft will fall until all of the released ACh has been split into acetate and choline.

It would be wasteful if the choline resulting from inactivation of ACh were lost and had to be replaced with fresh choline from inside the presynaptic cell. This potential waste is avoided because most of the choline is taken back up into the synaptic terminal, where it is reassembled into ACh by the enzyme choline acetyltransferase. Thus, both the vesicle membrane (the packaging material of the quantum) and the released neurotransmitter (the contents of the quantum) are effectively recycled by the presynaptic terminal.

Recording the Electrical Current Flowing Through a Single Acetylcholine-activated Ion Channel

Throughout our discussion of the membrane properties of excitable cells, we have made extensive use of the notion of ions crossing the membrane through specific pores or channels. For example, we saw that the effect of ACh on the muscle membrane is mediated via ion channels in the postsynaptic membrane that open in the presence of ACh. As discussed in Chapters 5 and 7, the flow of ions across the cell membrane constitutes a transmembrane electrical current that can be measured with electrical techniques like the voltage clamp. Recently, a new technique was developed by Neher and Sakmann to record transmembrane ionic currents, and the technique has sufficient resolution to measure the minute electrical current flowing through a single open ion channel. The technique is called the **patch clamp**, and it is illustrated in Figure 8-13.

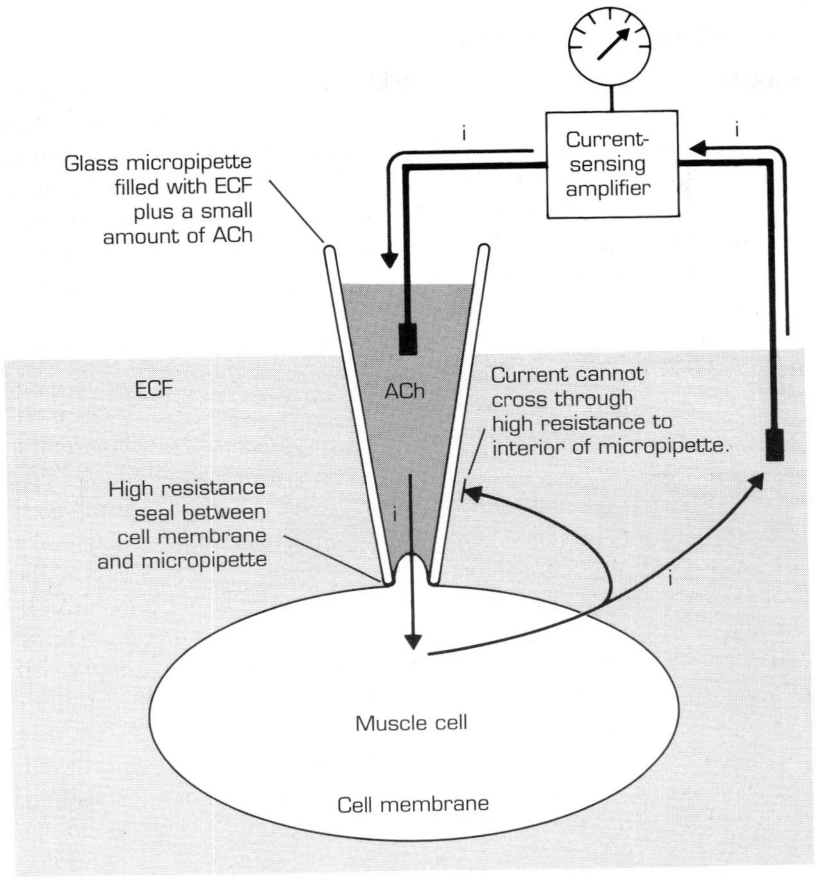

Figure 8-13 Schematic illustration of the procedure for recording the current through a single ACh-activated channel in a cell membrane. A micropipette with a tip diameter of 1–2 μm is placed on the external surface of the membrane. A tight electrical seal is made between the membrane and the glass of the micropipette, so that a resistance greater than 10^{10} ohms is imposed in the extracellular path for current flow through the channel. When a channel in the patch of membrane inside the micropipette opens, a current-sensing amplifier connected to the interior of the pipette detects the minute current flow.

The basic idea behind the patch clamp is to isolate electrically a small patch of cell membrane that contains only a few ion channels. The electrical isolation is achieved by placing a specially constructed miniature glass pipette in close contact with the membrane, so that a tight seal forms between the membrane and the glass. When one of the ion channels in the isolated patch opens, electrical current flows across the membrane; in the case of the ACh-activated channel that current would be a net inward (that is, depolarizing) current under normal conditions. We know from the basic properties of electricity that current must flow in a complete circuit. As shown in Figure 8-13, the return current path through the extracellular space is broken by the presence of the glass pipette; there is a high electrical resistance imposed by the seal between the cell membrane and the pipette. Under these conditions, the ionic current through the open channel is forced to complete its circuit through the current-sensing amplifier connected to the interior of the pipette. In order for the patch-clamp technique to achieve sufficient sensitivity to measure the current through a single channel, the electrical resistance between the interior of the patch pipette and the extracellular space must be greater than about 10^9 ohms, which is a very large resistance indeed. Fortunately for neurophysiologists, it is possible to achieve resistances greater than 10^{10} ohms.

Using the patch clamp, it is possible to record the current through ACh-activated channels of the postsynaptic membrane of muscle cells by placing a small amount of ACh (or structurally related compounds that are recognized by the receptors on the gate) inside the patch pipette. As shown in Figure 8-4, when the receptors are occupied, the gate opens and the channel allows ions to cross the membrane. Schematically, then, we might expect to record an electrical current like that shown in Figure 8-14a when the channel opens. There

(a)

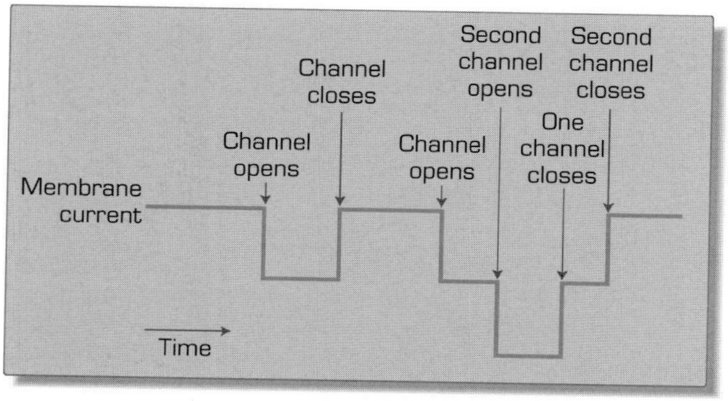

(b)

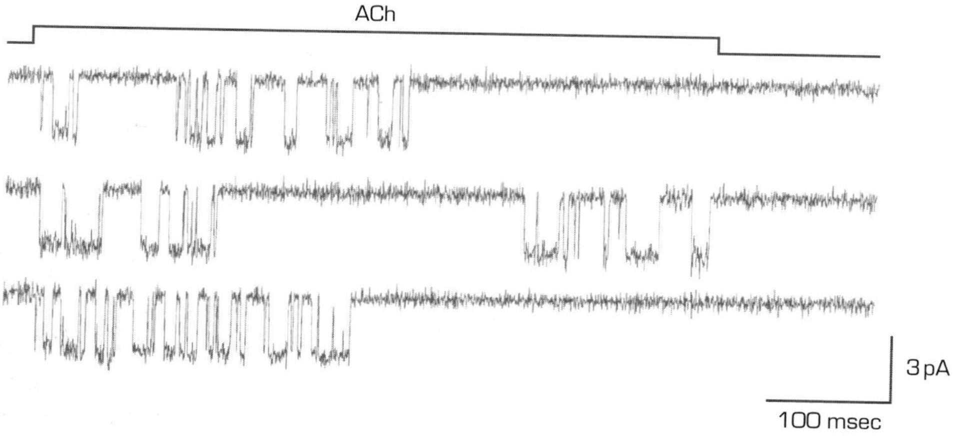

Figure 8-14 The current through single ACh-activated ion channels. (a) A schematic diagram of the current expected to flow through a single ACh-activated channel if the conducting state of the channel is controlled by a gate that is either completely open or completely closed. When ACh binds, the channel opens and a stepwise pulse of inward current flows through the channel. When ACh unbinds, the channel closes and the current abruptly disappears. (b) Actual recordings of currents flowing through single ACh-activated channels. (Data provided by D. Naranjo and P. Brehm of the State University of New York at Stony Brook.)

would be a rapid step of inward current that occurs as the gate opens, the current would be maintained at a constant level for as long as the channel is open, and the step would terminate when ACh unbinds from one of the receptor sites, causing the gate to close. If a second channel opens while the first is still open, the two currents simply add to produce a current twice as large as the single-channel current. This is also shown in Figure 8-14a.

Actual patch-clamp recordings of currents through single ACh-activated channels of human muscle cells are shown in Figure 8-14b. These records show that the currents through the channels are the rectangular events expected from the simple gating scheme of Figure 8-4. Experiments like that of Figure 8-14b confirm directly the view of ion permeation and channel gating that we have used to explain the electrical behavior of the membranes of excitable cells: the gated ion channels carrying electrical current across the plasma membrane are not just figments of the neurophysiologist's imagination. The development of the patch-clamp technique has led to a flurry of new information about ion channels of all types; for example, the currents flowing through single voltage-sensitive sodium and potassium channels that underlie the action potential (see Chapters 6 and 7) have also been observed using this technique.

Molecular Properties of the Acetylcholine-activated Channel

Techniques of molecular biology are being applied with great success to the study of ion channel function, particularly when combined with measurements of single-channel behavior using the patch-clamp technique just described.

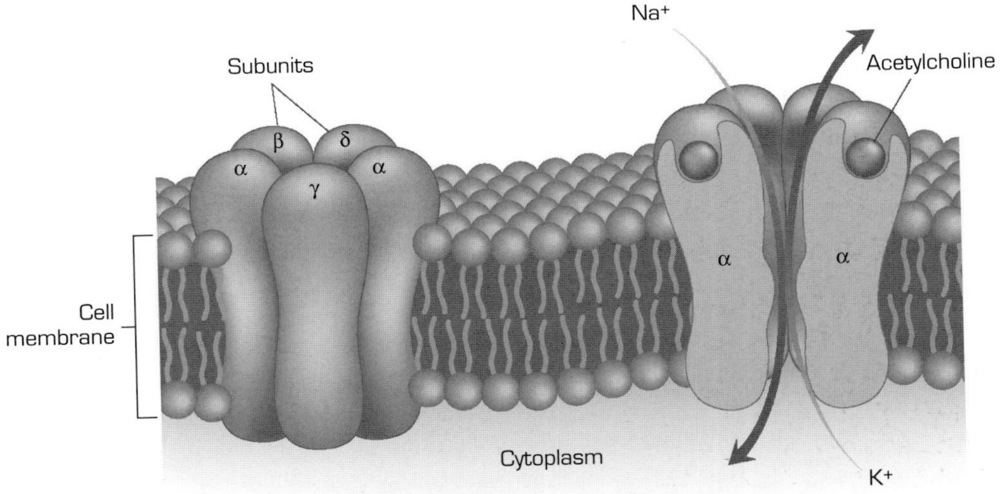

Figure 8-15 The subunit structure of the ACh-activated channel. The five subunits interact to form the gated ion channel of the end-plate membrane, with the pore at the center.

This has been especially true for the ACh-activated channel of the muscle end-plate. Biochemical studies have shown that this channel is formed by the aggregation of five individual protein subunits: two copies of an alpha-subunit, plus beta-, gamma-, and delta-subunits. The two ACh-binding sites have been located, one on each of the two alpha-subunits, thus accounting neatly for the fact that binding of two ACh molecules is required to open the channel (Figure 8-4). The subunits come together as shown in Figure 8-15 to form the ACh-activated channel, with parts of each subunit forming the aqueous pore at the center through which cations can cross the membrane. The genes encoding each of these subunits have been identified and analyzed, and the sequence of amino acids making up the protein has been deduced in each case by reading the genetic code from the pattern of nucleic acids in the DNA. This sequence of amino acids gives valuable structural information about the ACh-activated channel. But beyond that, molecular biological techniques can be used to assign functional roles to particular parts of the channel protein. This approach makes use of the fact that it is possible to make messenger RNA from the DNA sequence of the channel protein; when this mRNA is injected

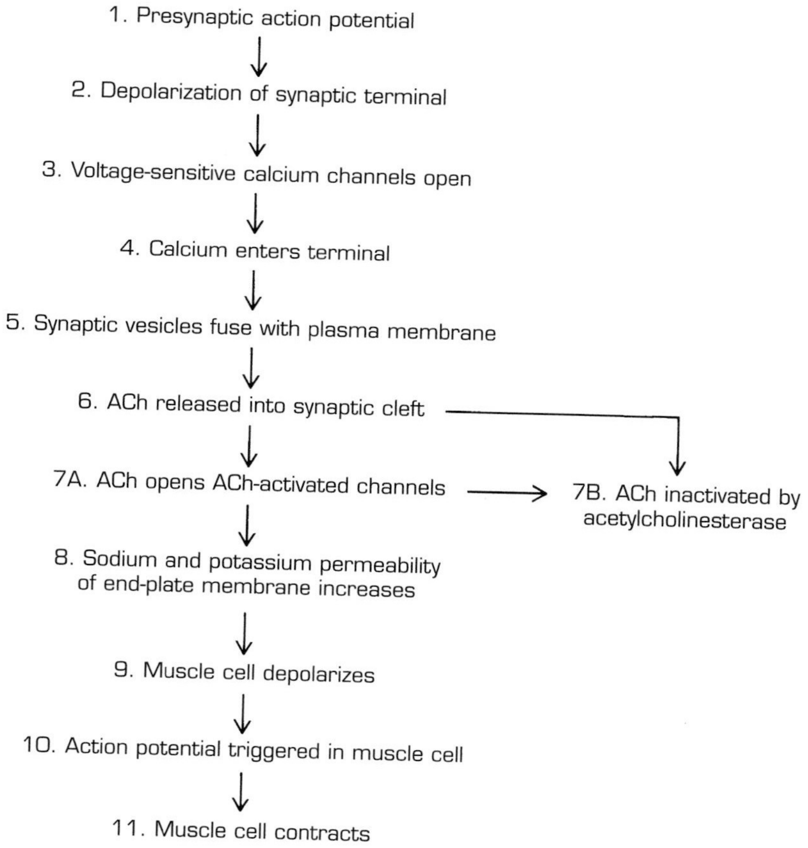
1. Presynaptic action potential
2. Depolarization of synaptic terminal
3. Voltage-sensitive calcium channels open
4. Calcium enters terminal
5. Synaptic vesicles fuse with plasma membrane
6. ACh released into synaptic cleft
7A. ACh opens ACh-activated channels
7B. ACh inactivated by acetylcholinesterase
8. Sodium and potassium permeability of end-plate membrane increases
9. Muscle cell depolarizes
10. Action potential triggered in muscle cell
11. Muscle cell contracts

Figure 8-16 A summary of the sequence of events during synaptic transmission at the neuromuscular junction.

into a cell that does not normally make ACh-activated channels (such as the egg cell of a frog), the cell's machinery of protein synthesis will read the message and make functional ACh-activated channels. The properties of these channels can then be examined using patch-clamp recording. Thus, by altering the mRNA, experimenters can make discrete changes in the channel protein and then see how the change affects the behavior of the channel. For example, in this way the parts of each subunit that probably make up the ion-conducting pore have been identified.

Summary

The sequence of events during synaptic transmission at the neuromuscular junction is summarized in Figure 8-16. The depolarization produced by an action potential in the synaptic terminal opens voltage-dependent calcium channels in the terminal membrane. Calcium ions enter the terminal down their concentration and electrical gradients, inducing synaptic vesicles filled with ACh to fuse with the plasma membrane facing the muscle cell. The ACh is thereby dumped into the synaptic cleft, and some of it diffuses to the muscle membrane and combines with specific receptors on ACh-activated channels in the muscle membrane. When ACh is bound, the channel opens and allows sodium and potassium ions to cross the membrane. This depolarizes the muscle membrane and triggers an all-or-none action potential in the muscle cell. The action of ACh is terminated by the enzyme acetylcholinesterase, which splits ACh into acetate and choline.

9 Synaptic Transmission in the Central Nervous System

Chemical synapses between neurons operate according to the same general principles as the synapse between a motor neuron and a muscle cell discussed in Chapter 8. In the patellar reflex, for example, presynaptic sensory neurons activate postsynaptic motor neurons through a sequence of events similar to those at the neuromuscular junction. However, despite the overall similarity between neuron-to-neuron synapses and neuron-to-muscle synapses, some important differences do exist. This chapter will consider some of those differences, as well as the similarities.

Excitatory and Inhibitory Synapses

At the neuromuscular junction, ACh depolarizes the muscle cell, causing it to fire an action potential. Synapses of this type are called **excitatory synapses** because the neurotransmitter brings the membrane potential of the postsynaptic cell toward the threshold for firing an action potential and thus tends to "excite" the postsynaptic cell. The synapse between the sensory neuron and the quadriceps motor neuron in the patellar reflex is an example of an excitatory synapse between two neurons. Synapses between neurons are not always excitatory, however. At **inhibitory synapses,** the postsynaptic effect of the neurotransmitter tends to prevent the postsynaptic cell from firing an action potential, by keeping the membrane potential of the postsynaptic cell more negative than the threshold potential. Thus, the postsynaptic cell is "inhibited" by the release of the inhibitory neurotransmitter. One major difference between synaptic transmission at the neuromuscular junction and synaptic transmission in the nervous system in general is that transmission at the neuromuscular junction is always excitatory, whereas transmission in the nervous system may be either excitatory or inhibitory.

We will return to a discussion of inhibitory synapses later in this chapter. At this point, the discussion will center on the properties of excitatory synaptic transmission between neurons in the nervous system.

Excitatory Synaptic Transmission Between Neurons

The synapse at the neuromuscular junction is unusual in one important aspect: a single action potential in the presynaptic motor neuron produces a sufficiently large depolarization in the postsynaptic muscle cell to trigger a postsynaptic action potential. Such a synapse is called a one-for-one synapse because one action potential appears in the output cell for each action potential in the input cell. Most synapses between neurons are not so strong, however. Instead, a single presynaptic action potential typically produces only a small depolarization of the postsynaptic cell. The synapse between a single stretch receptor sensory neuron and a quadriceps motor neuron is typical of this situation, as illustrated schematically in Figure 9-1.

Temporal and Spatial Summation of Synaptic Potentials

Figure 9-1a shows an experimental arrangement for recording the change in membrane potential of a motor neuron in response to action potentials in a single presynaptic sensory neuron. An intracellular microelectrode is placed inside the postsynaptic motor neuron, and presynaptic action potentials are triggered by electrical stimuli applied to the sensory nerve fiber. Figure 9-1b illustrates responses of the motor neuron to a single action potential in the sensory neuron and to a series of four action potentials. A single presynaptic action potential produces only a small depolarization of the motor neuron, called an **excitatory postsynaptic potential (e.p.s.p.)**. A single e.p.s.p. is typically much too small to reach threshold for triggering a postsynaptic action potential. Figure 9-2 shows a recording of an e.p.s.p. in a motor neuron produced by an action potential in a single sensory neuron. In this experiment, an intracellular electrode was placed inside the sensory fiber to record the presynaptic membrane potential and to inject depolarizing current that elicited an action potential in the presynaptic fiber (upper recording trace). A second intracellular microelectrode in the motor neuron recorded the change in membrane potential of the postsynaptic cell. Note that the single e.p.s.p. is only about 1 mV in amplitude, which is much smaller than the 10–20 mV depolarization required to reach threshold. Thus, summation of e.p.s.p.'s is required to trigger a postsynaptic action potential in the motor neuron.

If a second action potential arrives at the presynaptic terminal before the postsynaptic effect produced by the first action potential has dissipated, the second e.p.s.p. will sum with the first to produce a larger peak postsynaptic depolarization. As shown in Figure 9-1b, the e.p.s.p.'s produced by a rapid series of presynaptic action potentials can add up sufficiently to reach threshold for triggering a postsynaptic action potential. This kind of summation of the sequential postsynaptic effects of an individual presynaptic input is called **temporal summation**. Temporal summation is an important mechanism that allows even a weak excitatory synaptic input to stimulate an action potential in a postsynaptic cell.

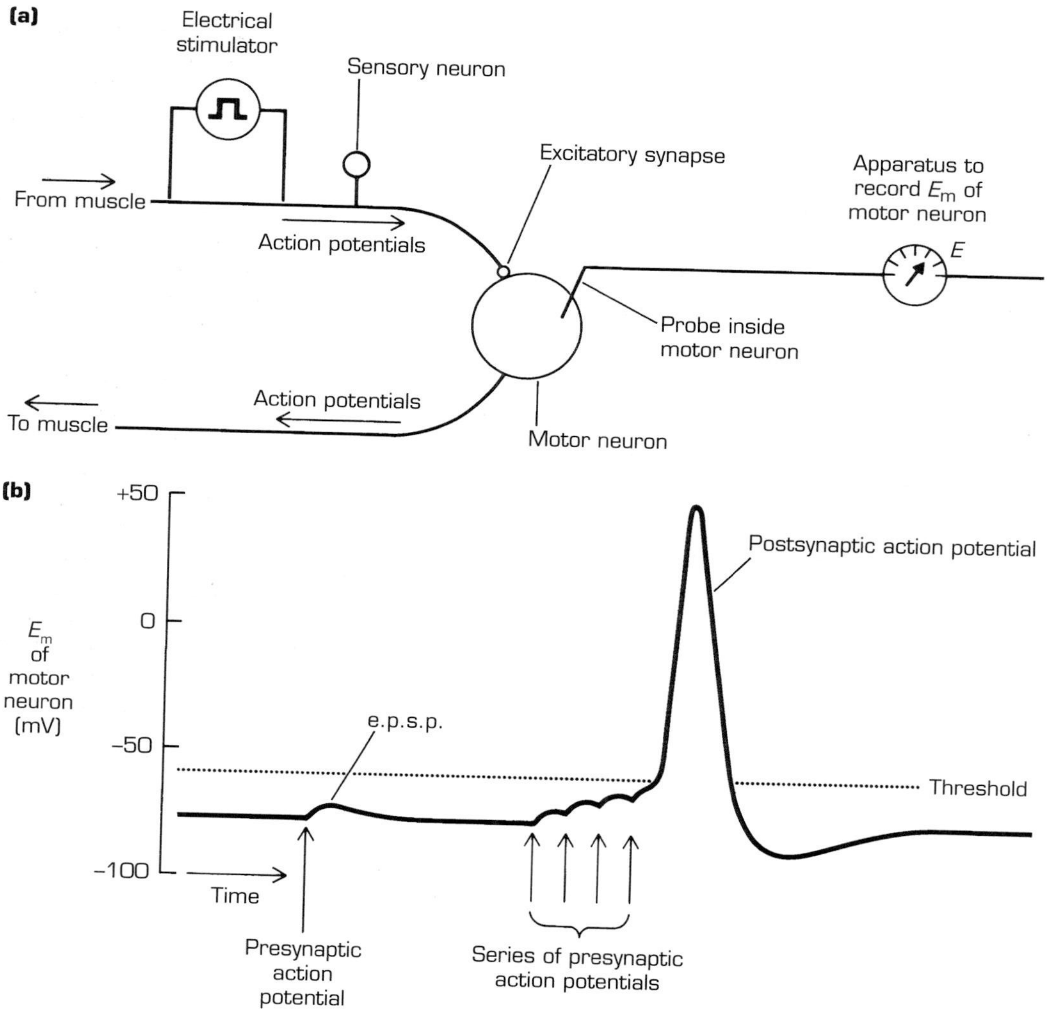

Figure 9-1 Synaptic transmission at an excitatory synapse between two neurons. (a) The experimental arrangement for examining transmission between a sensory and a motor neuron in the patellar reflex loop. (b) Responses of the postsynaptic motor neuron to action potentials in the presynaptic sensory neuron. At the upward arrows, action potentials are triggered in the presynaptic neuron by an electrical stimulus.

Temporal summation of e.p.s.p.'s is illustrated in the intracellular recordings shown in Figure 9-3, which were obtained from a motor neuron of the sympathetic nervous system. Each set of traces in the figure consists of superimposed responses to three postsynaptic stimuli. In each case, one stimulus fails to activate the presynaptic input and so produces no postsynaptic response (trace a), one stimulus produces a postsynaptic response that fails to

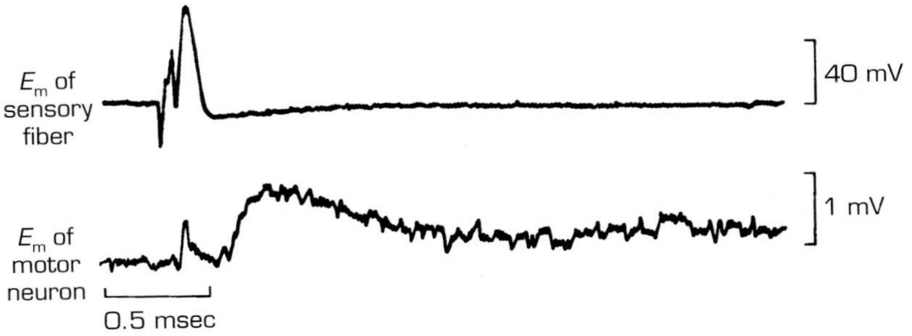

Figure 9-2 Simultaneous intracellular recordings from a single stretch-sensitive sensory nerve fiber and a motor neuron receiving synaptic input from the sensory fiber. The upper trace shows an action potential triggered in the sensory fiber by passing a depolarizing electrical current through the intracellular electrode. After a brief delay, a small excitatory postsynaptic potential was evoked in the postsynaptic motor neuron (lower trace). Note the different voltage scales for the two traces. (Data provided by W. Collins, M. Honig, and L. Mendell of the State University of New York at Stony Brook.)

reach threshold (trace b), and one stimulus produces a postsynaptic response that reaches threshold (trace c). Only if the successive e.p.s.p.'s summate sufficiently to reach threshold is an action potential triggered in the postsynaptic cell.

Another way that e.p.s.p.'s can sum to reach threshold is via the simultaneous firing of action potentials by several presynaptic neurons. A single neuron in the nervous system commonly receives synaptic inputs from hundreds or even thousands of presynaptic neurons. In the patellar reflex, for example, a single quadriceps motor neuron will receive excitatory synaptic connections from many stretch receptor sensory neurons, shown schematically in Figure 9-4a. An action potential in a single presynaptic cell produces only a small postsynaptic depolarization, as we have seen. If several presynaptic cells fire simultaneously, however, their postsynaptic effects sum together and can reach threshold (Figure 9-4b). This **spatial summation** of e.p.s.p.'s occurs when several spatially distinct synaptic inputs are active nearly simultaneously.

In the patellar reflex, both temporal and spatial summation are important in eliciting the reflexive response. In order to produce reflexive contraction of the quadriceps muscle, a tap to the patellar tendon must stretch the muscle sufficiently to fire a number of action potentials in each of a number of individual sensory neurons. Combined temporal summation of the effects of action potentials within the series and spatial summation of the effects of all of the individual sensory neurons ensure that postsynaptic motor neurons fire action potentials and trigger muscle contraction.

Some Possible Excitatory Neurotransmitters

The chemical neurotransmitter at the neuromuscular junction is ACh, as discussed in Chapter 8. Acetylcholine is also used as a neurotransmitter at some neuron-to-neuron synapses. In addition, many other substances act as neurotransmitters at excitatory synapses in the nervous system. The molecular structures of a representative sample of excitatory neurotransmitter substances are shown in Figure 9-5. Many excitatory neurotransmitters are relatively small molecules, often derived from amino acids by simple chemical

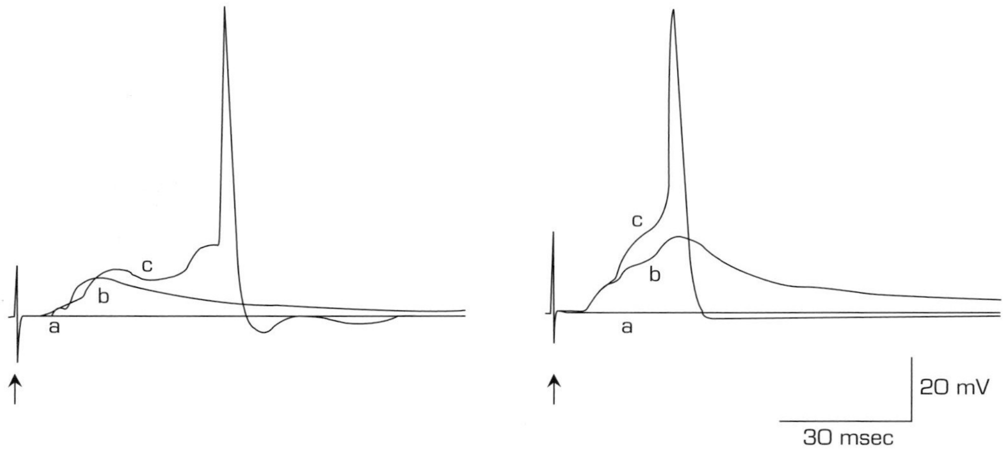

Figure 9-3 Intracellular recordings of e.p.s.p.'s in a neuron, showing summation of successive e.p.s.p.'s. Each set of traces shows three superimposed responses. The arrow indicates the electrical stimulus used to trigger action potentials in the presynaptic neurons that make excitatory synapses onto the recorded cell. Trace a in each set shows a stimulus that failed to trigger the presynaptic input. Trace b shows e.p.s.p.'s that failed to reach threshold. Trace c shows summated e.p.s.p.'s that reach threshold and produce a postsynaptic action potential. In this figure, the postsynaptic cell is a motor neuron of the sympathetic nervous system (which is described in Chapter 11). (Data provided by H.-S. Wang and D. McKinnon of the State University of New York at Stony Brook.)

modifications. Amino acids are more commonly thought of as the basic building blocks for the construction of proteins. In the nervous system, however, amino acids are also often used for cell-to-cell signaling in neurotransmission. For example, glutamate and aspartate are unmodified amino acids, norepinephrine and dopamine are derived from the amino acid tyrosine, and serotonin is derived from tryptamine. Glutamate is thought to be the excitatory transmitter at the synapse between the sensory and motor neurons in the patellar reflex.

Other neurotransmitters are more structurally complex than the small amino-acid derivatives. These substances—called **peptide neurotransmitters**, or more simply **neuropeptides**—are formed from a series of individual amino acids linked by peptide bonds, like a small piece of a protein molecule. Indeed, neuropeptides are synthesized by neurons as larger protein precursors, which are then processed proteolytically to release the embedded neuropeptide fragment. An example of a neuropeptide is substance P, whose amino-acid sequence is shown in Figure 9-5.

The list of excitatory neurotransmitters in Figure 9-5 is by no means exhaustive. As our knowledge of the brain grows, it is likely that new candidates will be added to the list.

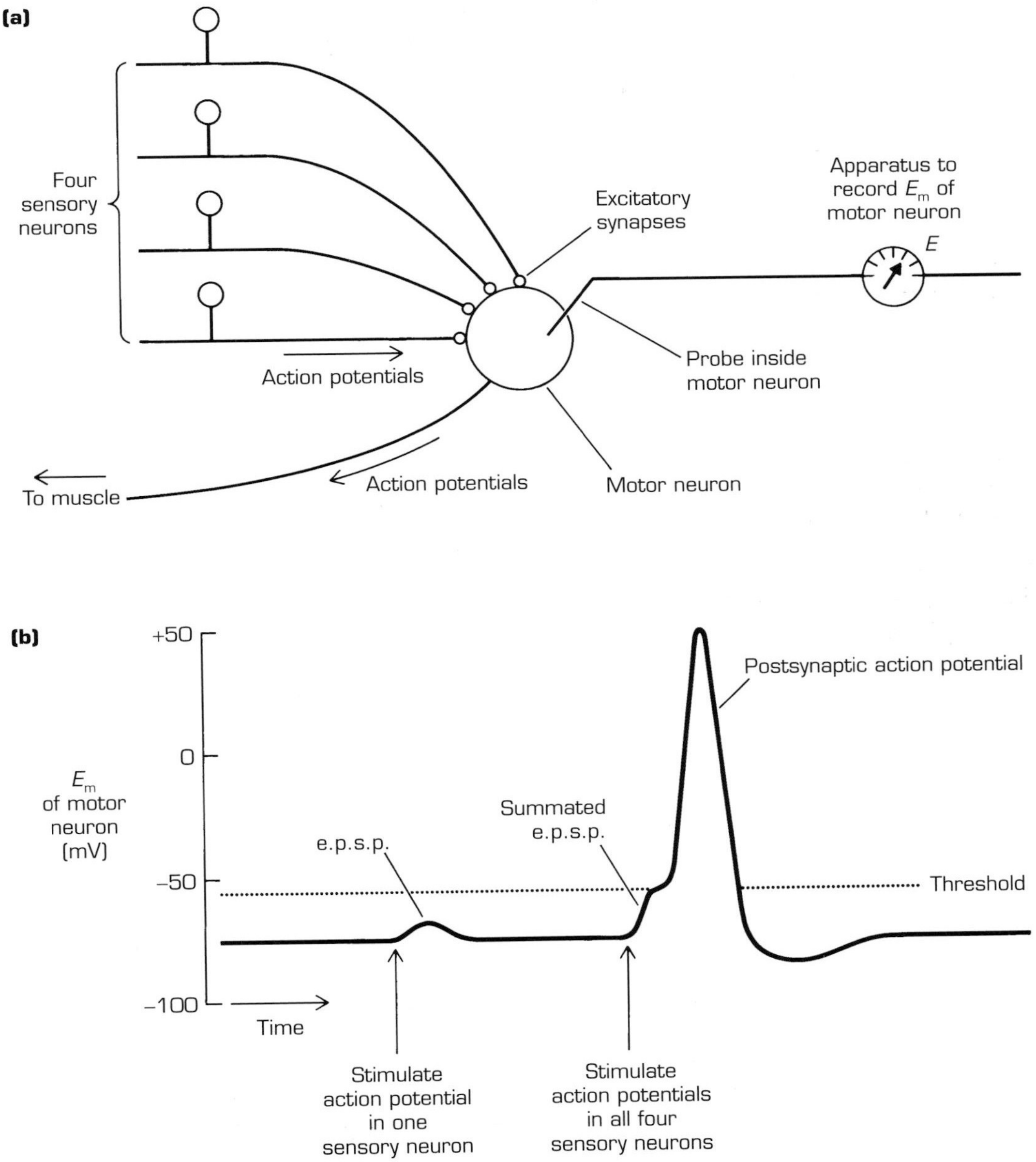

Figure 9-4 Spatial summation of excitatory inputs to a motor neuron. (a) A diagram of the synaptic circuitry and recording arrangement. (b) An illustration of synaptic responses in the postsynaptic motor neuron.

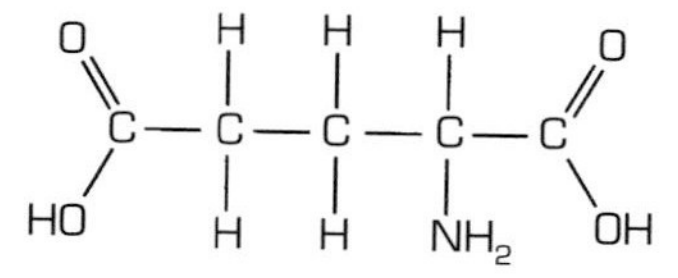

Glutamic acid
(Glutamate)

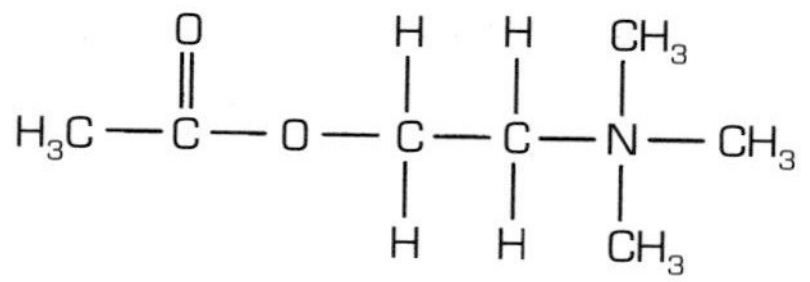

Acetylcholine

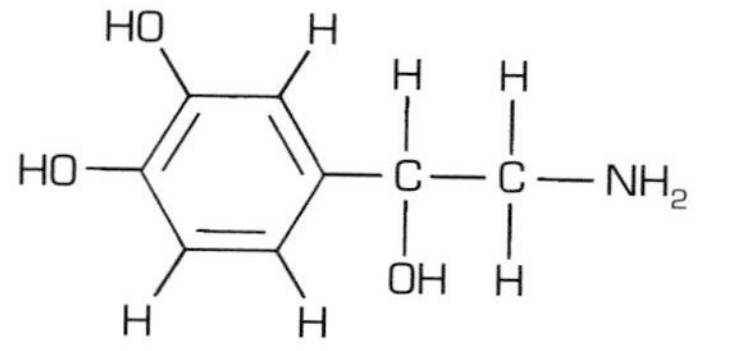

Norepinephrine

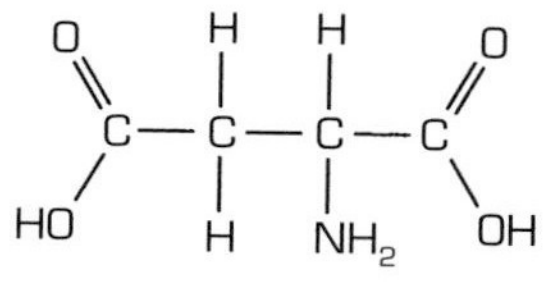

Aspartic acid
(Aspartate)

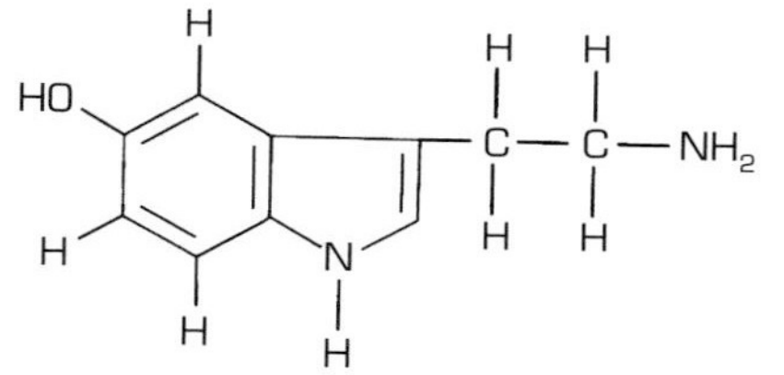

Serotonin
(5-hydroxytryptamine)

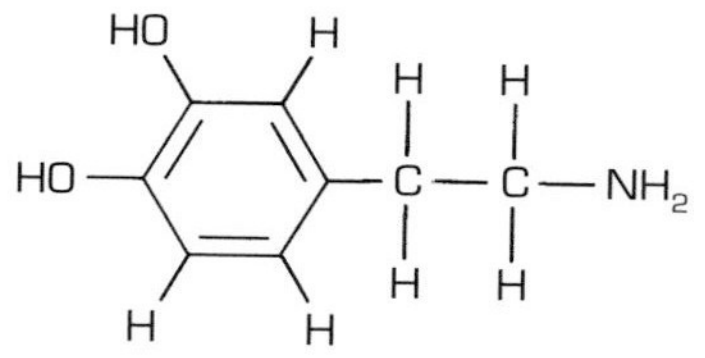

Dopamine

Arg-Pro-Lys-Pro-Gln-Gln- Phe-Phe-Gly-Leu-Met-NH_2

Substance P
(a string of 11 amino acids attached by peptide bonds)

Figure 9-5 Structures of some excitatory neurotransmitter substances in the nervous system.

Conductance-decrease Excitatory Postsynaptic Potentials

In most cases, the mechanism by which an excitatory neurotransmitter produces an e.p.s.p. in the postsynaptic cell is the same as that by which ACh depolarizes the muscle at the neuromuscular junction. That is, the transmitter opens channels in the postsynaptic membrane that are permeable to sodium and potassium ions. The altered balance of sodium and potassium permeability then depolarizes the postsynaptic cell, as described in Chapter 8. We saw in Chapter 5 that the membrane potential is controlled by the ratio of sodium to potassium permeability. Consequently, a depolarization might result from either an increase in sodium permeability or a decrease in potassium permeability. Indeed, at some synapses, the e.p.s.p. is produced by a reduction in postsynaptic potassium permeability. For instance, ACh produces a long-lasting depolarization of sympathetic ganglion neurons in the frog, caused by a decrease in the potassium permeability of the neuron. Acetylcholine closes a type of potassium channel in the neuron, so that outward potassium current declines and the resting inward sodium current exerts a greater influence on the membrane potential.

Inhibitory Synaptic Transmission

The Synapse between Sensory Neurons and Antagonist Motor Neurons in the Patellar Reflex

In the patellar reflex, muscles other than the quadriceps muscle must be taken into account for a more complete description, shown in Figure 9-6. Whereas the quadriceps muscle extends the knee joint, antagonistic muscles at the back of the thigh flex the knee joint. These flexor muscles also have a stretch reflex analogous to that of the quadriceps. That is, stretching the flexor muscle stimulates action potentials in stretch-sensitive sensory neurons, which then make excitatory synapses in the spinal cord onto motor neurons of the flexor muscle. When the patellar tendon is tapped, the quadriceps muscle reflexively contracts, causing the knee joint to extend (the "jerk" of the knee-jerk reflex).

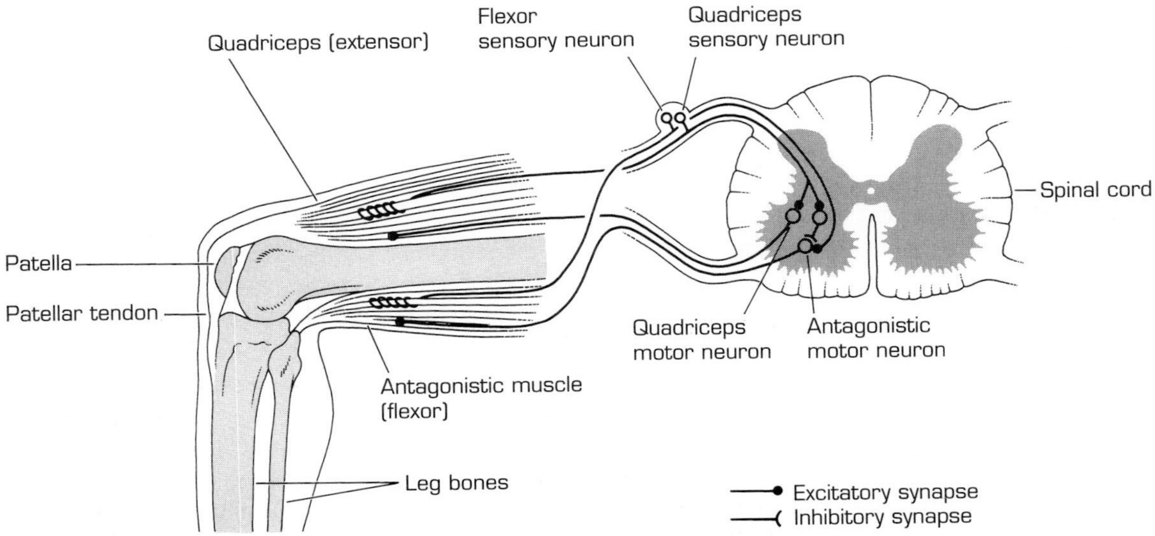

Figure 9-6 A revised diagram of the circuitry involved in stretch reflexes of thigh muscles.

Extension of the joint stretches the flexor muscles at the back of the thigh, which should then contract because of the action of their own stretch-reflex mechanism. The resulting flexion of the joint should again stretch the quadriceps and elicit reflexive extension, which should elicit another reflexive flexion, and so on. Thus, a single tap to the patellar tendon would send the knee joint into a series of oscillations that would continue until muscle exhaustion sets in.

Instead, tapping the patellar tendon elicits only a single knee jerk. What prevents the oscillatory response described above? The answer lies in the more elaborate neuronal circuitry diagrammed in Figure 9-6. The nerve fibers of the stretch-sensitive sensory neurons from the quadriceps muscle actually branch profusely when they enter the spinal cord and make synaptic connections with many kinds of neurons in addition to the quadriceps motor neurons. Among these other synaptic connections is an excitatory synapse onto neurons that in turn make an inhibitory synapse on the motor neurons of the antagonistic muscles. Thus, action potentials in quadriceps sensory neurons not only excite quadriceps motor neurons but also tend to prevent antagonistic motor neurons from being excited by the antagonistic sensory neurons by indirectly stimulating inhibitory inputs onto the antagonistic motor neurons.

Characteristics of Inhibitory Synaptic Transmission

We will now consider some of the properties of postsynaptic responses at an inhibitory synapse and then discuss the underlying mechanisms in the postsynaptic membrane. Figure 9-7 shows schematically an experimental arrangement to examine the inhibition of the antagonistic motor neuron in the patellar reflex. An intracellular microelectrode monitors the membrane potential of the motor neuron, while the inhibitory presynaptic neuron is stimulated electrically to fire action potentials.

Release of neurotransmitter at the inhibitory synapse follows the same basic scheme as at other chemical synapses: depolarization produced by the presynaptic action potential stimulates calcium entry through voltage-sensitive calcium channels, inducing synaptic vesicles containing neurotransmitter to fuse with the membrane and release their contents. On the postsynaptic side, however, the effect of the transmitter is very different from that of ACh at the neuromuscular junction, as shown in Figure 9-7b. An action potential in the presynaptic cell is followed by a transient increase in the postsynaptic membrane potential. When the membrane potential becomes more negative, the cell is said to be **hyperpolarized**. Because hyperpolarization moves the membrane potential away from the threshold for firing an action potential, it is less likely that an excitatory input will be able to trigger an action potential, and the postsynaptic cell is inhibited. The hyperpolarization of the postsynaptic cell caused by inhibitory neurotransmitter is called an **inhibitory postsynaptic potential** (i.p.s.p.).

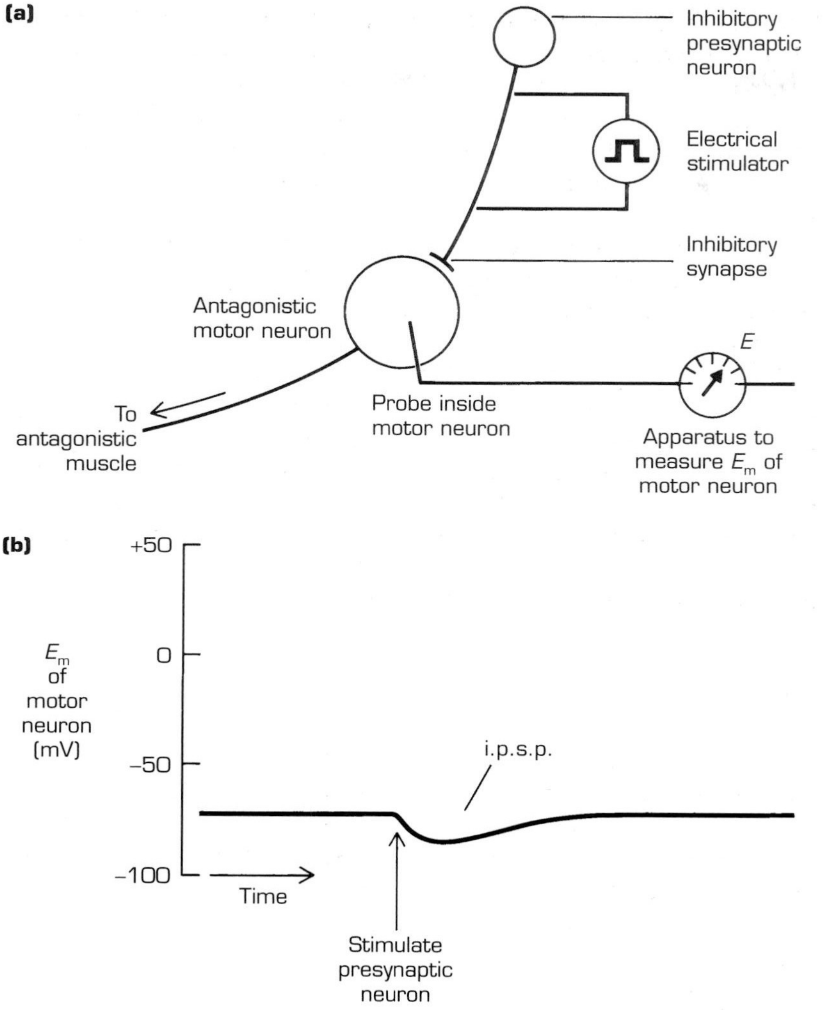

Figure 9-7 Inhibitory synaptic transmission between two neurons in the circuit of Figure 9-6. An action potential in the presynaptic neuron releases a neurotransmitter that hyperpolarizes the postsynaptic neuron. (a) A diagram of the synaptic circuitry and recording arrangement. (b) Postsynaptic response of the motor neuron.

Mechanism of Inhibition in the Postsynaptic Membrane

We have seen repeatedly that changes in membrane potential are produced by changes in ionic permeability of the plasma membrane. The i.p.s.p. is no different in this regard. When the permeability of the membrane to a particular ion increases, the membrane potential tends to move toward the equilibrium potential for that ion. What change in permeability might result in a hyperpolarizing response like an i.p.s.p.? One possible answer is illustrated in Figure 9-8. If potassium permeability of a cell membrane increases, the membrane potential would be expected to move toward E_K, which is about −85 mV for a typical mammalian cell (see Chapter 4). In this situation, p_{Na}/p_K would be smaller than usual, and the balance between potassium and sodium fluxes

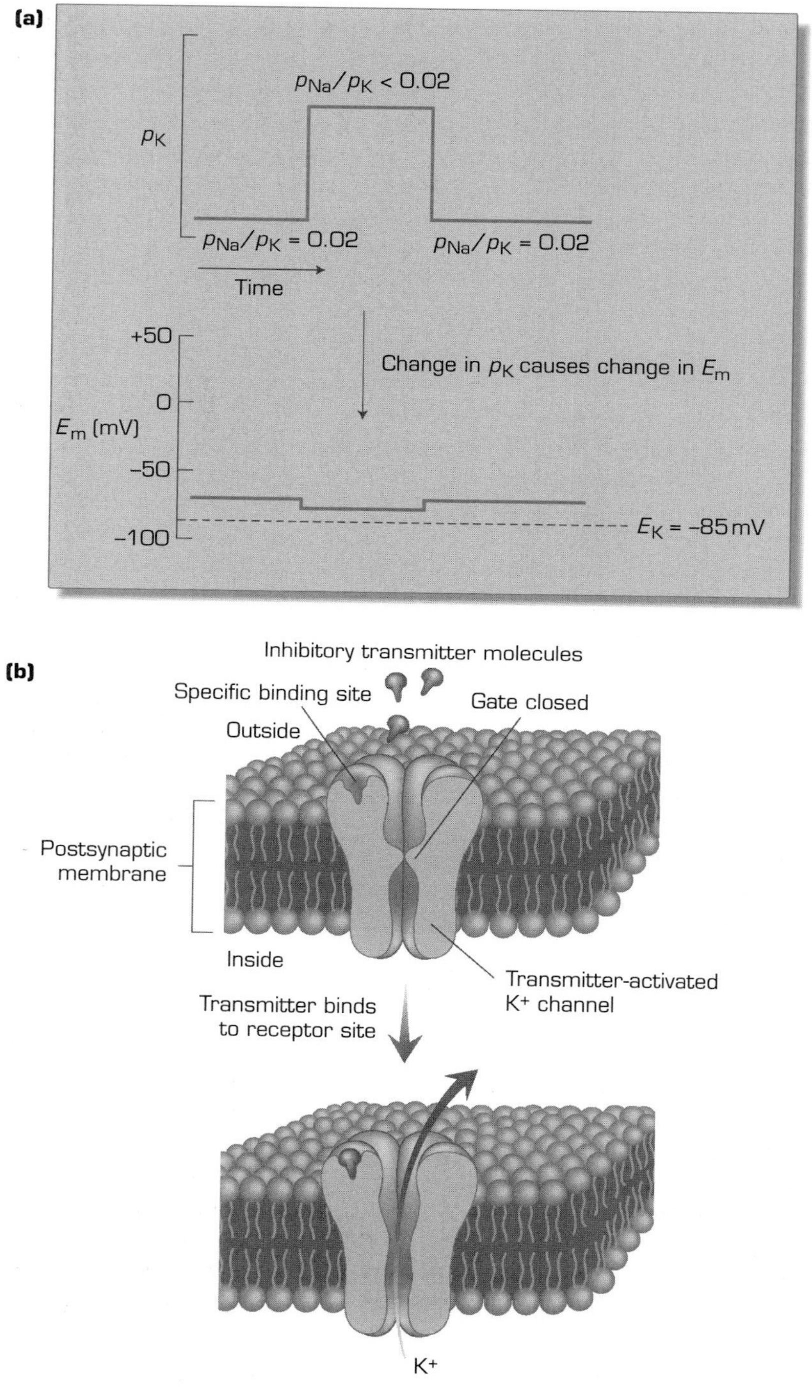

Figure 9-8 The mechanism by which increasing potassium permeability produces an inhibitory postsynaptic potential in a postsynaptic neuron. (a) The membrane potential moves toward the potassium equilibrium potential (E_K) when potassium permeability (p_K) increases. (b) At an inhibitory synapse, neurotransmitter molecules may act by opening potassium channels in the plasma membrane of a postsynaptic neuron. Efflux of potassium ions through the open channel then drives the membrane potential toward E_K.

would be struck closer to E_K. This is similar to the situation during the undershoot at the end of an action potential, when p_{Na}/p_K is transiently smaller than normal. As shown in Figure 9-8b, then, an inhibitory transmitter could hyperpolarize the postsynaptic cell by opening potassium channels in the postsynaptic membrane. As with ACh at the neuromuscular junction, the inhibitory transmitter might act by combining with specific binding sites associated with the gate on the channel. When the binding sites are occupied, the gate controlling movement through the channel opens, and potassium ions can move out of the cell, driving E_m closer to the potassium equilibrium potential.

At many inhibitory synapses, however, the transmitter-activated postsynaptic channels are not potassium channels. Instead, inhibitory neurotransmitters commonly open postsynaptic chloride channels, as illustrated schematically in Figure 9-9. In many neurons, chloride pumps in the plasma membrane maintain the chloride equilibrium potential, E_{Cl}, more negative than the resting membrane potential. An increase in chloride permeability will drive the membrane potential toward E_{Cl} and hyperpolarize the neuron. Thus, opening chloride channels can produce an i.p.s.p. in a postsynaptic cell.

In general, inhibition of a postsynaptic cell results when a neurotransmitter increases permeability to an ion whose equilibrium potential is more negative than the threshold potential for triggering an action potential. If the equilibrium potential for an ion is more negative than threshold, the ion will oppose any attempt to reach threshold as soon as the depolarization exceeds the ion's equilibrium potential. Thus, it is possible that inhibition could occur without any visible change in membrane potential from the resting level. For example, if the chloride equilibrium potential is equal to the resting potential, then opening a chloride channel would cause no change in membrane potential. However, if an excitatory input is activated, the size of the resulting e.p.s.p. would be reduced because of the enhanced ability of chloride ions to oppose depolarization.

Some Possible Inhibitory Neurotransmitters

Figure 9-10 shows the structures of some inhibitory neurotransmitters in the CNS. GABA and glycine are the most common transmitters at inhibitory synapses. Note that some of the molecules in Figure 9-10 also appeared in the list of excitatory neurotransmitters (Figure 9-5). A particular neurotransmitter substance may have an excitatory effect at one synapse but an inhibitory effect at another. Whether a neurotransmitter is excitatory or inhibitory at a particular synapse depends on the type of ion channel it opens in the postsynaptic membrane. If the transmitter-activated channel is a sodium or a sodium-potassium channel (as at the neuromuscular junction), an e.p.s.p. will result and the postsynaptic cell will be excited. If the transmitter-activated channel is a chloride or potassium channel, the postsynaptic cell will be inhibited. The same neurotransmitter could even have opposite effects at two different synapses on the same postsynaptic neuron.

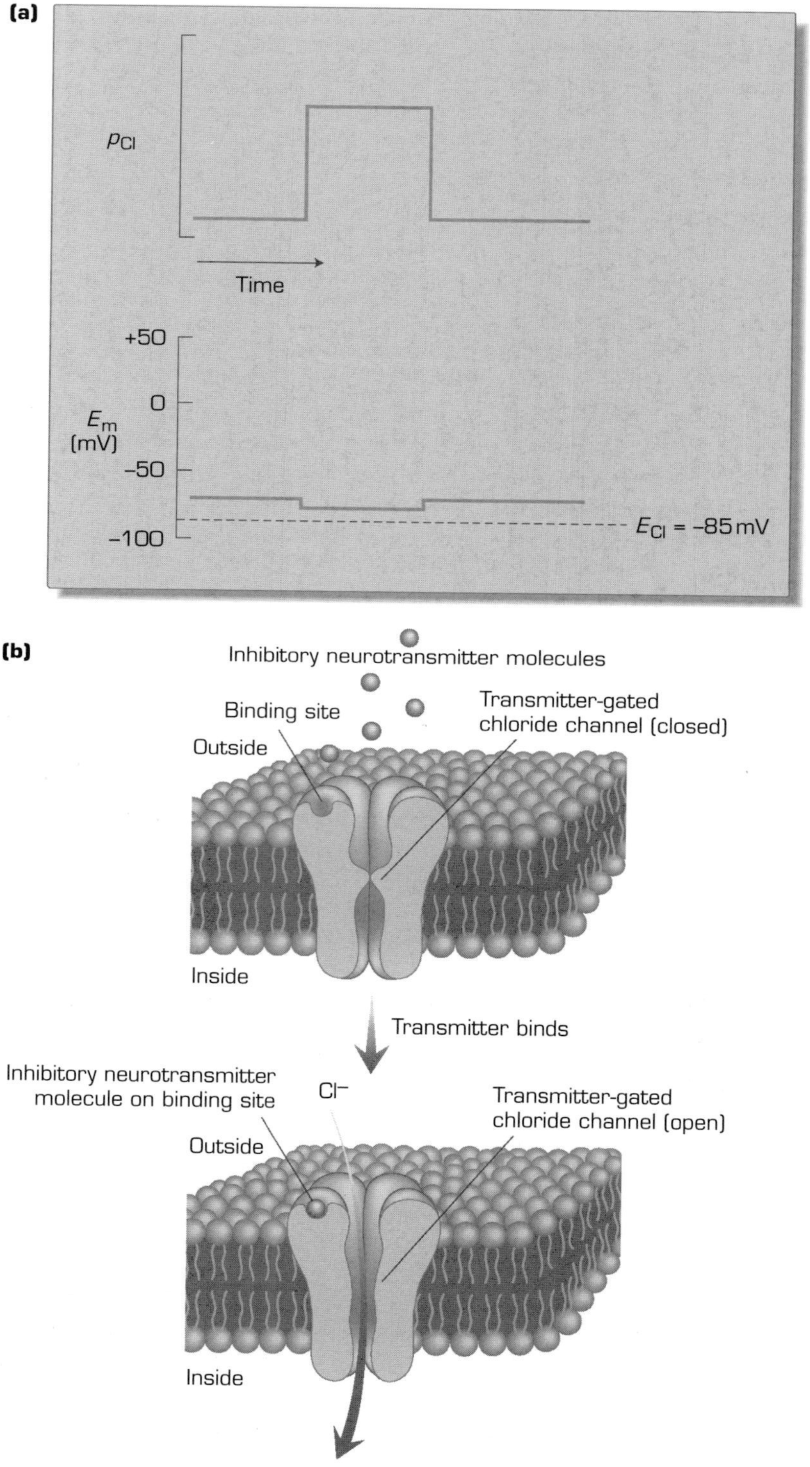

Figure 9-9 The mechanism by which increasing chloride permeability produces an inhibitory postsynaptic potential in a postsynaptic neuron. (a) The membrane potential moves toward the chloride equilibrium potential (E_{Cl}) when chloride permeability (p_{Cl}) increases. (b) At an inhibitory synapse, neurotransmitter molecules commonly act by opening chloride channels in the plasma membrane of a postsynaptic neuron. Chloride ions then enter the cell through the open channels to drive the membrane potential toward E_{Cl}. (Animation available at www.blackwellscience.com)

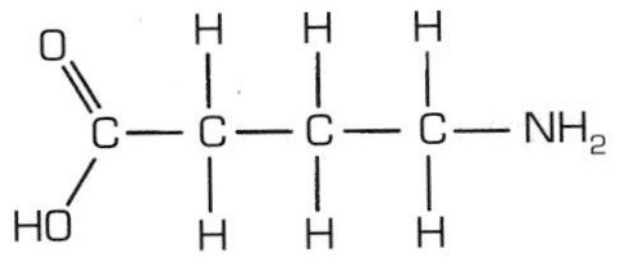

GABA
(γ-aminobutyric acid)

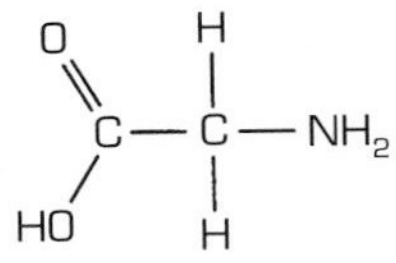

Glycine

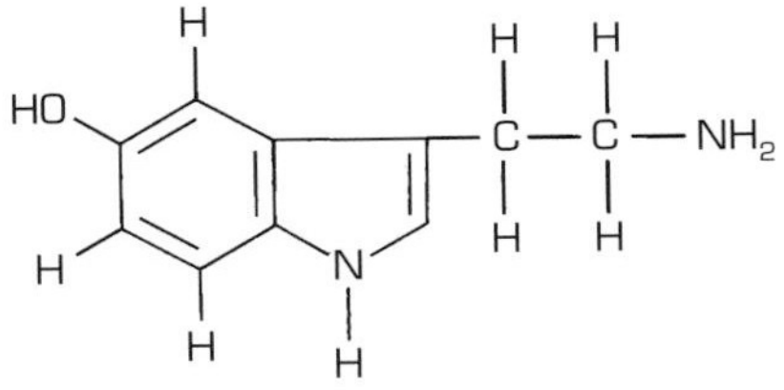

Serotonin
(5-hydroxytryptamine)

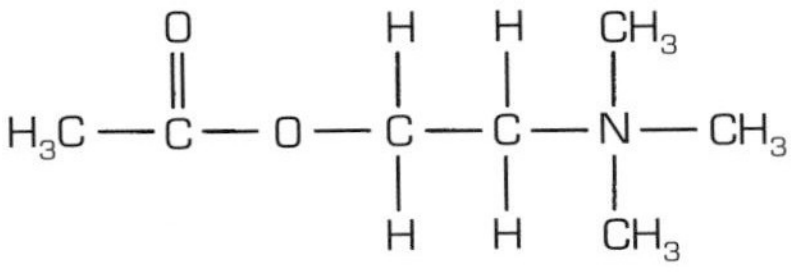

Acetylcholine

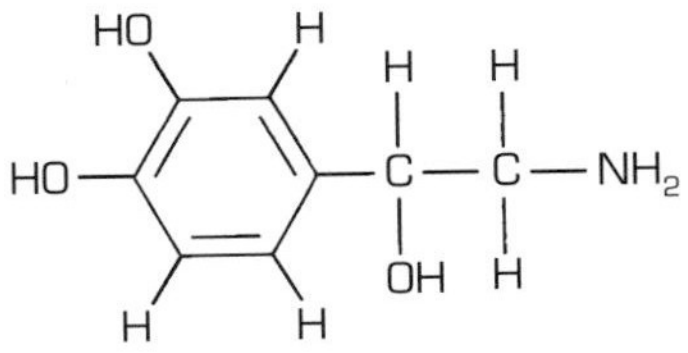

Norepinephrine

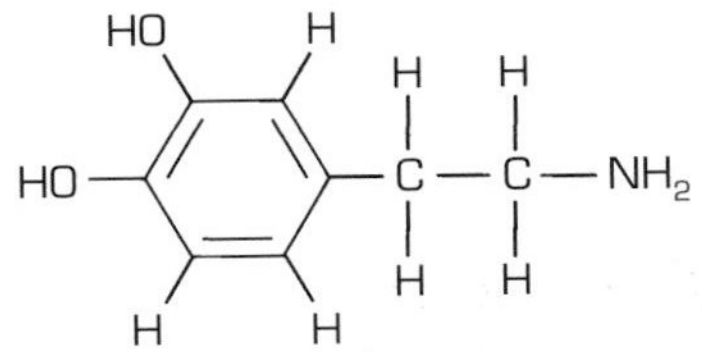

Dopamine

Tyr-Gly-Gly-Phe-Leu
Leucine enkephalin
(a series of five amino acids connected by peptide bonds)

Figure 9-10 Structures of some inhibitory neurotransmitter substances in the nervous system.

The Family of Neurotransmitter-gated Ion Channels

In Chapter 8, we learned that the ACh-gated channel is formed by the aggregation of several protein subunits. The other kinds of ion channels that underlie excitatory and inhibitory postsynaptic potentials have also been studied at the molecular level, and like the ACh-gated channel, these ion channels are formed by aggregates of individual subunits. Each type of subunit is encoded by a specific gene. The amino-acid sequences of subunits making up the neurotransmitter-gated channels are more or less similar. For example, GABA-activated channels are structurally similar to glycine-activated channels and ACh-activated channels. Glutamate-activated channels also are related to ACh-activated channels, although more distantly. Therefore, the genes encoding the subunits of the neurotransmitter-gated ion channels constitute a family of related genes, called the **ligand-gated ion channel family**. As the name implies, members of the family form ion channels that are opened by the binding of a chemical signal (the ligand) to a specific binding site on the channel.

Of course, there are also important functional differences among the members of this gene family. First, each channel type must be specifically activated by a particular neurotransmitter: a glutamate-activated channel is not activated by GABA, even though glutamate and GABA are structurally quite similar (Figures 9-5, 9-10). (In fact, GABA is formed enzymatically by modification of glutamate.) Thus, the part of the protein that forms the neurotransmitter binding site must be unique for each type of ligand-gated channel. Second, the ion-conducting pore differs among members of the ligand-gated ion channel family. Some ligand-gated channels form cation pores (e.g., glutamate-gated channels or ACh-gated channels), whereas others form anion pores (e.g., glycine-gated or GABA-gated channels). This difference in ionic selectivity reflects underlying differences in how the pore region is constructed. Differences in the pore region determine whether the effect of opening the channel is excitation or inhibition of the postsynaptic cell.

Neuronal Integration

In the nervous system, neurons receive both excitatory and inhibitory synaptic inputs. The decision of a postsynaptic neuron to fire an action potential is determined by only one factor: whether the threshold level of membrane potential has been reached. Reaching threshold is determined at any instant by the sum of all existing excitatory and inhibitory synaptic potentials. This process of summing up, or integrating, synaptic inputs is called **neuronal integration**.

Neuronal integration in the neural circuitry of the patellar reflex is shown in Figure 9-11. When the sensory neuron from the antagonistic muscle fires action potentials, e.p.s.p.'s are produced in the motor neurons that control the antagonist muscle. If there is sufficient temporal summation among the e.p.s.p.'s, an action potential is triggered (Figure 9-11b). If the inhibitory neuron is stimulated at the same time, however, the same series of excitatory inputs might be unable to reach threshold (Figure 9-11c). As shown in Figure 9-11d, this inhibitory effect can be overcome by increasing the strength of the excitatory input, which could be accomplished by increasing the number of presynaptic action potentials in the sensory neuron (temporal summation) or by increasing

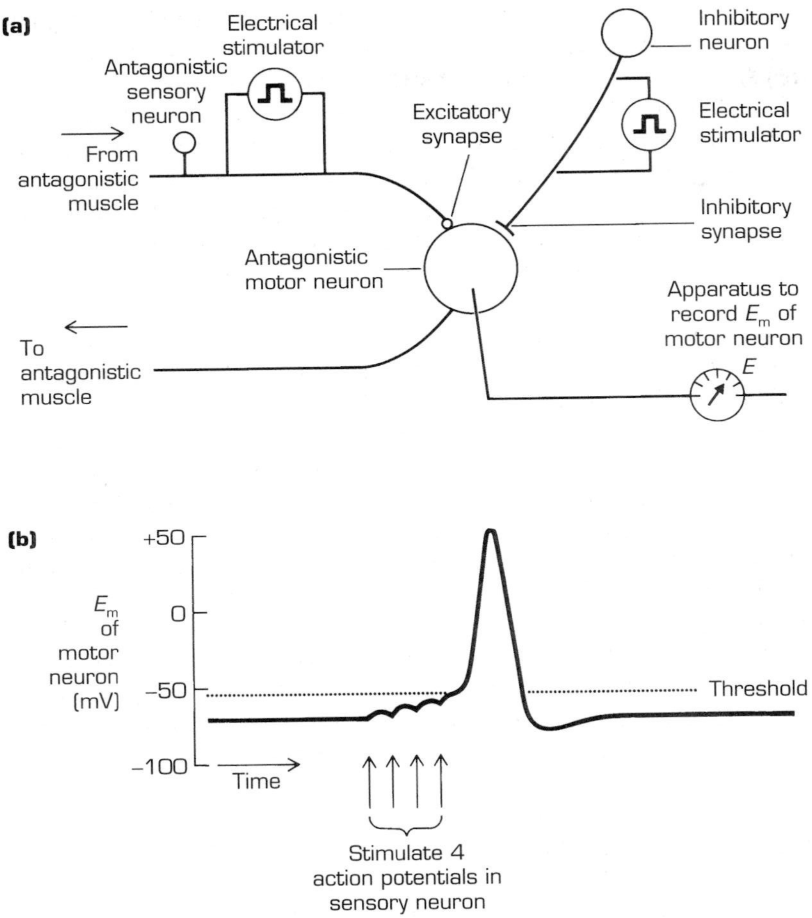

Figure 9-11 The integration of excitatory and inhibitory synaptic inputs by a postsynaptic neuron. (a) A schematic diagram of the experimental arrangement for the measurements shown in (b), (c), and (d). (b) Stimulation of the excitatory presynaptic neuron (the sensory neuron) produces a postsynaptic action potential if temporal summation is sufficient to reach threshold.

the number of sensory neurons activated (spatial summation). The balance between the excitatory and inhibitory inputs dictates whether a postsynaptic action potential is generated.

The information-processing capacity of a single neuron is considerable. A typical neuron receives hundreds or thousands of synapses from hundreds or thousands of other neurons and makes synaptic connections itself with an equal number of postsynaptic neurons. This capacity is increased still further by the widely varying weights of different synaptic inputs to a cell. Some synapses produce large changes in postsynaptic membrane potential, while others cause only tiny changes. Furthermore, the weight given a particular input might vary with time, as in the case of presynaptic inhibition. A network of some 10^{10} of these sophisticated units, like the human brain, has staggering information-processing ability.

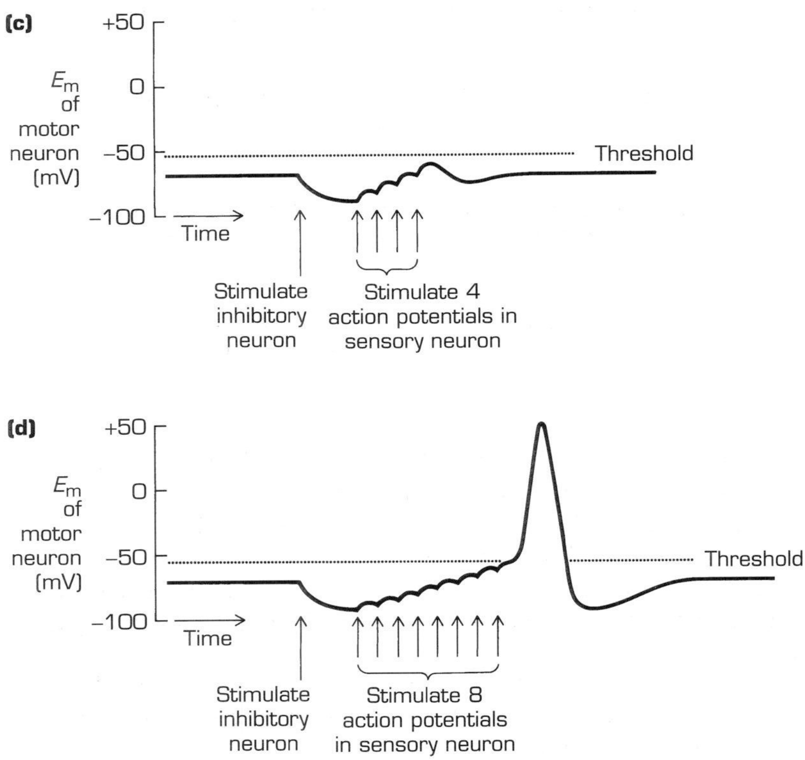

Figure 9-11 (cont'd) (c) Stimulation of the inhibitory presynaptic neuron prevents the excitatory inputs in (b) from reaching threshold. (d) The inhibitory effect can be overcome by increasing the amount of excitatory stimulation.

Indirect Actions of Neurotransmitters

The ligand-gated ion channels provide a direct linkage between neurotransmitter binding and the change in postsynaptic ion permeability. The binding site for neurotransmitter molecules is part of the ion channel protein. In addition, however, postsynaptic effects of neurotransmitters often involve an indirect linkage, in which neurotransmitter binding and the change in ion permeability are carried out by distinct protein molecules. Separation of neurotransmitter binding and the postsynaptic response allows a single neurotransmitter substance to have diverse effects on a postsynaptic neuron—closing one type of ion channel while opening others, affecting the metabolism of the postsynaptic cell as well as its membrane permeability, or altering gene expression.

The basic scheme for indirect actions of neurotransmitters is shown in Figure 9-12. Neurotransmitter molecules bind to a postsynaptic receptor molecule, as with ligand-gated ion channels. The receptor molecule is not itself an ion channel. Instead, the activated receptor molecule stimulates or inhibits production of an internal substance, called a second messenger (the neurotransmitter

1. Neurotransmitter is released by presynaptic neuron

↓

2. Neurotransmitter combines with specific receptor in membrane of postsynaptic neuron

↓

3. Combination of neurotransmitter with receptor leads to intracellular release or production of a second messenger

↓

4. Second messenger interacts (directly or indirectly) with ion channel, causing it to open or close

Figure 9-12 Overview of the indirect linkage of a neurotransmitter to activity of an ion channel via an intracellular second messenger in the postsynaptic cell.

being the first messenger), that alters the state of the postsynaptic cell. Common second messenger molecules include:

- Cyclic AMP (cyclic adenosine monophosphate), which is produced from ATP by the enzyme adenylyl cyclase.
- Cyclic GMP (cyclic guanosine monophosphate), which is produced from GTP (the guanine nucleotide equivalent of ATP) by the enzyme guanylyl cyclase.
- The dual second messengers diacylglycerol and inositol trisphosphate, both of which are produced from a particular kind of membrane lipid molecule by the enzyme phospholipase C.
- Arachidonic acid, which is produced from membrane lipid molecules by the enzyme phospholipase A.

Second messenger substances have a variety of effects in postsynaptic cells. Excitation results if the second messenger promotes opening of sodium channels or closing of potassium channels. Conversely, if the second messenger results in opening of potassium or chloride channels, or closing of sodium channels, then inhibition results.

How are the second messenger and the target ion channel linked? In some cases, the second messenger molecule directly binds to the ion channel, causing it to open or close. For example, in photoreceptor cells of the retina cyclic GMP directly opens sodium channels in the plasma membrane. In other instances, the second messenger acts indirectly, by activating an enzyme that then affects the ion channel. For example, cyclic AMP activates an enzyme called cyclic-AMP-dependent protein kinase (or protein kinase A). Protein kinase A phosphorylates proteins, by attaching inorganic phosphate to specific amino acids in the protein. Phosphorylation is a common biochemical mechanism by which protein function is altered, including ion channels. For example, phosphorylation of voltage-activated calcium channels is necessary for normal operation of the channel. Thus, a neurotransmitter might indirectly affect calcium channels in a postsynaptic cell by altering the level of cyclic AMP and hence altering phosphorylation of the channels.

How is the activated neurotransmitter receptor molecule linked to enzymes that alter second messenger levels? Once again, the linkage is indirect and involves a protein called a GTP-binding protein (or G-protein). In its inactive state, GDP is bound to the G-protein. The neurotransmitter receptor molecule catalyzes the replacement of GDP by GTP on the G-protein and thus activates the G-protein. The activated G-protein then stimulates the enzyme that produces the second messenger (adenylyl cyclase in the case of cyclic AMP, for example).

Numerous varieties of G-proteins have been identified, each with specific effects on specific target enzymes. Some G-proteins stimulate the activity of the target enzyme, while others inhibit it. Thus, activation of one type of neurotransmitter receptor molecule might increase the level of a second messenger, whereas activation of a different receptor molecule might decrease the level of the second messenger, depending on the type of G-protein to which the receptor is coupled. In addition to acting via second messengers, activated G-proteins may sometimes serve as a messenger that directly activates ion channels.

The indirect actions of neurotransmitters are summarized in Figure 9-13. This sequence can be envisioned as an enzymatic cascade, in which an

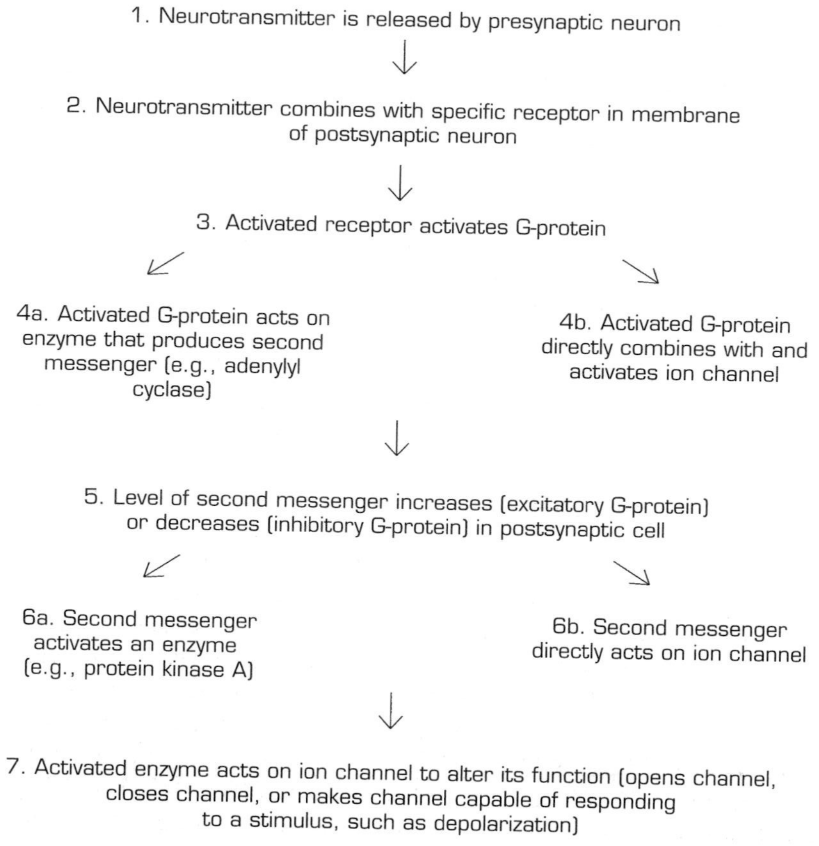

1. Neurotransmitter is released by presynaptic neuron
2. Neurotransmitter combines with specific receptor in membrane of postsynaptic neuron
3. Activated receptor activates G-protein
4a. Activated G-protein acts on enzyme that produces second messenger (e.g., adenylyl cyclase)
4b. Activated G-protein directly combines with and activates ion channel
5. Level of second messenger increases (excitatory G-protein) or decreases (inhibitory G-protein) in postsynaptic cell
6a. Second messenger activates an enzyme (e.g., protein kinase A)
6b. Second messenger directly acts on ion channel
7. Activated enzyme acts on ion channel to alter its function (opens channel, closes channel, or makes channel capable of responding to a stimulus, such as depolarization)

Figure 9-13 The sequence of events in the indirect action of a neurotransmitter on membrane permeability of a postsynaptic cell. Ion channel activity may be altered by G-proteins, by second messengers, or by second-messenger-dependent enzymes. (Animation available at www.blackwellscience.com)

activated neurotransmitter receptor acts as an enzyme to activate G-protein, which in turn activates an enzyme that produces a second messenger. The second messenger then activates another enzyme that affects ion channel operation. In this sequence, an ion channel might be affected at three different points:

- Activated G-protein might bind to and activate an ion channel.
- The second messenger might directly bind to the channel.
- An enzyme, such as a protein kinase, that depends on the presence of the second messenger might act on the ion channel.

In all cases, the net excitatory or inhibitory effect of the neurotransmitter depends on the type of ion channel affected in the postsynaptic membrane and on whether the ion channel is opened or closed by the indirect action of the neurotransmitter.

Presynaptic Inhibition and Facilitation

Inhibition in the nervous system is sometimes accomplished indirectly by targeting excitatory presynaptic terminals, instead of the postsynaptic cell. This type of inhibition, called **presynaptic inhibition**, is illustrated schematically in Figure 9-14. The inhibitory terminal makes synaptic contact with an excitatory synaptic terminal, which in turn contacts the cell to be inhibited. Inhibition is produced by decreasing the release of excitatory neurotransmitter by the excitatory synaptic terminal. The electronmicrograph in Figure 9-14b shows a synaptic arrangement that might give rise to presynaptic inhibition.

Presynaptic inhibition often involves reduced calcium influx into the excitatory terminal during a presynaptic action potential. Reduced calcium influx during presynaptic inhibition results in some cases from decreased size or duration of the depolarization during the presynaptic action potential, which could be accomplished by activating potassium channels in the terminal. Smaller depolarization opens fewer voltage-sensitive calcium channels, and less calcium enters the excitatory terminal. In other cases, presynaptic inhibition involves reduced opening of voltage-sensitive calcium channels, possibly by decreased phosphorylation of the channels.

A synapse onto a synapse, such as the arrangement shown in Figure 9-14, might also facilitate rather than inhibit the release of neurotransmitter from the excitatory terminal. Presynaptic facilitation is known to occur, for example in the nervous system of a sea slug, *Aplysia*. The neurotransmitter serotonin increases the effectiveness of an excitatory synaptic connection from presynaptic sensory neurons onto postsynaptic motor neurons in *Aplysia*. The mechanism of the facilitation of synaptic transmission by serotonin is illustrated in Figure 9-15. Serotonin activates receptors that stimulate G-proteins in the synaptic terminal. One target of the G-proteins is adenylyl cyclase, which increases the concentration of cyclic AMP inside the synaptic terminal. This rise in cyclic AMP enhances neurotransmitter release in two ways. First, cyclic

(a)

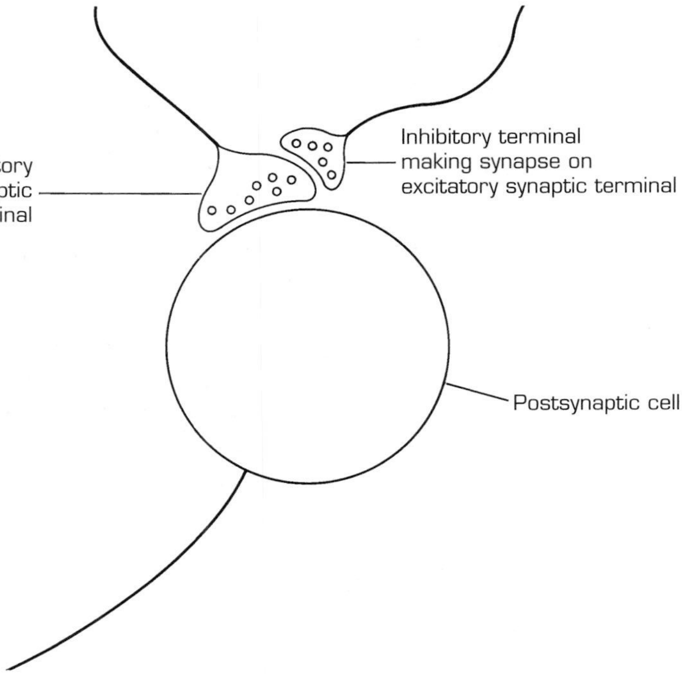

(b)

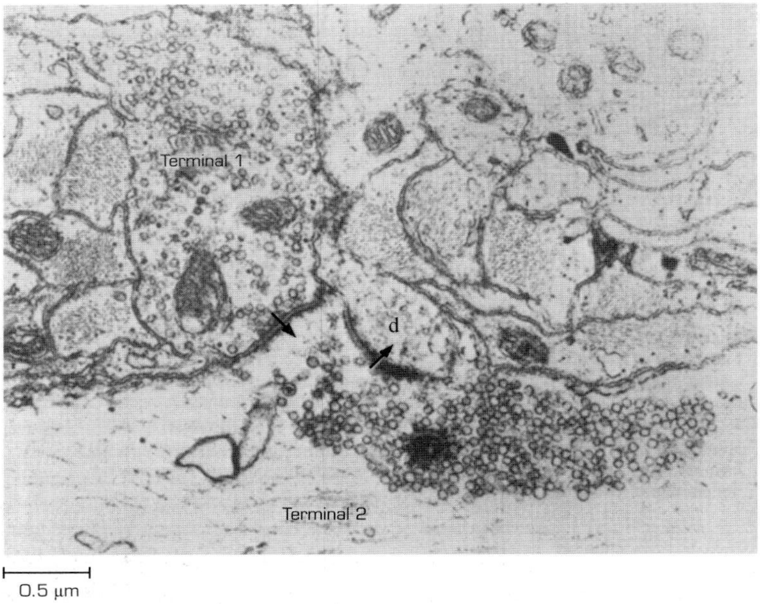

Figure 9-14 (a) Schematic arrangement for presynaptic inhibition in the nervous system. The inhibitory terminal makes synaptic contact with another synaptic terminal, rather than directly with the postsynaptic cell. (b) Electronmicrograph showing a synapse (terminal 1) in the vertebrate central nervous system onto an axon (terminal 2) that in turn makes a synapse onto a third neuronal process (labeled "d" for dendrite). The arrows show the direction of synaptic transmission from terminal 1 to terminal 2 and from terminal 2 to the dendrite. Note the accumulations of synaptic vesicles in terminals 1 and 2. (Part (b) reproduced with permission from W. O. Wickelgren, Physiological and anatomical characteristics of reticulospinal neurones in lamprey. *J. Physiol.* 1977; 270:89–114.)

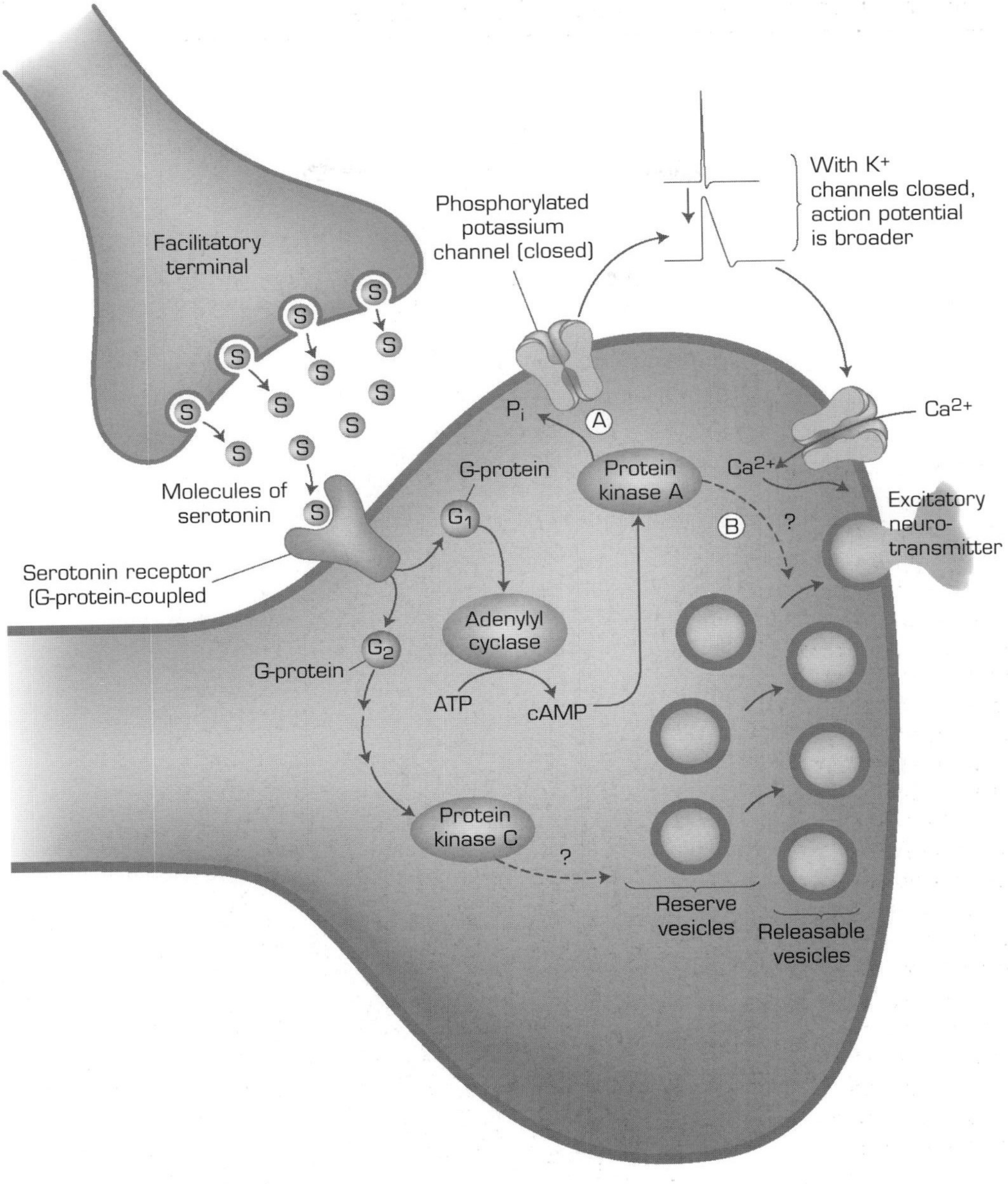

Figure 9-15 A model for presynaptic changes associated with sensitization of the gill withdrawal reflex in *Aplysia.* The synaptic terminals of the facilitatory interneuron release the neurotransmitter serotonin, which combines with serotonin receptors in the membrane of the excitatory synaptic terminals of the gill withdrawal circuit. The activated receptor stimulates two G-proteins: one increases intracellular cyclic AMP via adenylyl cyclase, and a second activates protein kinase C. Cyclic AMP stimulates protein kinase A, which in turn phosphorylates and closes potassium channels. Reduced potassium permeability broadens the presynaptic action potential and enhances calcium influx through voltage-dependent calcium channels. In addition, protein kinase A and possibly protein kinase C may promote movement of synaptic vesicles from reserve pools to releasable pools, thereby potentiating transmitter release.

AMP activates protein kinase A, which phosphorylates potassium channels (pathway A in Figure 9-15). The phosphorylated channels do not open during depolarization, which slows action potential repolarization and allows voltage-activated calcium channels to remain open for a longer time. Thus, a single action potential releases a greater amount of neurotransmitter. Second, the number of synaptic vesicles available to be released by a presynaptic action potential increases in response to cAMP. This effect may be generated by the movement of vesicles from a reserve group to the active zones, where they can fuse with the plasma membrane in response to calcium influx (pathway B in Figure 9-15). The molecular mechanism of this second action of cAMP remains unknown.

Evidence suggests that protein kinase C (PKC) may also be involved in enhancement of neurotransmitter release during sensitization. During sensitization, serotonin receptors indirectly activate PKC via a pathway initiated by a different subclass of G-proteins (Figure 9-15). Activation of PKC closes potassium channels and broadens action potentials, which potentiates calcium influx as described above. In addition, PKC may increase the pool of releasable synaptic vesicles at active zones.

Synaptic Plasticity

Short-term Changes in Synaptic Strength

The size of the postsynaptic response produced by a particular synaptic input in the nervous system is not fixed but instead varies, depending on the past history of activity at that particular synapse. This variation in the strength of a synaptic connection is called **synaptic plasticity**. The presynaptic facilitation of transmission produced by serotonin in *Aplysia* sensory neurons described in the preceding paragraph is one example of synaptic plasticity. In addition to enhancement of synaptic strength, synaptic plasticity sometimes involves a decline in the effectiveness of a synaptic connection. Some presynaptic factors that generate synaptic depression are summarized in Figure 9-16. During a series of presynaptic action potentials, the pool of synaptic vesicles available for release may become depleted, causing the amount of neurotransmitter released to decline with time (Figure 9-16a). In addition, accumulation of calcium inside the presynaptic terminal during a series of action potentials can depress further neurotransmitter release by closing calcium channels (Figure 9-16b; also see Chapter 12). Calcium-dependent inactivation of calcium channels reduces the amount of calcium entering during a presynaptic action potential and thus decreases the amount of neurotransmitter released. Accumulation of calcium in the presynaptic terminal can also open calcium-activated potassium channels (see Chapter 6), which would depress neurotransmitter release by hyperpolarizing the terminal and promoting rapid repolarization following an action potential.

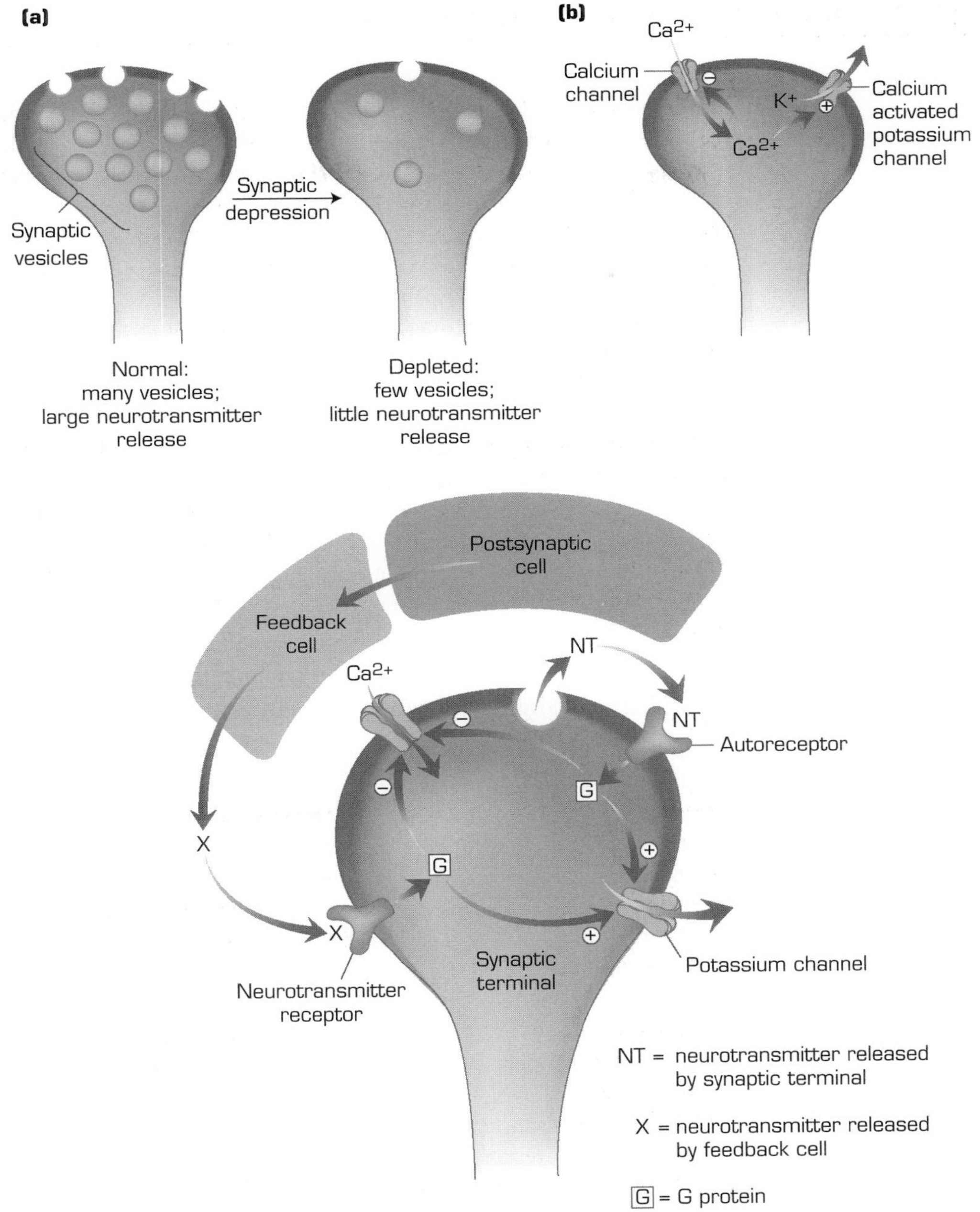

Figure 9-16 Three mechanisms for synaptic depression. (a) After repetitive presynaptic action potentials, the pool of releasable synaptic vesicles can become depleted, leaving fewer vesicles available to respond to subsequent action potentials. (b) Accumulation of calcium ions inside the terminal can inactivate calcium channels (negative sign), or activate calcium-sensitive potassium channels (plus sign). (c) The neurotransmitter molecules (NT) released by a synaptic terminal bind to autoreceptors on the surface of the terminal. The activated autoreceptors then activate G-proteins, leading to closure of voltage-dependent calcium channels (negative sign) or opening of potassium channels (plus sign). In addition, the postsynaptic cell contacted by the synaptic terminal can feed back either directly or indirectly and release a different neurotransmitter (X), which alters calcium and/or potassium channel opening.

Feedback mechanisms are also thought to play a role in synaptic depression (Figure 9-16c). The neurotransmitter released by previous action potentials feeds back, either directly or indirectly, onto the releasing terminal and influences the release of transmitter by subsequent action potentials. Indirect feedback can occur via presynaptic inhibition, described previously in this chapter. Direct feedback can occur via **autoreceptors** in the plasma membrane of the presynaptic terminal. Autoreceptors are activated by neurotransmitter released from the synaptic terminal on which they are located. Because they are usually located in parts of the synaptic terminal at a distance from the synaptic cleft, autoreceptors are activated only when enough neurotransmitter is released to spill out of the synaptic cleft and reach the surrounding parts of the extracellular space. Autoreceptors are usually members of the G-protein-coupled family of receptors, linked indirectly to ion channels via intracellular second messengers. In some cases, the activated autoreceptors reduce neurotransmitter release by closing calcium channels, which reduces the amount of calcium entering during a presynaptic action potential. In other cases, they are linked to the opening of potassium channels, which hyperpolarizes the terminal and speeds repolarization during a presynaptic action potential.

Long-term Changes in Synaptic Strength

Short-term changes in synaptic strength affect neurotransmitter release on a time-scale of seconds to minutes after a burst of activity. In addition, neuronal activity can lead to longer-term aftereffects that alter neurotransmitter release on a time-scale of hours or days. Such long-lasting changes require different cellular mechanisms from those that underlie short-term synaptic enhancement and depression. In this section, we will examine a particularly well studied example of these long-term changes: **long-term potentiation** (abbreviated **LTP**). As the name implies, LTP involves enhancement of synaptic strength lasting a week or more. Although LTP occurs at a variety of sites in the nervous system, we will concentrate on LTP in synaptic connections in a brain region called the **hippocampus**, which is involved in the formation of new memories.

In LTP, a burst of high-frequency activity in a presynaptic input enhances subsequent postsynaptic excitatory responses. Activity in one synapse can affect subsequent responses evoked by another synapse on the same postsynaptic cell (i.e., the potentiation is **heterosynaptic**), provided the synapses are active at approximately the same time (i.e., the potentiation is **associative**). Thus, a weakly stimulated synapse that is active contemporaneously with strong stimulation of the postsynaptic cell becomes potentiated. LTP is initiated in active synapses whenever the synaptic activity is paired with depolarization of the postsynaptic cell. If the postsynaptic neuron is depolarized by injecting positive current into the cell through a microelectrode, LTP is triggered in synaptic responses to the presynaptic cells that were active (even at a low rate) during the artificial depolarization. Synapses that were silent during the postsynaptic depolarization are not potentiated.

How does depolarization of the postsynaptic cell affect subsequent synaptic responses, and why does the potentiation affect only those synapses that are active during the depolarization? To answer these questions, we must first examine the anatomical arrangement of the excitatory synapses in the hippocampus and the properties of the postsynaptic receptor molecules that detect the neurotransmitter, glutamate, released by the presynaptic terminals. As with many other excitatory synapses in the central nervous system, the synaptic terminals contact the dendrites of hippocampal neurons at short, hair-like protuberances called **dendritic spines**. At high magnification, each spine is seen to consist of a knob-like swelling connected via a thin neck of cytoplasm to the main branch of the dendrite, as shown schematically in Figure 9-17. The thin connecting neck allows each spine to behave as a separate intracellular compartment, within which biochemical events can occur in isolation from the rest of the cell. Thus, internal signals can be generated in one spine without spreading to and affecting other spines on the dendrite. Each spine receives input from a single excitatory synaptic terminal. The combination of one terminal and one spine forms a functional synaptic unit that can be modulated separately from the other units on the dendrite of a single neuron. This structural organization may play a central role in the ability of LTP to selectively enhance transmission at active synapses, leaving inactive synapses unaffected.

Two types of glutamate receptors, called NMDA receptors and AMPA receptors, are located in the postsynaptic membrane of the dendritic spine. Both receptor types are glutamate-gated cation channels that have about equal permeability to sodium and potassium, but in addition, NMDA receptors permit influx of calcium ions while AMPA receptors do not. Another important difference between the two receptor types is that AMPA receptors open when glutamate binds, regardless of the membrane potential, whereas NMDA receptors require both glutamate and depolarization to open. NMDA receptors are blocked by external magnesium ions at negative membrane potentials. Block of the channel by magnesium, a divalent cation like calcium, is relieved during depolarization, allowing sodium, potassium, and calcium to move through the open channel. Influx of calcium through the open NMDA channel is the actual trigger for LTP, which explains why both activity and postsynaptic depolarization is required to initiate LTP. Activity provides glutamate and depolarization relieves block by magnesium, both of which are necessary to open NMDA channels and permit calcium to enter the dendritic spine.

The increase in internal calcium has multiple targets in the dendritic spine, summarized in Figure 9-18. LTP reflects, at least in part, an increase in neurotransmitter release from the presynaptic terminal, which raises the question of how an increase in postsynaptic calcium can influence presynaptic events. A **retrograde messenger** is required to transmit information from the dendritic spine to the synaptic terminal. Among the cellular targets for elevated calcium in the dendritic spine is nitric oxide synthase, which is an enzyme that produces **nitric oxide** (NO). NO is membrane permeant and can diffuse from the dendritic spine to the presynaptic terminal, where it might potentiate transmitter

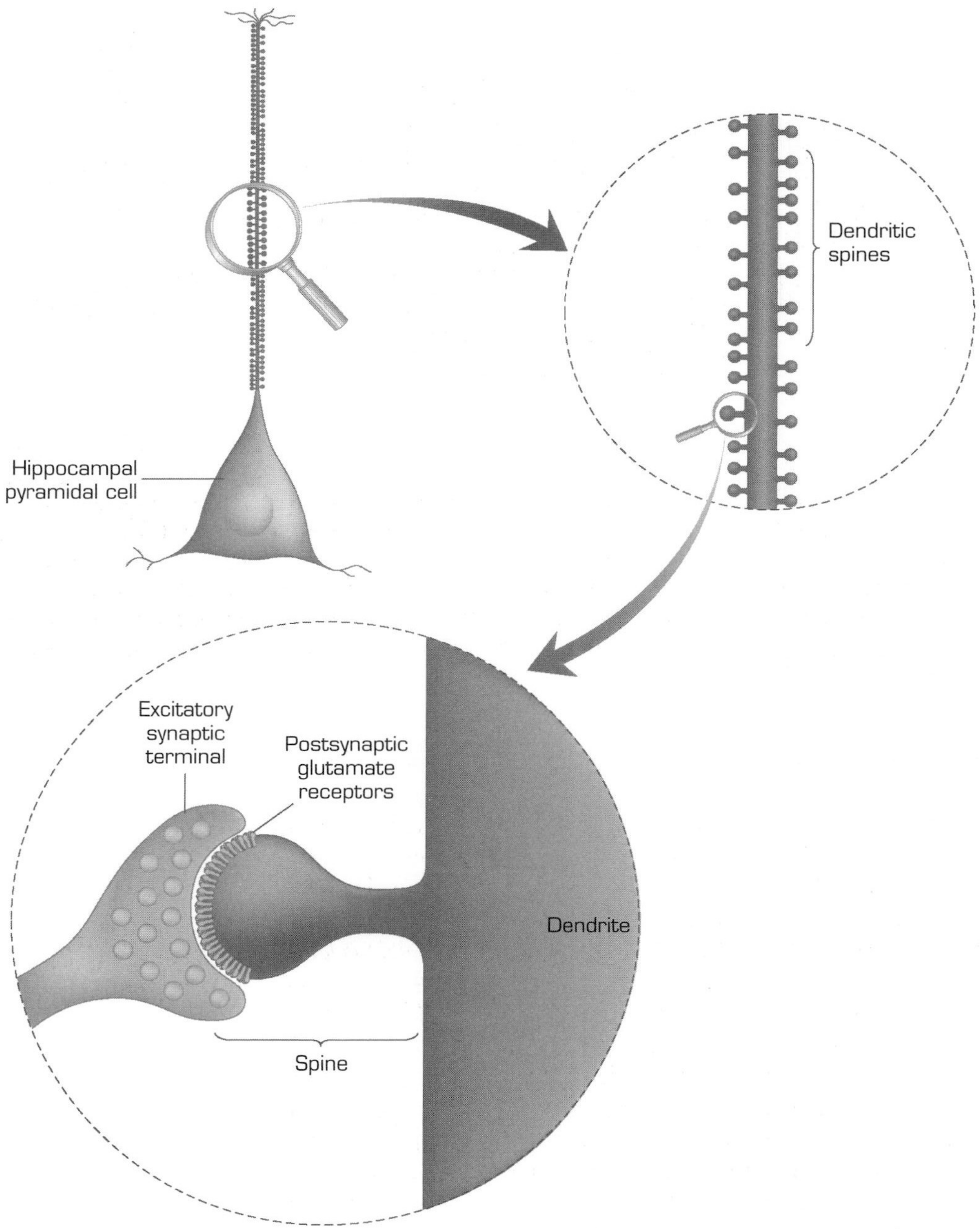

Figure 9-17 Excitatory synapses onto hippocampal pyramidal cells are made onto spike-like protrusions of the dendrites, called dendritic spines.

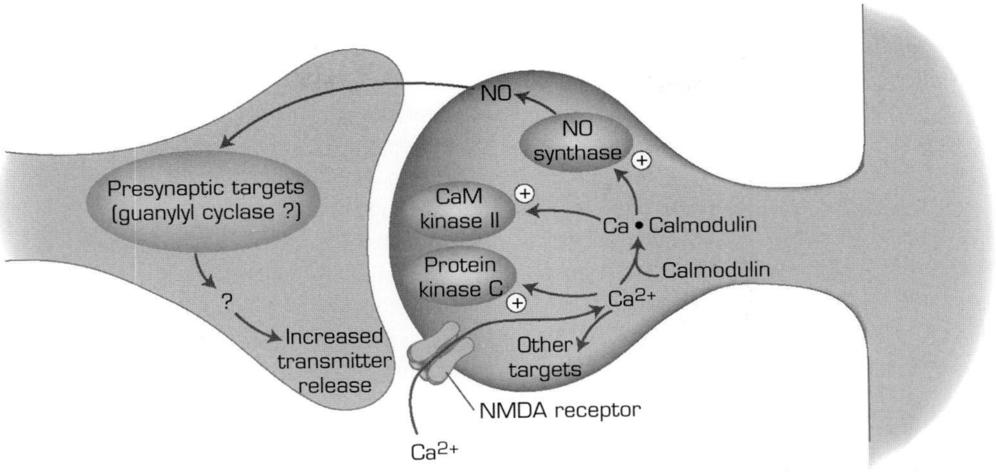

Figure 9-18 Elevated intracellular calcium activates several cellular signals in dendritic spines. Calcium influx through NMDA receptors increases intracellular calcium, which binds to calmodulin. Calcium/calmodulin then activates two enzymes: nitric oxide synthase (NO synthase) and calcium/calmodulin-dependent protein kinase II (CaM kinase II). Calcium also activates protein kinase C. NO synthase produces nitric oxide (NO) from arginine, and the membrane permeant messenger is thought to diffuse to the presynaptic terminal. NO then interacts with cellular signaling pathways, possibly including guanylyl cyclase, to potentiate transmitter release.

release by activating guanylyl cyclase, the synthetic enzyme for the second messenger cyclic GMP.

How might elevated calcium in a spine trigger postsynaptic factors that might also contribute to LTP? Several possible mechanisms for enhanced postsynaptic sensitivity to glutamate have been suggested. Figure 9-18 illustrates some other cellular targets for calcium in the dendritic spine, including two different kinases: **protein kinase C (PKC)** and **calcium/calmodulin-dependent kinase II (CaM kinase II)**. When activated by elevated calcium, these enzymes phosphorylate specific target proteins in the postsynaptic cell. Phosphorylation is a cellular signal often used to activate or inactivate various kinds of proteins. In the case of LTP, the targets for phosphorylation by calcium-dependent kinases have not been established. Phosphorylation may increase the number of functional postsynaptic glutamate receptors, either because phosphorylation allows the channels to open in response to glutamate or because phosphorylation allows channels to attach to the cytoskeleton, anchoring them at the appropriate position in the postsynaptic membrane. Increased glutamate sensitivity might also arise from insertion of additional AMPA receptors into the postsynaptic membrane. All of these factors can potentiate e.p.s.p.'s by making more glutamate receptors available in the postsynaptic membrane to respond to glutamate released by the presynaptic terminal.

The excitatory synapses in the hippocampus demonstrate **long-term depression** (**LTD**), as well as LTP. If a synaptic input is activated at a low rate for a few minutes without strong activity in other synapses, the size of the e.p.s.p. elicited by that synaptic input diminishes and remains at this lower level for many hours. In this regard, LTD can be considered the opposite of LTP. In LTP, the effectiveness of a weakly stimulated synaptic input is enhanced when paired with strong activation of other pathways. In LTD, the effectiveness of a weakly stimulated synapse becomes reduced if its activation occurs in the absence of strong stimulation in other synaptic inputs. If LTP is induced at a particular synapse, it can subsequently be reversed by LTD. This fact suggests that LTD represents an erasure mechanism for LTP in the hippocampus: unless activation of a synaptic input is *consistently* strongly activated or paired with strong activation of other inputs, potentiation of that input is reversed by LTD.

Summary

Chemical synapses between neurons in the nervous system are similar to the synapse at the neuromuscular junction in the following ways:

- Neurotransmitter molecules are stored in the synaptic terminal in membrane-bound synaptic vesicles.
- Influx of external calcium ions into the terminal triggers release of neurotransmitter.
- Synaptic vesicles fuse with the plasma membrane of the terminal to release their neurotransmitter content.
- Neurotransmitter molecules combine with specific postsynaptic receptors' molecules and open ion channels in the postsynaptic membrane.

Nervous system synapses differ from the neuromuscular junction in the following ways:

- At most synapses, a single presynaptic action potential produces only a small change in postsynaptic membrane potential. By contrast, a single presynaptic action potential at the neuromuscular junction produces a large depolarization of the muscle cell and triggers a postsynaptic action potential.
- Synapses between neurons can be either excitatory or inhibitory.
- Acetylcholine is the neurotransmitter at the neuromuscular junction, but many different neurotransmitter substances (including ACh) are released at synapses in the nervous system.
- A skeletal muscle cell receives synaptic input from only one neuron—a single motor neuron. A neuron in the nervous system may receive synaptic connections from thousands of different neurons. The output of a neuron depends on the integration of all the inhibitory and excitatory inputs active at a given instant.
- At the neuromuscular junction, ACh directly opens channels by combining with postsynaptic binding sites that are part of the channel protein. In other

parts of the nervous system, a neurotransmitter may directly bind to an ion channel or may indirectly affect ion channels via an internal second messenger in the postsynaptic cell.

• Synapses in the central nervous system usually undergo short-term or long-term changes in synaptic strength, and this synaptic plasticity plays a role in the complex information-processing capacity of the brain.

Cellular Physiology of Muscle Cells

part III

Part III of this book describes the second major type of excitable cell: muscle cells. These cells are specialized to produce movement when they are electrically stimulated. Because muscle cells produce visible movements, their actions are the most obvious external manifestation of the activity of the nervous system. The point of interaction between the nervous and muscular systems—the neuromuscular junction—was the central focus for the discussion of chemical synaptic transmission in Chapter 8. In the first chapter of Part III, Chapter 10, we return to the neuromuscular junction and examine the sequence of events linking an action potential in the postsynaptic muscle cell to mechanical contraction. This linkage is the process called **excitation–contraction coupling**, and the explanation of this process in terms of underlying molecular mechanisms stands as one of the major accomplishments of cellular physiology. Chapter 11 then discusses how the nervous system controls the strength of contraction of an entire skeletal muscle by regulating the twitch contractions of the individual muscle cells making up the muscle. Chapter 12 considers the important electrical differences between the muscle cells of skeletal muscles and the heart. These electrical differences underlie the ability of the heart to produce the rhythmic, coordinated contractions necessary to pump blood through the body. The control of the heart by the autonomic nervous system is also considered in Chapter 12.

Excitation–Contraction Coupling in Skeletal Muscle

10

Throughout Part II of the book, we used the patellar reflex as an example system to explore the cellular signals underlying nervous system function. The final stage of the patellar reflex is the contraction of the quadriceps muscle brought about by the activity of the motor neurons making excitatory synaptic connections with that muscle. The arrival of an action potential in the synaptic terminal of the presynaptic motor neuron causes release of the chemical neurotransmitter, ACh. The ACh in turn depolarizes the end-plate region of the postsynaptic muscle cell, initiating an action potential in the muscle cell. This action potential propagates along the long, thin muscle cell just as the neuron action potential propagates along nerve fibers. The muscle action potential serves as the trigger for contraction of the muscle cell. This chapter will examine the events that intervene between the occurrence of the action potential in the plasma membrane of the muscle cell and the activation of the contraction: the process of excitation–contraction coupling. Then we will move on in Chapter 11 to look at how the motor nervous system is organized to integrate the twitch-like contractions of individual muscle fibers into the smooth and graded contractions of a muscle as a whole.

To begin, it will be useful to examine the structure of the muscle cells at the level of both the light and electron microscopes. We will then consider the molecular makeup of the contractile apparatus and discuss the biochemical mechanisms that control the action of that apparatus. The chapter will conclude with a discussion of how the action potential of the muscle cell is coupled to the contractile machinery to produce the muscle contraction.

The Three Types of Muscle

There are two general classes of muscle in the body: striated and smooth. Both are named for the characteristic appearance of the individual cells making up the muscle tissue when viewed through a microscope. Striated muscle cells

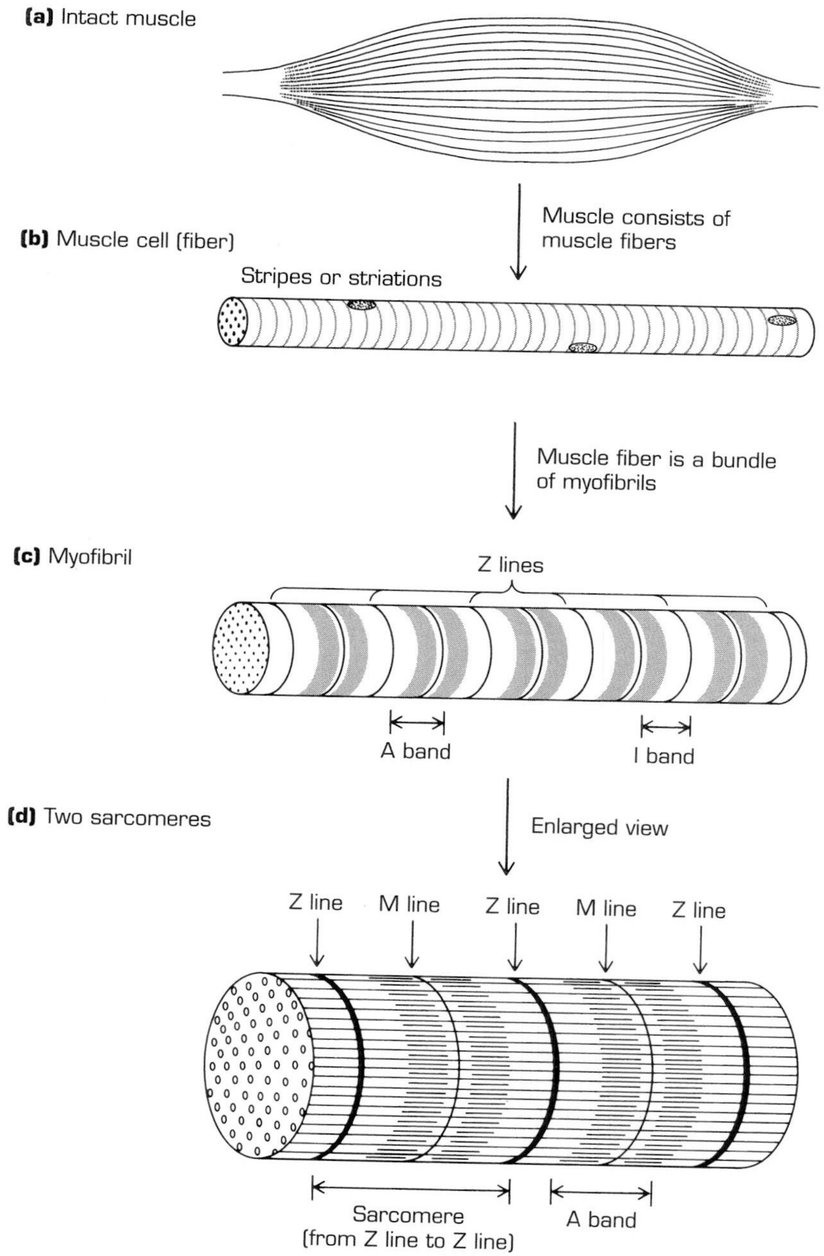

Figure 10-1 Microscopic structure of skeletal muscle. Muscle components are viewed at increasing magnification in (a) through (d).

(Figure 10-1) exhibit closely spaced, crosswise stripes (striations). Smooth muscle cells have no striations and have a smooth appearance. Smooth muscle is found in the gut, blood vessels, the uterus, and other locations where contractions are usually slow and sustained. The muscles that move and support the skeletal framework of the body—the skeletal muscles—are made up of striated muscle cells. This chapter will focus on the structure and properties of skeletal muscle cells.

The cells that make up the muscle of the heart are also striated, like skeletal muscle. Because the membranes of cardiac muscle cells are electrically quite different from those of skeletal muscle cells, cardiac muscle is usually regarded as a distinct class of muscle in its own right. The characteristics of cardiac muscle will be discussed in Chapter 12.

Structure of Skeletal Muscle

Figure 10-1 shows the structure of a typical mammalian skeletal muscle at progressively greater magnification. To the naked eye, an intact muscle appears to be vaguely striped longitudinally, as in Figure 10-1a. Upon closer inspection, the muscle is made up of bundles of individual cells: the muscle cells or muscle fibers (Figure 10-1b). In mammalian muscle, the individual cells are about 50 μm in diameter and are typically as long as the whole muscle. Thus, muscle cells are long, thin fibers similar in shape to neuronal axons. The end-plate region, where synaptic input from the motor neuron is located, is only a few microns in length. Therefore, a rapidly propagating action potential—like that of a nerve cell—is required in skeletal muscle cells to transmit the depolarization initiated at the end-plate along the entire length of the muscle fiber.

Individual muscle cells consist of bundles of still smaller fibers called myofibrils. The plasma membrane of a single muscle cell encloses many myofibrils. At the level of the myofibrils, the structural basis of the crosswise striations of skeletal muscle cells becomes apparent. As shown in Figure 10-1c, myofibrils exhibit a repeating pattern of crosswise light and dark stripes: the A band, I band, and Z line. The I band is a predominantly light region with the dark Z line at its center, while the A band is a darker region separating two I bands of the repeating pattern. At still higher magnification, the A band can be seen to have its own internal structure (Figure 10-1d); two darker areas at the outer edges of the A band are separated by a lighter region with a faint dark line, called the M line, at the center. The basic unit of the repeating striation pattern of a myofibril is called a sarcomere, which is defined as extending from one Z line to the next—that is, from the center of one I band to the center of the next I band.

Changes in Striation Pattern on Contraction

When a muscle cell contracts, the relationship among the stripes changes. This change can best be appreciated at the level of the electron microscope, as shown diagrammatically in Figure 10-2. Through the electron microscope, a myofibril can be seen to consist of two kinds of longitudinally oriented filaments, called thick and thin filaments. Both the thick and thin filaments are arrayed in parallel groups. As shown in Figure 10-2, the Z line corresponds to the position where the thin filaments of one sarcomere join onto those of the neighboring

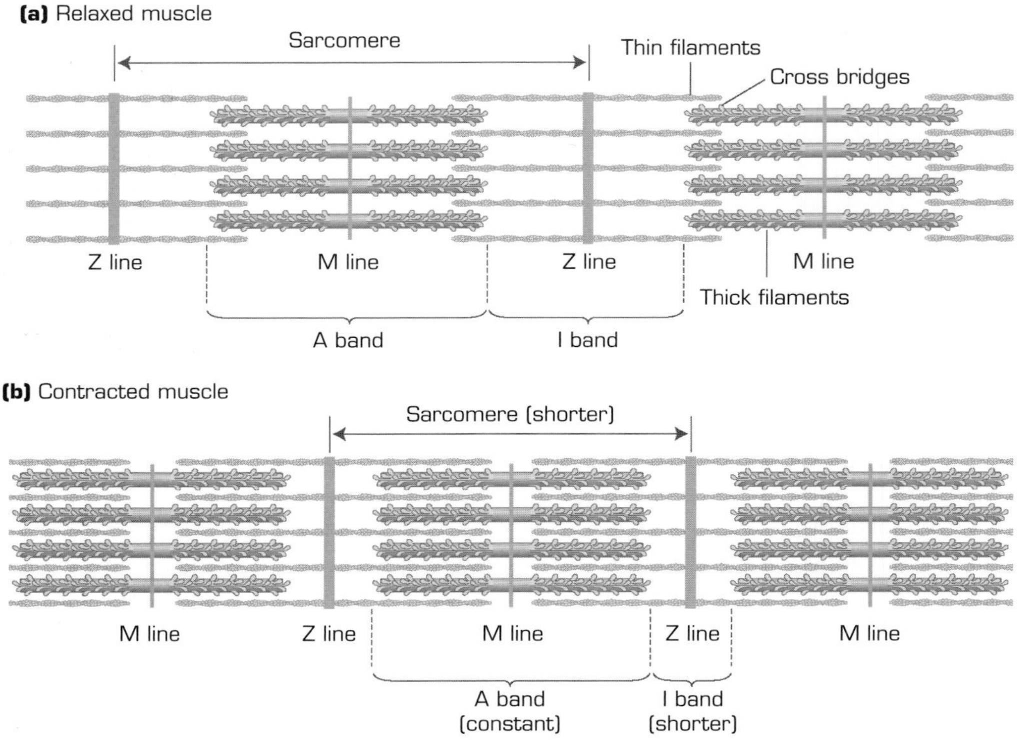

Figure 10-2 Schematic representation of the relationships between thick and thin filaments of a myofibril in a relaxed muscle (a) and a contracted muscle (b). In a contracted muscle, the sarcomere is shorter, because the degree of overlap of thick and thin filaments is greater.

sarcomere and where cross-connections are made among the parallel thin filaments. The thick filaments within a sarcomere are joined to each other at the M line. It is clear from comparing Figure 10-2a with Figure 10-1d that the lighter I band corresponds to the region occupied only by thin filaments, and the darker A band corresponds to the spatial extent of the thick filaments. The darker regions at the two edges of the A band correspond to the region of overlap of the thick and thin filaments. The thick filaments bear thin fibers that appear to link to the thin filaments in the region of overlap, forming cross-bridges between the thick and thin filaments.

Upon contraction, the length of each sarcomere shortens—that is, the distance between successive Z lines diminishes. However, the width of the A band is unaffected by contraction; thus, only the I band becomes thinner during a contraction. In terms of the thick and thin filaments, this observation can be explained by the sliding filament hypothesis, which is illustrated in Figure 10-2b. Neither the thick nor the thin filaments change in length during a contraction; rather, shortening occurs because the filaments slide with respect to one another, so that the region of overlap increases. In order to understand

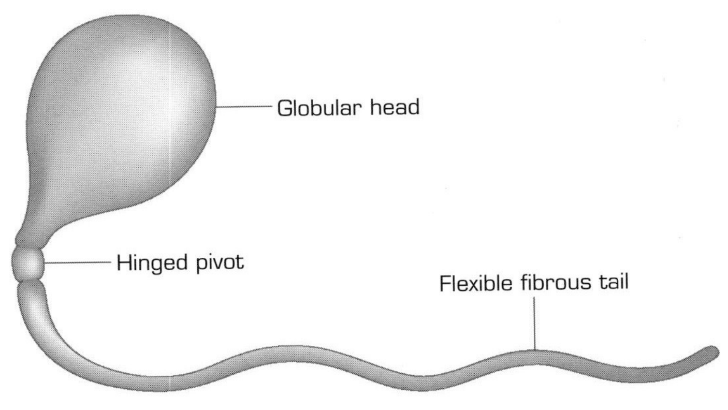

Figure 10-3 The overall structure of a single molecule of the thick filament protein, myosin. The flexible fibrous tail is connected to the globular head region via a hinged point. The globular head includes a region that can bind and split a molecule of ATP.

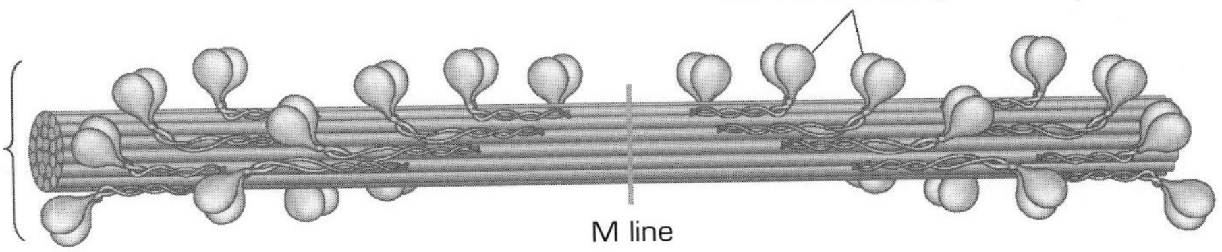

Figure 10-4 The structure of a thick filament. The fibrous tails of individual myosin molecules polymerize to form the backbone of the filament. The globular heads radiate out perpendicular to the long axis of the filament to form the cross-bridges to the thin filament. The myosin molecules reverse orientation at the M line, at the midpoint of the filament.

how the sliding occurs, it will be necessary to examine the molecular makeup of the thick and thin filaments.

Molecular Composition of Filaments

The thick filaments are aggregates of a protein called myosin, which consists of a long fibrous "tail" connected to a globular "head" region, as shown schematically in Figure 10-3. The fibrous tails tend to aggregate into long filaments, with the heads projecting off to the side. Figure 10-4 shows a generally accepted view of how myosin molecules are arranged in the thick filaments of a sarcomere. The aggregated tails form the backbone of the thick filament, and the globular heads form the cross-bridges that connect with adjacent thin filaments.

The globular head of myosin contains a region that can bind ATP and split one of the high-energy phosphate bonds of the ATP, releasing the stored chemical energy. That is, myosin acts as an ATPase. The energy provided by the ATP is transferred to the myosin molecule, which is transformed into an "energized" state. This sequence can be summarized as follows:

$$\text{Myosin} + \text{ATP} \rightarrow \text{Myosin}\cdot\text{ATP} \rightarrow \text{Energized Myosin}\cdot\text{ADP} + P_i$$

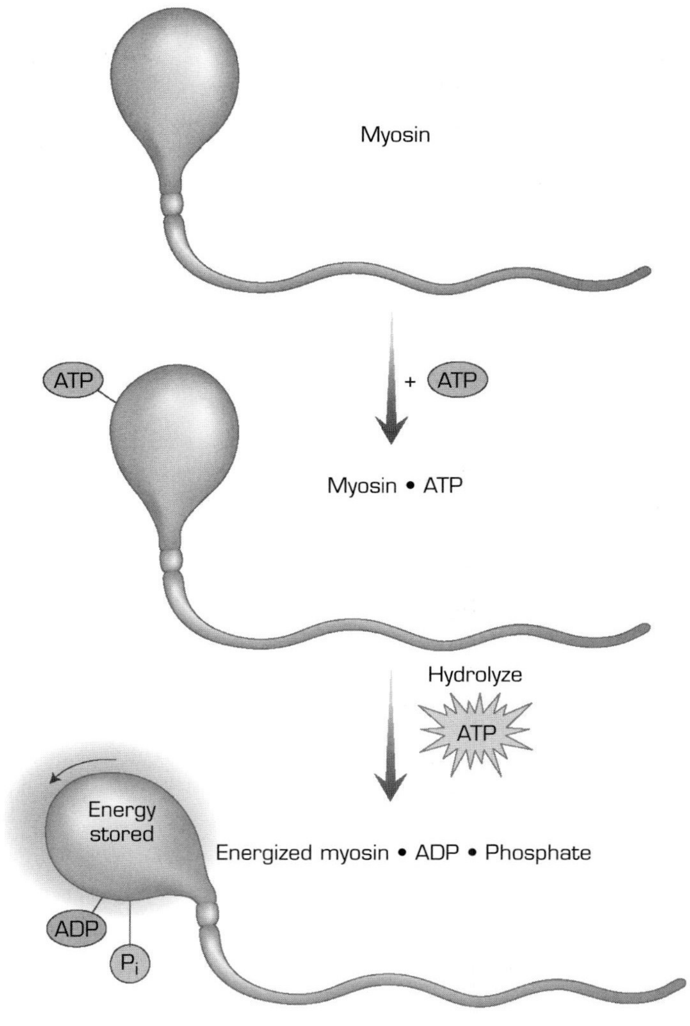

Figure 10-5 Myosin is an ATPase. ATP binds to the globular head of myosin, which catalyzes hydrolysis of ATP to ADP + inorganic phosphate (P_i). Energy released by ATP hydrolysis is stored in myosin, which is transformed into an energized form. The transition from the resting to the energized state of myosin involves rotation of the globular head around its flexible attachment site to the fibrous tail.

Here, the dot indicates that two molecules are bound together, as in an enzyme–substrate complex. To make a mechanical analogy, the globular head behaves as though it is attached to the fibrous tail at a hinged connection point. The energy released by ATP causes the head to pivot about the hinge into the energized state, as drawn schematically in Figure 10-5. To continue with mechanical analogies, this can be thought of as cocking the spring-loaded hammer of a cap pistol. As we will see shortly, the energy stored in this energized form of myosin is the energy that fuels the sliding of the filaments during contraction.

The thin filaments within a myofibril are largely made up of the protein actin. Thin filaments also contain two other kinds of protein molecule called troponin and tropomyosin, whose roles in contraction will be discussed a little later; for the present we will concentrate on the actin molecules. Actin is a globular protein that polymerizes to form long chains; thus, the thin filament can be thought of as a long string of actin molecules, like a pearl necklace. (Actually, each thin filament consists of two actin chains entwined about each other in a helix, but for a conceptual understanding of the sliding filament hypothesis it is not necessary to keep this in mind.) Each actin molecule in the chain contains a binding site that can combine with a specific site on the globular head of a myosin molecule. This is the site of attachment of the crossbridge on the thin filament.

Interaction between Myosin and Actin

When actin combines with energized myosin, the stored energy in the myosin molecule is released. This causes the myosin molecule to return to its resting state, and the globular head pivots about its hinged attachment point to the thick filament. The pivoting motion requires that the thick and thin filaments move longitudinally with respect to each other. This mechanical analog is illustrated schematically in Figure 10-6. The exact nature of the chemical changes in a myosin molecule during the transition from resting to energized state and back is unknown at present; the sliding filament hypothesis, however, requires that there be some chemical equivalent of the hinged arrangement shown in Figure 10-6.

How is the bond between actin and myosin broken so that a new cycle of sliding can be initiated? In the scheme presented so far, each myosin molecule on the thick filament could interact only one time with an actin molecule on the thin filament, and the total excursion of sliding would be restricted to that produced by a single pivoting of the globular head. In order to produce the large movements of the filaments that actually occur, it is necessary that the attachment of the cross-bridges is broken so that the cycle of myosin energization, binding to actin, and movement can be repeated. The full cycle that allows this to occur is summarized in Figure 10-7. When energized myosin binds to actin and releases its stored energy, the ADP bound to the ATPase site of the globular head is released. This allows a new molecule of ATP to bind to the myosin. When this happens, the bond between actin and myosin is broken, possibly because of structural changes in the globular head induced by the interaction between ATP and myosin. The new ATP molecule can then be split by myosin to regenerate the energized form, which is then free to interact with another actin molecule on the chain making up the thin filament. Note that there are two roles for ATP in this scheme: to provide the energy to "cock" myosin for movement, and to break the interaction between actin and myosin after movement has occurred. If there is no ATP present, actin and myosin get stuck together and a rigid muscle results (as in rigor mortis).

Each of the many myosin heads on an individual thick filament independently goes through repetitive cycles of energization by ATP, binding to actin,

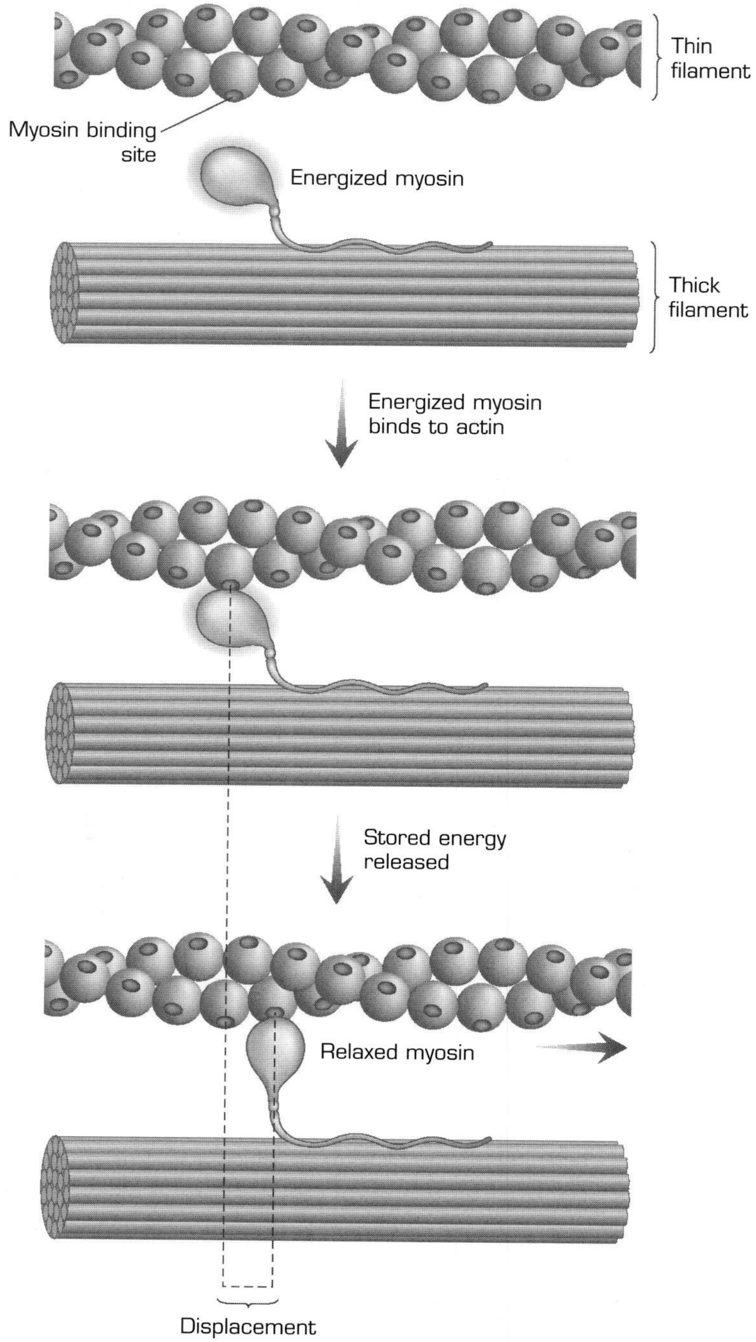

Figure 10-6 A schematic representation of the mechanism of filament sliding during contraction of a myofibril. The globular head of energized myosin binds to a specific binding site on actin, and the energy stored in myosin is released. The resulting relaxation of the myosin molecule entails rotation of the globular head, which induces longitudinal sliding of the filaments.

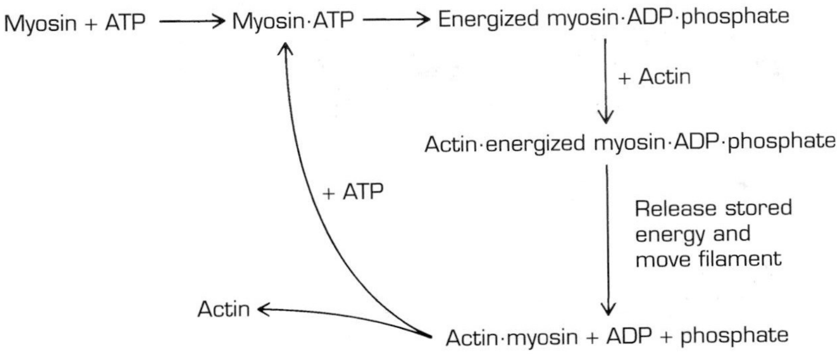

Figure 10-7 The cycle of cross-bridge formation and dissociation between myosin and actin during filament sliding.

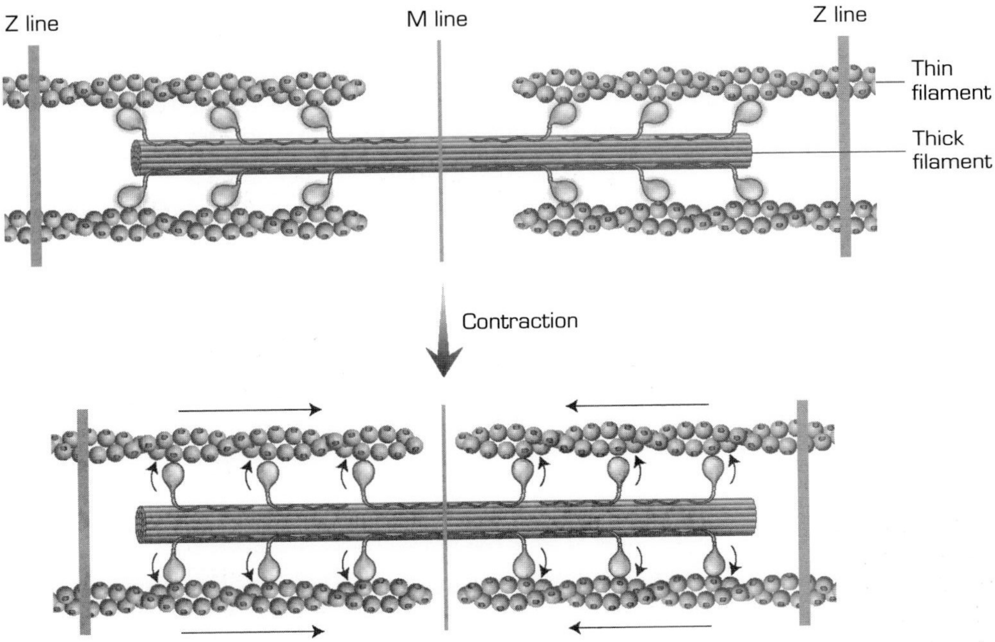

Figure 10-8 The mechanism of sarcomere shortening during contraction. For clarity, the myosin heads are shown acting in concert, although in reality they behave independently.

releasing stored energy to produce sliding, and detachment from actin. Each cycle results in the splitting of one molecule of ATP to ADP and inorganic phosphate. Note from Figure 10-4 that the orientation of the myosin heads reverses at the midpoint of the thick filament, the M line. This is the proper orientation to pull both Z lines at the boundary of a sarcomere toward the center (Figure 10-8). The thin filaments attached to the left Z line will be pulled to the right by the cyclical pivoting of the myosin cross-bridges. Similarly, the thin filaments attached to the right Z line will be pulled to the left. Thus, each sarcomere in each myofibril shortens, and the whole muscle shortens.

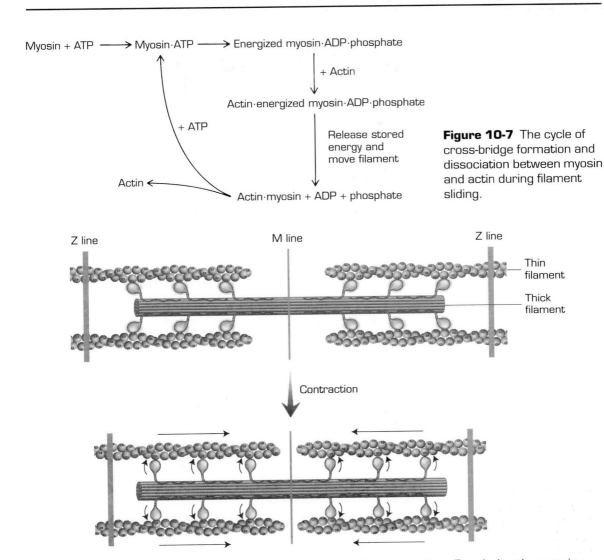

Regulation of Contraction

In the scheme summarized in Figure 10-7, there is no mechanism to control the interaction between actin and myosin. That is, as long as ATP is present, we would expect every muscle in the body to be in a perpetual state of maximum contraction. This section will examine the molecular mechanisms that prevent the interaction of actin with myosin except when a contraction is triggered by an action potential in the muscle cell membrane.

Recall that thin filaments also contain the proteins troponin and tropomyosin. These proteins are responsible for regulating the interaction between individual myosin and actin molecules in the thick and thin filaments. The regulatory scheme is summarized by the diagrams in Figure 10-9. In the resting muscle, tropomyosin is in a position on the thin filament that allows it to effectively cover the myosin binding site on actin. Myosin's access to the binding site is blocked by the tropomyosin. The position of tropomyosin on the actin polymer is in turn regulated by troponin. In the resting state, troponin locks tropomyosin in the blocking position. Thus, tropomyosin acts like a trapdoor

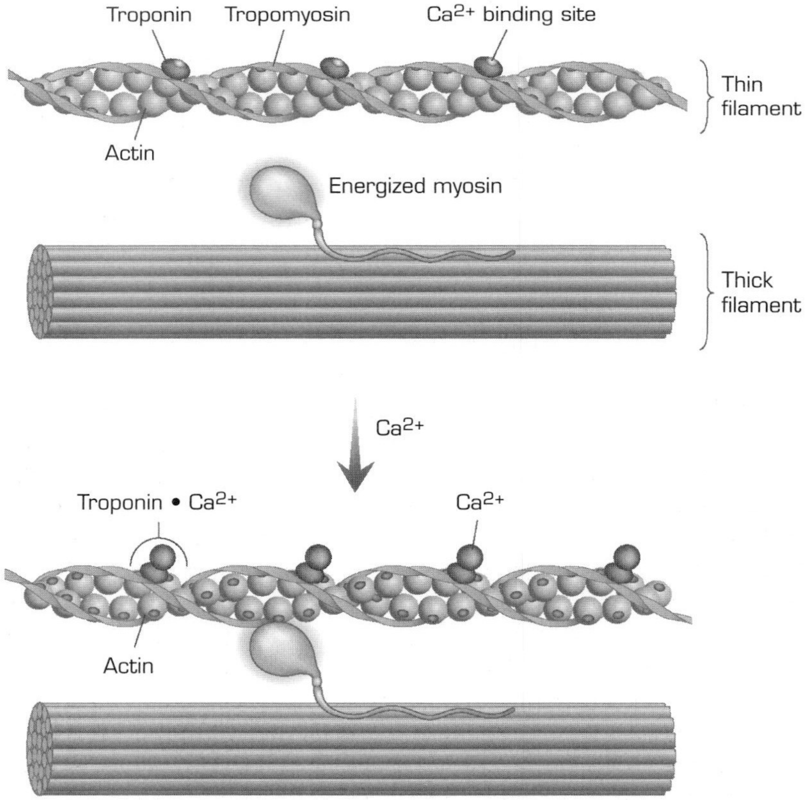

Figure 10-9 The regulation of the interaction between actin and myosin by calcium ions, troponin, and tropomyosin. In the absence of calcium ions, tropomyosin blocks access to the myosin-binding site of actin (upper diagram). In the absence of calcium ions, troponin locks tropomyosin in the blocking position. When calcium binds to troponin, the positions of troponin and tropomyosin are altered on the thin filament, and myosin then has access to its binding site on actin. The cycle of filament sliding is then free to begin. (Animation available at www.blackwellscience.com)

covering the myosin binding site, and troponin acts like a lock to keep the door from opening.

What is the trigger that causes tropomyosin to reveal the myosin binding sites on actin? The signal that initiates contraction is the binding of calcium ions to troponin. Each troponin molecule contains a specific binding site for a single calcium ion. Normally, the concentration of calcium inside the cell is very low, and the binding site is not occupied. It is in this state that troponin locks tropomyosin in the blocking position. When an action potential occurs in the muscle cell plasma membrane, however, the concentration of calcium ions in the intracellular fluid rises dramatically, and calcium binds to troponin. When this happens, there is a structural change in the troponin molecule, and the interaction between troponin and tropomyosin is altered in such a way that tropomyosin uncovers the myosin binding site on actin. The cycle of events depicted in Figure 10-7 is then allowed to occur, and the filaments slide past each other.

The Sarcoplasmic Reticulum

Where does the calcium come from to trigger the interaction of actin and myosin underlying the sliding of the filaments? Recall from Chapter 8 that a rise in internal calcium is also responsible for the release of chemical transmitter during synaptic transmission, and that in that case the calcium enters the cell from the ECF through voltage-sensitive calcium channels in the plasma membrane. In the case of skeletal muscle, however, the calcium does not come from outside the cell; rather, the calcium is injected into the intracellular fluid from a separate intracellular compartment called the sarcoplasmic reticulum. The sarcoplasmic reticulum is an intracellular sack that surrounds the myofibrils of a muscle cell. This sack forms a separate intracellular compartment, bounded by its own membrane that is not continuous with the plasma membrane of the muscle cell.

The concentration of calcium ions inside the sarcoplasmic reticulum is much higher than it is in the rest of the space inside the cell; in fact, it is probably higher than the concentration of calcium in the ECF. This accumulation of calcium inside the sarcoplasmic reticulum is accomplished by a calcium pump in the membrane of the sarcoplasmic reticulum. Like the sodium pump of the plasma membrane, this calcium pump uses metabolic energy in the form of ATP to transport calcium ions across the membrane against a large concentration gradient; in this case, the pump moves calcium ions into the sarcoplasmic reticulum. To initiate a contraction, a puff of calcium ions is released from the sarcoplasmic reticulum, which is strategically located surrounding the contractile apparatus of the myofibrils. The action of the released calcium is terminated as the ions are pumped back into the sarcoplasmic reticulum by the calcium pump. Here, then, is a third role for ATP in the contraction process: ATP, as the fuel for the calcium pump, is responsible for terminating a contraction as well as for energizing myosin and breaking the bond between actin and myosin.

Calcium is released from the sarcoplasmic reticulum via calcium-selective ion channels, which are located in the sarcoplasmic reticulum membrane. These calcium channels are quite different from the voltage-dependent calcium channels we have encountered previously in our discussion of synaptic transmission. Rather than being activated by depolarization, as are the voltage-dependent calcium channels, these calcium channels in the sarcoplasmic reticulum are activated by an increase in cytoplasmic calcium concentration. For this reason, they are referred to as calcium-induced calcium release channels. If calcium is released from the sarcoplasmic reticulum via channels that are themselves activated by an increase in calcium, then the calcium release process exhibits positive feedback reminiscent of the rising phase of the action potential (where depolarization opens sodium channels, which in turn produce further depolarization). This positive feedback ensures that the calcium release is large and rapid, producing fast and complete activation of the contraction mechanism.

The Transverse Tubule System

How does an action potential in the plasma membrane of a muscle cell trigger release of calcium from the sarcoplasmic reticulum, whose membrane is separate from the plasma membrane? The crucial aspect of the action potential in triggering contraction is depolarization of the plasma membrane. However, to affect the sarcoplasmic reticulum surrounding myofibrils deep within the muscle cell, the depolarization produced by the action potential at the outer surface of the cell must somehow be transmitted to the interior of the muscle cell. To accomplish this task, the plasma membrane of the muscle cell forms periodic infoldings, called transverse tubules, that extend into the depths of the muscle fiber (Figure 10-10). The long fingers of plasma membrane projecting into the cell provide a path for depolarization resulting from an action potential in the surface membrane to influence events in the interior of the cell.

In most species, the transverse tubules are located at the boundary between the A band and the I band. This location represents the edge of overlap between the thick and thin filaments in a resting muscle fiber, and it makes sense that calcium release should be triggered first at this position at the leading edge of filament sliding. Where the transverse tubules encounter the sarcoplasmic reticulum, the membranes come into close apposition to form a structure called a triad. This is presumably the point of interaction between the depolarizing signal and the membrane of the sarcoplasmic reticulum. Note, however, that although the membranes are close together at a triad, they do not touch. Because the membranes are not in continuity, the depolarization produced during the action potential cannot spread directly to the sarcoplasmic reticulum. Therefore, some indirect signal is required to link depolarization of the transverse tubule to calcium release by the sarcoplasmic reticulum.

Because calcium is released from the sarcoplasmic reticulum through calcium-induced calcium release channels, it is natural to suppose that the link between transverse tubules and sarcoplasmic reticulum is mediated by an

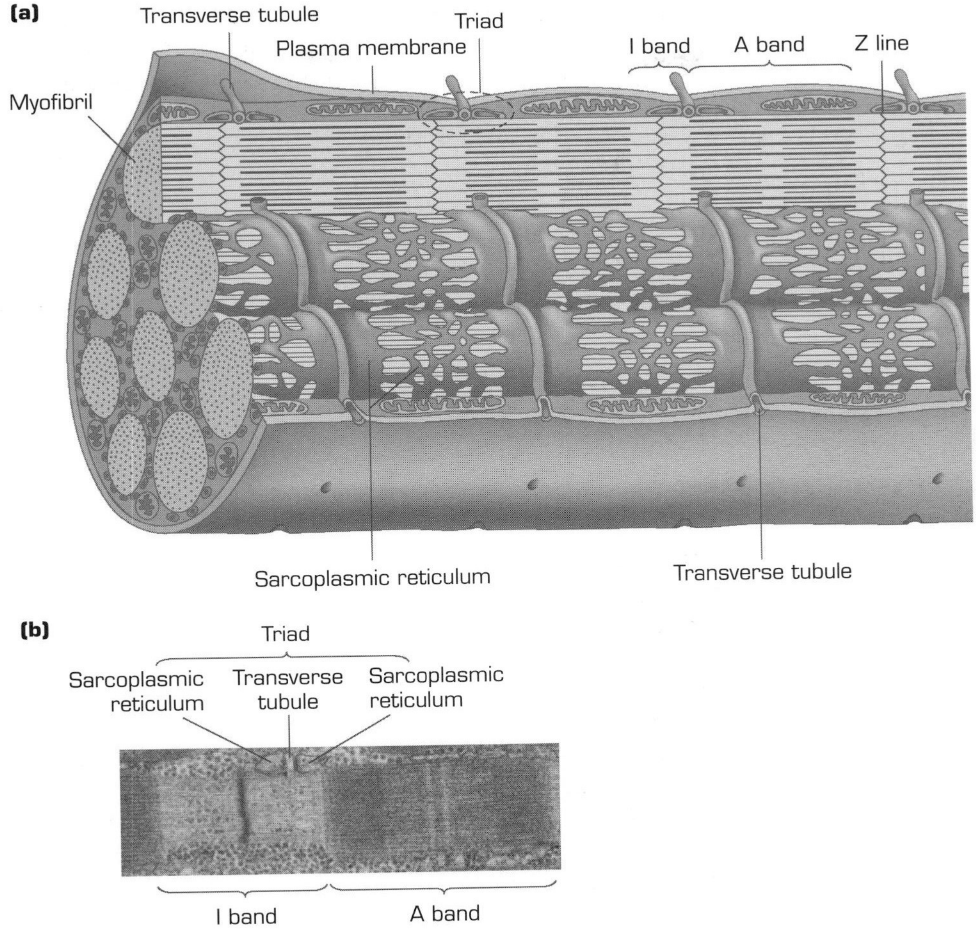

Figure 10-10 The sarcoplasmic reticulum and transverse tubules. (a) The transverse tubules are invaginations of the plasma membrane of the muscle cell. Depolarization during an action potential can spread along the transverse tubules to the interior of the fiber. The sarcoplasmic reticulum is an intracellular compartment surrounding each myofibril in the muscle cell. Calcium ions that trigger contraction are released from the sarcoplasmic reticulum. The membranes of the transverse tubules and the sarcoplasmic reticulum come close together at the triad. Depolarization of the membrane of the tubules triggers calcium release from the sarcoplasmic reticulum. (b) The triad near a single myofibril, viewed through the electron microscope. (Electron micrograph provided by B. Walcott of the State University of New York at Stony Brook.)

influx of calcium ions from the extracellular space. Indeed, the membrane of the transverse tubules contain voltage-dependent calcium channels that are opened by depolarization, and these calcium channels are required for the initiation of contraction. In cardiac muscle, influx of calcium from the extracellular fluid via these depolarization-activated calcium channels is in fact required to initiate the muscle contraction, so that calcium ions through the tranverse tubule calcium channels do indeed trigger the calcium release from the sarcoplasmic reticulum in cardiac muscle cells. However, in skeletal muscle, calcium influx via the calcium channels of the transverse tubules is not required to trigger contraction, and some other linkage mechanism is required. The mechanism is not yet understood in detail, but it is thought that the calcium channels of the transverse tubule act as voltage sensors to detect the depolarization produced by the action potential and that there is a direct physical link—extending through the intracellular space—connecting single calcium-induced calcium-release channels in the sarcoplasmic reticulum membrane to single voltage-dependent calcium channels in the immediately adjacent membrane of the transverse tubule. Through this physical link, a conformation change in the transverse-tubule calcium channel upon depolarization is thought to induce a conformation change in the calcium-induced calcium-release channels. The sarcoplasmic reticulum channel then opens, locally releasing calcium and initiating the explosive calcium-induced release of calcium from the sarcoplasmic reticulum as a whole.

Summary

The sequence of events leading to contraction of a skeletal muscle fiber following stimulation of its motor neuron can be summarized as follows:

1. Acetylcholine released from the presynaptic terminal depolarizes the end-plate region of the muscle fiber.

2. The depolarization initiates an all-or-none action potential in the muscle fiber, and the action potential propagates along the entire length of the fiber.

3. Depolarization produced by the action potential spreads to the interior of the fiber along the transverse tubule system.

4. Depolarization of the transverse tubules causes release of calcium ions by the sarcoplasmic reticulum.

5. Released calcium ions bind to troponin molecules on the thin filaments.

6. When calcium combines with troponin, tropomyosin uncovers the myosin-binding site of actin.

7. Globular heads of myosin molecules, which have been energized by splitting a high-energy phosphate bond of ATP, are then free to bind to actin.

8. The stored energy of the activated myosin is released to propel the thick and thin filaments past each other. The spent ADP is released from myosin at this point.

9. A new ATP binds to myosin, releasing its attachment to the actin molecule.

10. The new ATP is split to re-energize myosin and return the contraction cycle to step 7 above.

11. Contraction is maintained as long as internal calcium concentration is elevated. The calcium concentration falls as calcium ions are taken back into the sarcoplasmic reticulum via an ATP-dependent calcium pump.

Neural Control of Muscle Contraction

11

Up to this point, we have been concerned with the physiology of muscle at the level of single muscle fibers. We have come to some understanding of the mechanisms involved in the linkage between an action potential in a presynaptic motor neuron and the contraction of the postsynaptic muscle cell. We saw that the motor neuron depolarizes the muscle fiber, causing an action potential that, in turn, triggers the all-or-none contraction of the fiber. At this point, we will step back a bit from this cellular perspective and look at the functioning of a skeletal muscle as a whole. We will consider how the all-or-none twitches of single muscle fibers are integrated into the smooth, graded movements we know our muscles are capable of making.

The Motor Unit

A single motor neuron makes synaptic contact with a number of muscle fibers. The actual number varies considerably from one muscle to another and from one motor neuron to another within the same muscle; a single motor neuron may contact as few as 10–20 muscle fibers or more than 1000. However, in mammals, a single muscle fiber normally receives synaptic contact from only one motor neuron. Therefore, a single motor neuron and the muscle fibers to which it is connected form a basic unit of motor organization called the **motor unit**. A schematic diagram of the organization of a motor unit is shown in Figure 11-1. Recall from Chapter 8 that the synapse between a motor neuron and a muscle fiber is a one-for-one synapse—that is, a single presynaptic action potential produces a single postsynaptic action potential and hence a single twitch of the muscle cell. This means, then, that all the muscle cells in a motor unit contract together and that the fundamental unit of contraction of the whole muscle will not be the contraction of a single muscle fiber, but the contraction produced by all the muscle cells in a motor unit.

Gradation in the overall strength with which a particular muscle contracts is under control of the nervous system. There are two basic ways the nervous system uses to accomplish this task: (1) variation in the total number of motor neurons activated, and hence in the total number of motor units contracting;

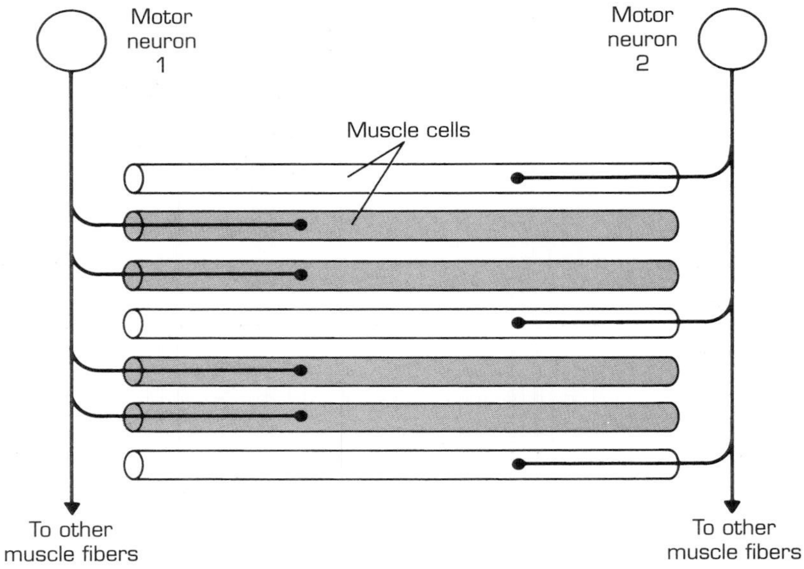

Figure 11-1 Schematic illustration of the innervation of a small number of muscle fibers in a muscle. The shaded muscle fibers form part of the motor unit of motor neuron I and the unshaded fibers form part of the motor unit of motor neuron 2.

and (2) variation in the frequency of action potentials in the motor neuron of a single motor unit. The greater the number of motor units activated, the greater the strength of contraction; similarly, within limits, the greater the rate of action potentials within a motor unit, the greater the strength of the resulting summed contraction. We will consider each of these mechanisms in turn below.

The Mechanics of Contraction

When the nerve controlling a muscle is stimulated, the resulting action potentials in the muscle fibers set up the sliding interaction between the filaments of the individual myofibrils in the muscle. This sliding generates a force that tends to make the muscle fibers, and therefore the muscle as a whole, shorten. Whether or not the muscle actually shortens, however, depends on the load attached to the muscle. While we might attempt to order the muscles in our arms to lift an automobile, it is unlikely that the muscles would be able to shorten against such a load. The force developed in an activated muscle is called the muscle **tension**, and only if the tension is great enough to equal the weight of the load will the muscle shorten and lift the load.

We can distinguish between two kinds of response to activation of a muscle. If the muscle tension is less than the load, the contraction is said to be **isometric** ("same length") because the length of the muscle does not change even though the tension increases. That is, the force exerted on the load by the muscle is not sufficient to move the load, so the muscle cannot shorten. An isometric contraction is diagrammed in Figure 11-2a. In the figure, an isolated muscle

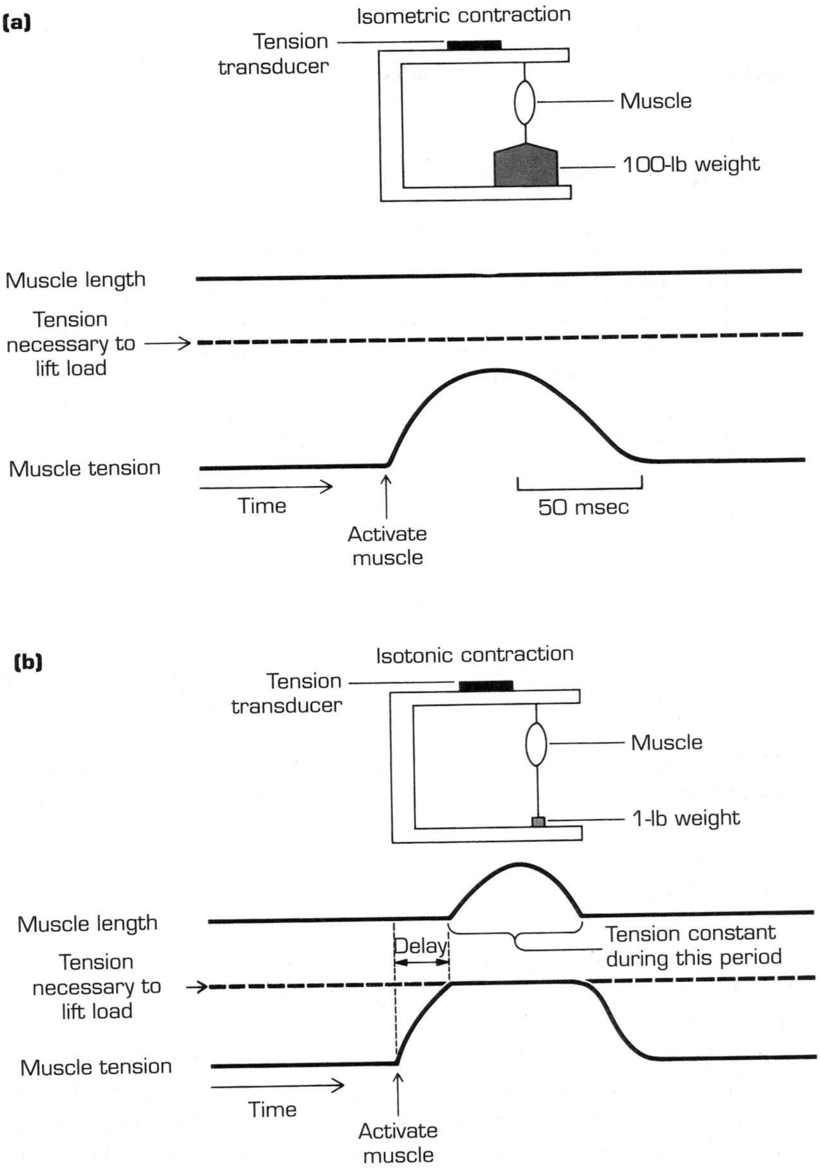

Figure 11-2 Measurements of muscle length and muscle tension during (a) isometric and (b) isotonic contractions. At the upward arrow, the nerve innervating the muscle is stimulated, causing activation of the muscle fibers.

is attached to a load it cannot lift. When the muscle is activated, the resulting tension is registered by a strain gauge that measures the miniscule flexing of the rigid strut to which the muscle is attached. A single activation of the muscle triggers a transient increase in tension lasting typically about 0.1 sec. You can easily feel the tension developed in an isometric contraction by placing your palms together with your arms flexed in front of your chest and pushing with both hands, one against the other.

If the tension is great enough to overcome the weight of the load, the contraction is said to be **isotonic** ("same tension") because the tension remains constant once it reaches the level necessary to move the load. This situation is diagrammed in Figure 11-2B. The strain gauge again records the increase in tension, as with the isometric contraction. When the tension reaches the level necessary to lift the load, it levels off and the muscle begins to shorten as the load is lifted. During the change in muscle length, the tension remains constant and equal to the weight of the load. This is because it is this weight—hanging from the muscle and support strut—that determines the flexing measured by the strain gauge. Thus, while the muscle is changing length the contraction is isotonic. In an isotonic contraction, the force developed by the sliding filaments in the myofibrils making up the muscle produces work in the form of moving the load through space.

One difference between isometric and isotonic contractions can be seen in the different delays between muscle activation and the occurrence of a measurable change in either muscle tension (isometric) or muscle length (isotonic). The tension begins to rise within a few milliseconds, the time required for the effect of the excitation–contraction process discussed in Chapter 10 to take hold. However, if muscle length is measured instead there is a pronounced delay between activation of the muscle and beginning of shortening. This delay is the time required for the tension to rise to the point where the load is lifted, which will depend on the size of the load. Thus, with light loads the shortening begins quickly, but with heavier loads the onset of shortening is progressively delayed. Finally, with sufficiently heavy loads there is no shortening at all and the contraction becomes isometric. In addition, with heavier loads the duration of shortening will be less and the maximum speed of shortening will be slower. In a sense, the measurement of tension during an isometric contraction gives a more direct view of the contractile state of the muscle; for this reason, subsequent examples in this chapter will be of isometric contractions.

The Relationship Between Isometric Tension and Muscle Length

At this point, it is worth considering how the magnitude of the isometric tension developed by a muscle depends on the muscle length at which the tension is measured; this will allow us to relate the tension of the muscle as a whole to the microscopic contractile apparatus within each muscle fiber and to the sliding filament hypothesis discussed in Chapter 10. Suppose the experiment diagrammed in Figure 11-2a were repeated at a variety of lengths of the muscle, as set by varying the distance between the upper support bar and the weight. We would find that as the muscle is stretched beyond its normal resting length, the tension developed upon stimulating the muscle would fall off rapidly, falling to zero at about 175% of resting length. This is shown in Figure 11-3a, in which peak isometric tension is plotted against muscle length (expressed as a percentage of the resting length of the unstimulated muscle). Such behavior can be easily understood in terms of the underlying state of the thick and

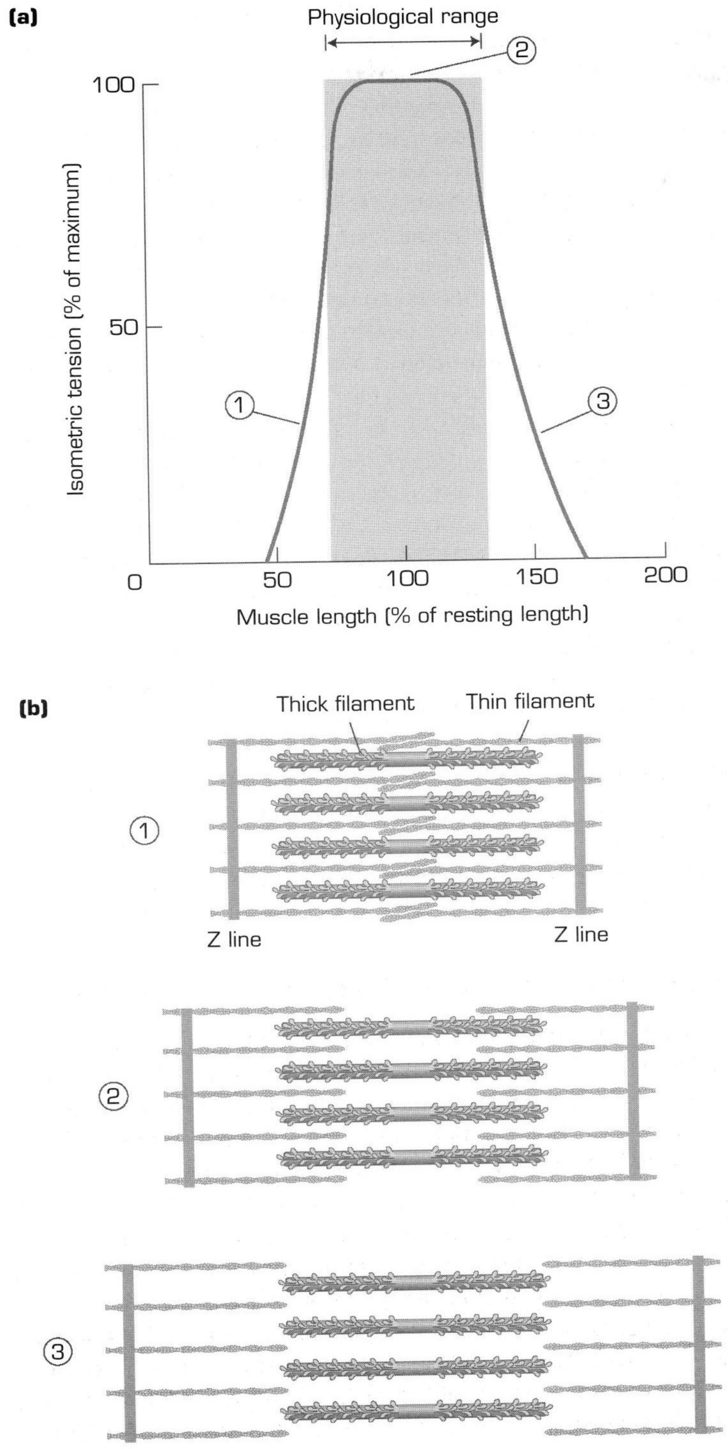

Figure 11-3 The relationship between muscle length and strength of isometric tension. (a) The graph shows the relation between the isometric tension produced when a muscle is stimulated, expressed as a percentage of the maximum observed, and the length of the muscle at the time it is stimulated, expressed as a percentage of resting length. The shaded area shows the range of muscle length over which the muscle would actually operate in the body. (b) The diagrams show the states of the thick and thin filaments within a sarcomere at each of the three numbered positions marked in (a).

thin filaments making up each sarcomere. As the distance between Z lines increases, the degree of overlap between thick and thin filaments declines, and thus the number of myosin head groups available to form cross-bridges is reduced. Finally, with sufficient stretch, there is no overlap, as shown in Figure 11-3b (number 3), and there can be no tension developed.

If the muscle is artificially shortened, isometric tension is also reduced, falling to zero at about 50% of resting length. This effect occurs as the distance between successive Z lines becomes sufficiently short that there is overlap between the thin filaments attached to neighboring Z lines. This overlap distorts the necessary spatial relation between thin and thick filaments required for cross-bridge attachments to form, so that once again there are fewer cross-bridges available to develop tension, as shown in Figure 11-3b (number 1). In addition to this geometric effect, other factors (such as reduced coupling between depolarization of the membrane and release of calcium from the sarcoplasmic reticulum) might contribute to the reduced peak tension at short muscle lengths. The fall-off of tension with both increasing and decreasing length means that there is an optimal range of length for development of tension; in this optimal range, there is maximal overlap of thin filaments with the cross-bridges of the thick filament (Figure 11-3b, number 2). If a muscle is to operate at maximum efficiency, the range of length over which it is required to develop force when in actual use in the body should be close to this optimal range. This is indeed the case; as a skeletal muscle operates, its length remains within about 30% of the optimal length (the shaded region in Figure 11-3a). In order to ensure that this range is not exceeded, precise arrangements of muscle-fiber length, tendon length and attachment sites, and joint geometry have evolved that are appropriate for the functional task of each muscle.

Control of Muscle Tension by the Nervous System

Recruitment of Motor Neurons

A single muscle typically receives inputs from hundreds of motor neurons. Thus, tension in the muscle can be increased by increasing the number of these motor neurons that are firing action potentials; the tension produced by activating individual motor units sums to produce the total tension in the muscle. A simplified example is shown in Figure 11-4. The increase in the number of active motor neurons is called **recruitment** of motor neurons and is an important physiological means of controlling muscle tension. When motor neurons are recruited into action during naturally occurring motor behavior, such as locomotion or lifting loads, the order of recruitment is determined by the size of the motor unit. As the tension in a muscle is increased, starting from the relaxed state, motor units containing a small number of muscle fibers are the first to be recruited; larger motor units are recruited later. Thus, when there is little

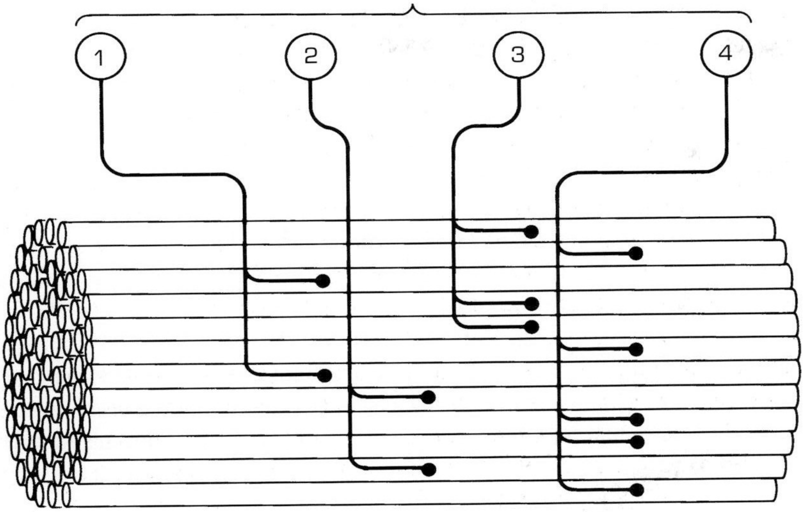

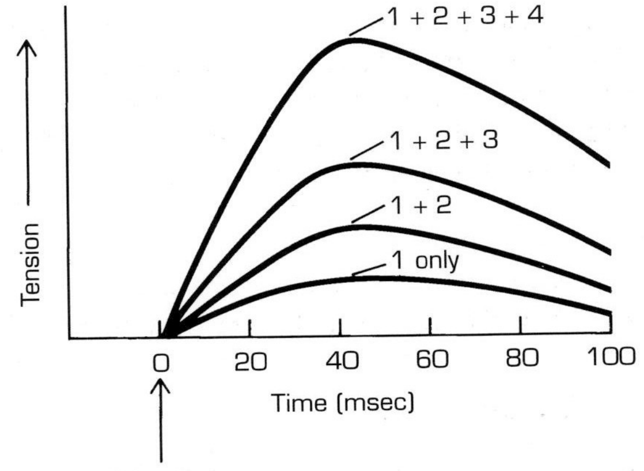

Figure 11-4 A simple muscle consisting of four motor units of varying size. The graph (bottom) shows isometric tension in response to simultaneous action potentials (at the arrow) in various combinations of the motor neurons.

activity in the pool of motor neurons controlling a muscle and the tension in the muscle is low, small motor units are recruited to produce an increase in tension. This insures that the added increments of tension are small and prevents large jerky increases in tension when the tension is small. As tension increases, however, further increases in tension must be larger in order to make a significant difference; thus, larger motor units are added, resulting in larger increments of tension when the background tension is already high. This

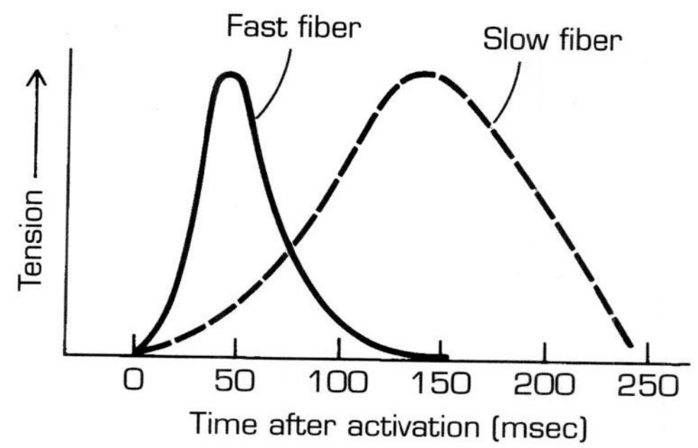

Figure 11-5 Comparison of the speed of development of isometric tension in fast and slow muscle fibers.

behavior is referred to as the **size principle** in motor neuron recruitment. In Figure 11-4, for example, it would be expected that tension would be increased by adding the smaller motor units (numbers 1 and 2) first and the largest unit (number 4) last.

Fast and Slow Muscle Fibers

The time delay between the occurrence of the muscle fiber action potential and the peak of the resulting tension is not constant across all muscle fibers. The delay to peak tension can be as little as 10 or as many as 200 msec. In general, muscle fibers can be grouped into two classes—fast and slow—on the basis of this speed. Samples of isometric contractions in fast and slow fibers are shown in Figure 11-5. Both slow and fast fibers are found together in most muscles, but slow fibers predominate in muscles that must maintain steady contraction, such as those involved in keeping us standing upright. Fast muscle fibers are more common in muscles that require rapid contraction, such as those involved in jumping and running. The fastest muscle fibers are those of muscles that move the eyes in rapid jumps, like those your eyes are making as you scan the words on this page.

Temporal Summation of Contractions Within a Single Motor Unit

When motor neurons are activated during naturally occurring movement, they do not typically fire just a single action potential, as has been the case in all our examples so far. Rather, action potentials tend to occur in bursts of several or as steady discharges at a relatively constant frequency. It is not uncommon for action potentials within a burst to be separated by only 10 msec or less. Under normal conditions, each of these many action potentials in a motor neuron will produce a corresponding action potential in each of the motor unit's muscle fibers. Because the tension resulting from a single action potential typically lasts for many tens or hundreds of milliseconds, there is considerable opportunity for summation of the effects of succeeding muscle action potentials in a series, as illustrated in Figure 11-6. Such temporal summation of individual

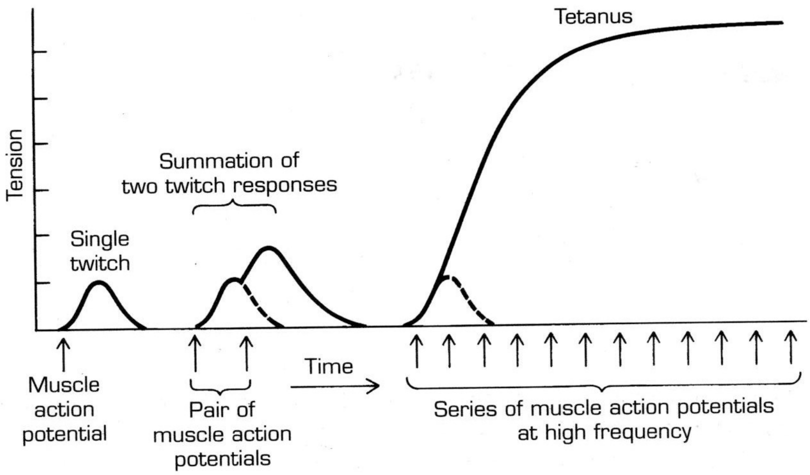

Figure 11-6 Isometric tension in response to a series of action potentials in a muscle. The dashed lines show the expected response if only the first action potential of a series occurred.

twitches is a major way in which the nervous system controls tension in skeletal muscles.

The amount of summation within a burst of muscle action potentials depends on the frequency of action potentials: the higher the frequency, the greater the resulting summed tension. However, as shown in Figure 11-6, when the frequency is sufficiently high the individual tension responses of the muscle fuse together into a plateau of tension. Further increase in frequency beyond this point does not increase tension: the muscle has reached its maximum response and cannot develop further tension. This plateau state is called **tetanus**. As expected from the examples shown in Figure 11-5, the frequency of stimulation required to produce tetanus varies considerably depending on whether slow or fast fibers are involved. For fast fibers, a frequency of more than 100 action potentials per second may be required, while for slow fibers a frequency of 20 per second may suffice.

Asynchronous Activation of Motor Units During Maintained Contraction

As anyone who has done prolonged physical labor or exercised vigorously can attest, muscle contractions cannot be maintained indefinitely; muscles fatigue and must be rested. Thus, the state of tetanus in Figure 11-6 could not be maintained in a single motor unit for very long without allowing the muscle fibers in the motor unit to relax. However, some muscles—such as those involved in maintaining body posture—are required to contract for prolonged periods. What mechanism helps prevent muscle fatigue during such prolonged contractions? During maintained tension in a muscle, all the motor neurons to the muscle are not active at the same time. The activity of the motor units occurs in bursts separated by quiet periods, and the activity of different motor units is

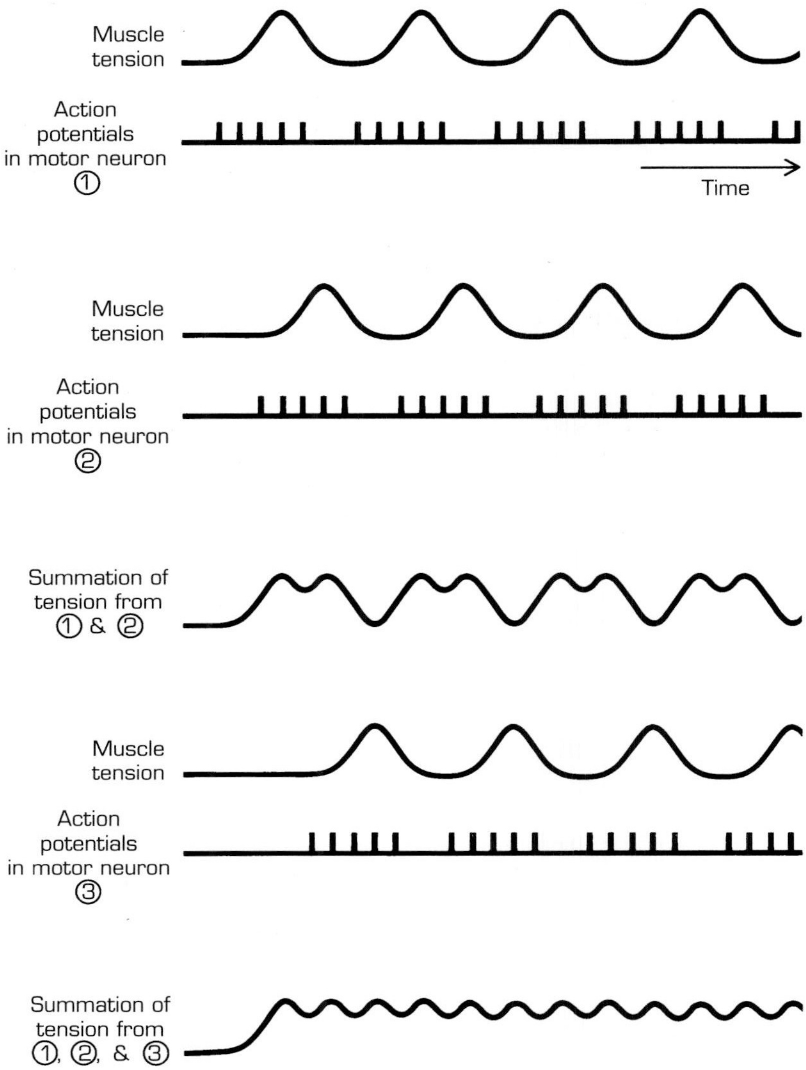

Figure 11-7 Summation of muscle tension during asynchronous activation of three motor units in a muscle.

staggered in time. An example of this kind of asynchronous activity during steady contraction is shown in Figure 11-7. Notice that the summation of the tensions produced by the activity of only three motor units, each active only half the time, can produce a reasonably smooth, steady tension. With hundreds of motor units available in many muscles, a much smoother and larger steady tension could be maintained with less effort on the part of any one motor unit. Thus, asynchronous activation of motor neurons to a muscle allows a prolonged contraction with reduced fatigue of individual motor units in the muscle.

Summary

The basic unit of contraction of a skeletal muscle is the contraction of the group of muscle fibers making up a single motor unit, which consists of a single motor neuron and all the muscle fibers receiving synaptic connections from that neuron. Whenever the motor neuron fires an action potential, all the muscle fibers in that motor unit twitch together. The magnitude of the contraction generated by activation of a motor unit depends on the number of muscle fibers in that motor unit. The number of fibers in a unit, and hence the magnitude of the tension produced by activating the unit, varies considerably among the set of motor neurons innervating a particular muscle.

The type of contraction produced by activation of a whole muscle depends on the load against which the muscle is contracting. If the load is too great for the muscle to move, the length of the muscle does not change during the contraction, which is then called an isometric contraction. If the tension is sufficient to overcome the weight of the load, the contraction will be accompanied by a shortening of the muscle. During the shortening, the tension in the muscle remains constant and equal to the weight of the load. Such a contraction is called isotonic. The overall tension developed by a muscle depends on how many motor units are activated and on the frequency of action potentials within a motor unit. Increasing muscle tension by increasing the number of active motor neurons is called motor neuron recruitment. When the frequency of action potentials within a motor unit is increased, the resulting muscle tension increases until a steady plateau state, called tetanus, is reached. Normally, during a maintained contraction all the motor neurons of a muscle are not active simultaneously; rather, the activity of individual motor neurons is restricted to periodic bursts that occur asynchronously among the pool of motor neurons controlling a muscle. This helps reduce fatigue in the muscle by allowing individual motor units to rest periodically during a maintained contraction.

12 Cardiac Muscle: The Autonomic Nervous System

The motor functions we have described so far in this part of the book have been concerned with the control of skeletal muscles. These are the muscles that produce overt movements of the body and give rise to the observable external actions that we normally think of as the "behavior" of an animal. However, even in an animal that appears to an external observer to be quiescent, the nervous system is quite busy coordinating many ongoing motor actions that are as important for survival as skeletal muscle movements. These motor activities include such things as regulating digestion, maintaining the proper glucose balance in the blood, regulating heart rate, and so on. The part of the nervous system that controls these functions is called the **autonomic nervous system**. The motor targets of the autonomic nervous system include gland cells, cardiac muscle cells, and smooth muscle cells such as those found in the gut. To distinguish it from the autonomic nervous system, the parts of the nervous system we have been discussing up to this point—whose motor targets are the skeletal muscles—are collectively called the **somatic nervous system**.

In addition to the differences in their target cells, there are other differences between the autonomic and somatic nervous systems. As we have seen, in the somatic nervous system, the cell bodies of the motor neurons are located within the central nervous system, either in the spinal cord or in the nuclei of cranial nerves in the brainstem. By contrast, the cell bodies of the motor neurons in the autonomic nervous system are located outside the central nervous system altogether, in a system of **autonomic ganglia** distributed throughout the body. The central nervous system controls these autonomic ganglia by way of output neurons called **preganglionic neurons**, which are located in the spinal cord and brainstem. This arrangement is illustrated in Figure 12-1. The motor neurons in the autonomic ganglia are also called **postganglionic neurons**. The axons of the preganglionic neurons entering the ganglia are referred to as the **preganglionic fibers**, while the axons of the autonomic motor neurons carrying the output to the target cells are called the **postganglionic fibers**. Thus, in the

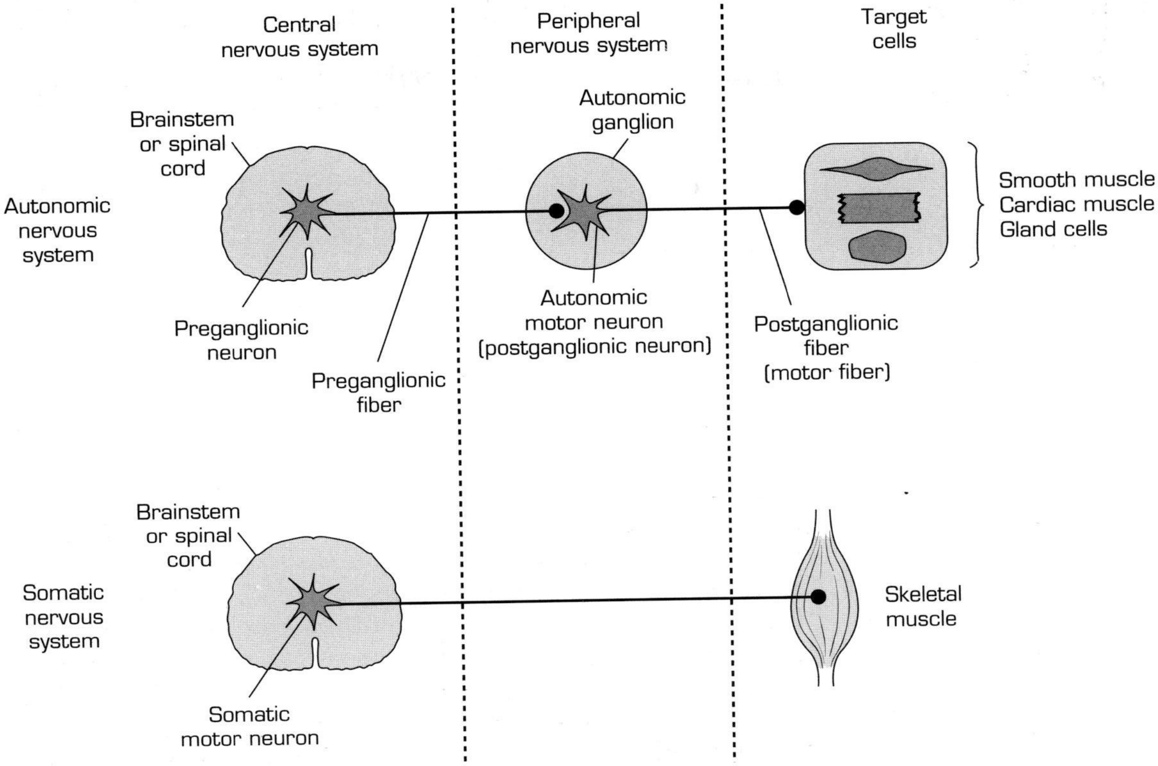

Figure 12-1 Differences between autonomic and somatic nervous systems. In the autonomic nervous system, the motor neurons are located outside the central nervous system, in autonomic ganglia. The motor neurons contact smooth muscle cells, cardiac muscle cells, and gland cells. The central nervous system controls the ganglia via preganglionic neurons. In the somatic nervous system, the motor neurons are located within the central nervous system and contact skeletal muscle cells.

somatic nervous system, the motor commands exiting from the central nervous system go directly to the target cells, while in the autonomic nervous system, the motor commands from the central nervous system are relayed via an additional synaptic connection in the peripheral nervous system.

The autonomic and somatic nervous systems also differ in the effects that the motor neurons have on the target cells. In Chapter 8, we discussed in detail the synaptic interaction between motor neurons and skeletal muscle cells at the neuromuscular junction. All of the somatic motor neurons release ACh as their neurotransmitter, and the effect on the skeletal muscle cells is always excitatory: contraction is stimulated. In the autonomic nervous system, however, some motor neurons release ACh and other motor neurons release the neurotransmitter norepinephrine (see Chapter 9), instead of ACh. Further, an autonomic motor neuron may either excite or inhibit its target cell. In general, if norepinephrine excites the target cells, then ACh inhibits them, and vice versa. For example, norepinephrine increases the rate of beating of the heart, while ACh decreases the heart rate, as we will examine in detail shortly.

The norepinephrine-releasing and ACh-releasing motor neurons are organized into anatomically distinct divisions of the autonomic nervous system, called the **sympathetic division** (norepinephrine-releasing) and the **parasympathetic division** (ACh-releasing). The ganglia containing the sympathetic motor neurons are called sympathetic ganglia, and those containing parasympathetic motor neurons are called parasympathetic ganglia. Most of the sympathetic ganglia are arrayed parallel to the spinal cord, one ganglion on each side just outside the vertebral column. There is one pair of these **paravertebral ganglia** for each vertebral segment. The ganglia are interconnected by thick, longitudinal bundles of axons containing the preganglionic fibers exiting from the spinal cord. Because of these connectives, the paravertebral ganglia form two long chains parallel to the spinal column, sometimes referred to as the **sympathetic chains**. In addition to the paravertebral ganglia that make up the chains, there are also sympathetic ganglia called the **prevertebral ganglia**, located within the abdomen.

The parasympathetic ganglia are distributed more diffusely throughout the body and tend to be located closer to their target organs. In some cases, the parasympathetic ganglia are actually located within the target organ itself. This is the case, for example, in the heart. Because the sympathetic ganglia are located predominantly near the central nervous system while the parasympathetic ganglia are located mostly near to their target organs, the preganglionic fibers of the sympathetic nervous system are usually much shorter than the preganglionic fibers of the parasympathetic nervous system, which must extend all the way from the central nervous system to the near vicinity of the target organ in order to reach the postganglionic neurons. Conversely, the postganglionic fibers are typically much longer in the sympathetic nervous system than in the parasympathetic nervous system.

Most target organs receive inputs from both the sympathetic and parasympathetic divisions of the autonomic nervous system. As noted above, the sympathetic and parasympathetic inputs produce opposing effects on the target. In general, excitation of the sympathetic nervous system has the overall effect of placing the organism in "emergency mode," ready for vigorous activity. The parasympathetic nervous system has the opposite effect of placing the organism in a "vegetative mode." For example, sympathetic activity increases the heart rate and blood pressure, diverts blood flow from the skin and viscera to the skeletal muscles, and reduces intestinal motility, all appropriate preparations for rapid reaction to an external threat. Parasympathetic activity, on the other hand, decreases heart rate and blood pressure, and promotes blood circulation to the gut and intestinal motility. All of these actions are appropriate for resting and digesting, in the absence of any threatening situation in the environment.

Autonomic Control of the Heart

To see how the motor neurons of the sympathetic and parasympathetic divisions exert their actions on target cells, it will be useful to examine a particular example in detail. The example we will explore is the neural control of the heart. The heart is made up of muscle cells, which are in some ways similar to the skeletal muscle cells we learned about in Chapter 10. However, there are some important differences, which we must understand before we can examine the effects that the sympathetic and parasympathetic neurons have on the heart. Thus, we will first discuss the electrical and mechanical properties of the heart muscle, and then return to the modulation of those properties by norepinephrine and ACh, which are the neurotransmitters released by the sympathetic and parasympathetic inputs to the heart, respectively.

The Pattern of Cardiac Contraction

Cardiac muscle cells contain a contractile apparatus like that of other striated muscle, being made up of bundles of myofilaments with a microscopic structure like that discussed in Chapter 10. Unlike other striated muscles in the body, the heart muscle is specialized to produce a rhythmic and coordinated contraction in order to drive the blood efficiently through the blood vessels. The heart has a number of tasks to accomplish in order to carry out its role in providing oxygen to the cells of the body. It must receive the oxygen-poor blood returning from the body tissues via the venous circulation and send that blood to the lungs for oxygenation. It must also receive the oxygenated blood from the lungs and send it out through the arterial circulation to the rest of the body. Carrying out these tasks requires precise timing of the contractions of the various heart chambers; otherwise, the flow of oxygenated blood will not occur efficiently or will cease altogether—with disastrous consequences. What is the normal timing sequence of the heart contractions underlying the coordinated pumping of the blood?

A schematic diagram of the flow of blood through a human heart during a single contraction cycle is shown in Figure 12-2. Humans, like all other mammals, have a four-chambered heart, consisting of the left and right **atria** and the left and right **ventricles**. The two atria can be thought of as the receiving chambers, or "priming" pumps, of the heart, while the two ventricles are the "power" pumps of the circulatory system. The right atrium receives the blood returning from the body through the veins, and the left atrium receives the freshly oxygenated blood from the lungs. During the phase of the heartbeat when the atria are filling with blood, the valves connecting the atria with the ventricles are closed, preventing flow of blood into the ventricles. When the atria have filled with blood, they contract and the increase in pressure opens the valves leading to the ventricles and drives the collected blood into the ventricles. At this point, the muscle of the ventricles is relaxed, and the valves

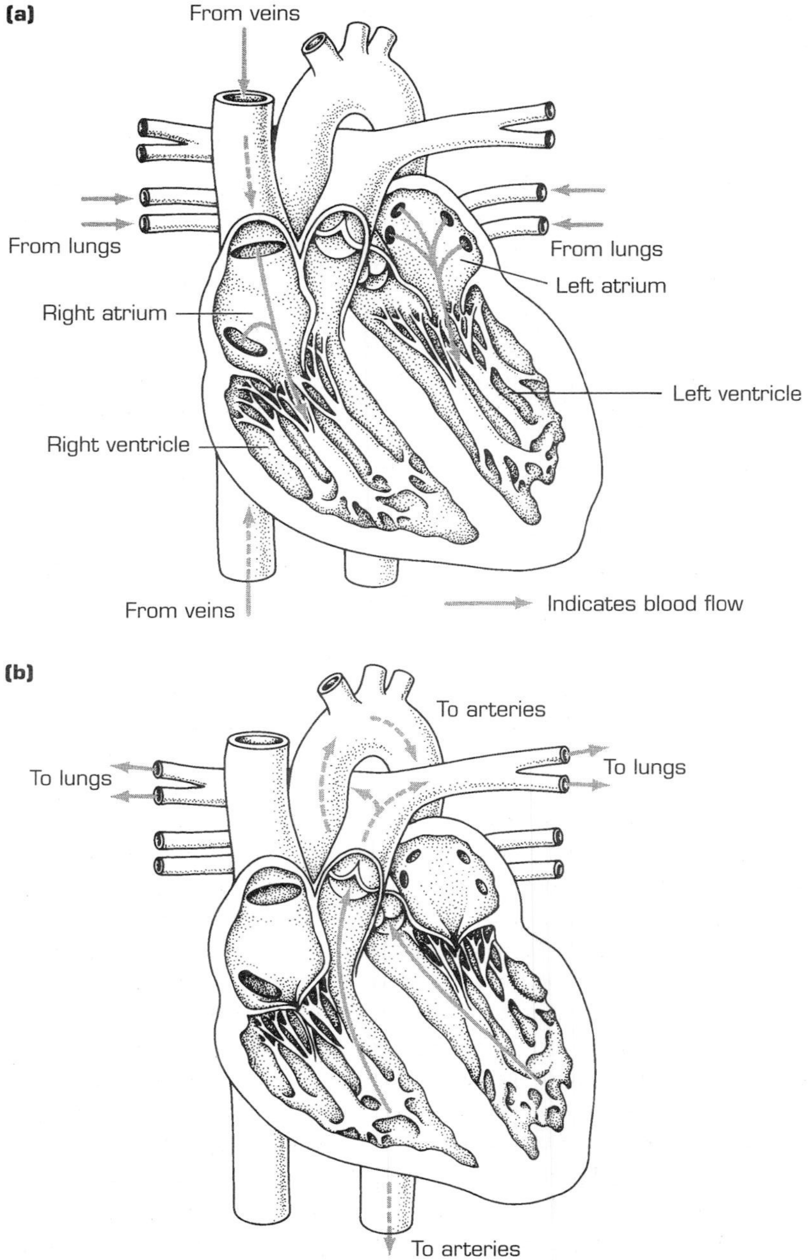

Figure 12-2 Schematic drawings of the state of the heart valves and the direction of blood flow during two stages in a single heartbeat. (a) The atria are contracting and the ventricles are filling with blood. (b) The valves between the atria and ventricles are closed and the ventricles are contracting, forcing the blood from the right ventricle to the lungs and from the left ventricle to the arteries supplying the rest of the body.

connecting the ventricles to the vessels leaving the heart are closed. When the ventricles have filled with blood, they contract, opening these valves and delivering the power stroke to drive the blood out to the lungs and to the rest of the body, as shown in Figure 12-2b. Thus, during a normal heartbeat the two atria contract together, followed after a delay by the simultaneous contraction of the two ventricles.

Coordination of Contraction Across Cardiac Muscle Fibers

In order for the contraction of a heart chamber to be able to propel the expulsion of fluid, all the individual muscle fibers making up the walls of that chamber must contract together. It is this unified contraction that constricts the cavity of the chamber and drives out the blood into the blood vessels of the circulation. In skeletal muscles, an action potential in one muscle fiber is confined to that fiber and does not influence neighboring fibers; therefore, contraction is restricted to the particular fiber undergoing an action potential. In cardiac muscle, however, the situation is quite different. When an action potential is generated in a cardiac muscle fiber, it causes action potentials in the neighboring fibers, which in turn set up action potentials in their neighbors, and so on. Thus, the excitation spreads rapidly out through all the muscle fibers of the chamber. This insures that all the fibers contract together.

What characteristic of cardiac muscle fibers allows the action potential to spread from one fiber to another? The answer can be seen by looking at the microscopic structure of the cells of cardiac muscle, shown schematically in Figure 12-3. At the ends of each cardiac cell, the plasma membranes of neighboring cells come into close contact at specialized structures called **intercalated disks**. The contact at this point is sufficiently close that electrical current flowing inside one fiber can cross directly into the interior of the next fiber; in electrical terms, it is as though the neighboring cells form one larger cell. Recall from Chapter 6 that an electrical current flowing along the interior of a fiber has at each point two paths to choose from: across the plasma membrane or continuing along the interior of the fiber. The amount of current taking each path at a particular point depends on the relative resistances of the two paths; the higher the resistance, the smaller the amount of current taking that path. Normally, at the point where one cell ends and the next begins, there is little opportunity for current to flow from one cell to the other because the current would have to flow out across one cell membrane and in across the other in order to do so; this is a high resistance path because current must cross two membranes. At the specialized structure of the intercalated disk, however, the resistance to current flow across the two membranes is low, so that the path to the interior of the neighboring cell is favored. This means that depolarizing current injected into one cell during the occurrence of an action potential can spread directly into neighboring cells, setting up an action potential in those cells as well. The low resistance path from one cell to another is through membrane structures called **gap junctions**. These structures consist of arrays of small pores directly connecting the interiors of the joined cells. The pores are formed by pairs of protein molecules, one in each cell, that attach to each other and bridge the small extracellular gap between the two cell membranes (Figure 12-3). The pores at the center of each of these **gap junction channels** are

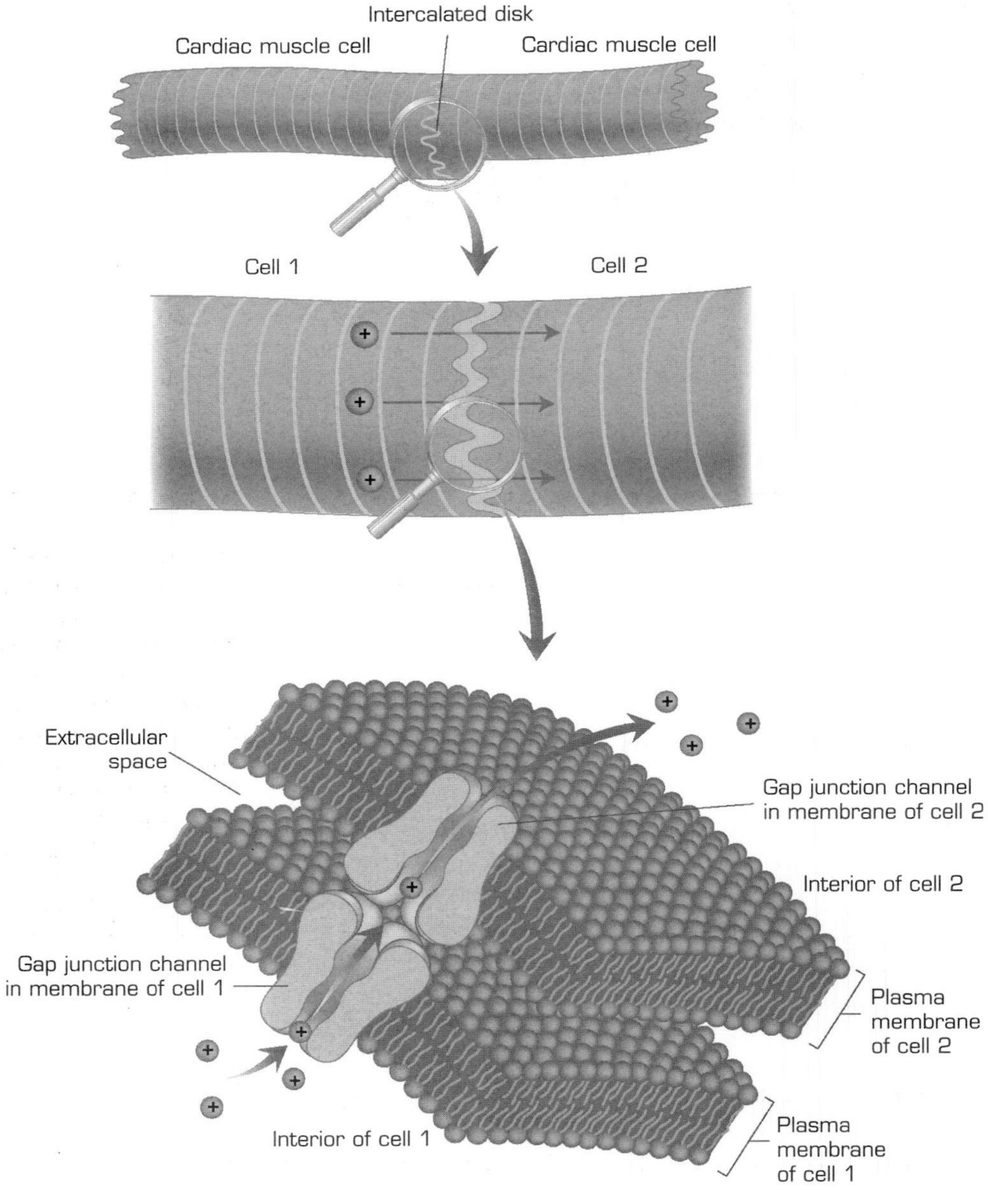

Figure 12-3 Electrical current can flow from one cardiac muscle cell to another through specialized membrane junctions located in a region of contact called the intercalated disk. The current flows through pores formed by pairs of gap junction channels that bridge the extracellular space at the intercalated disk.

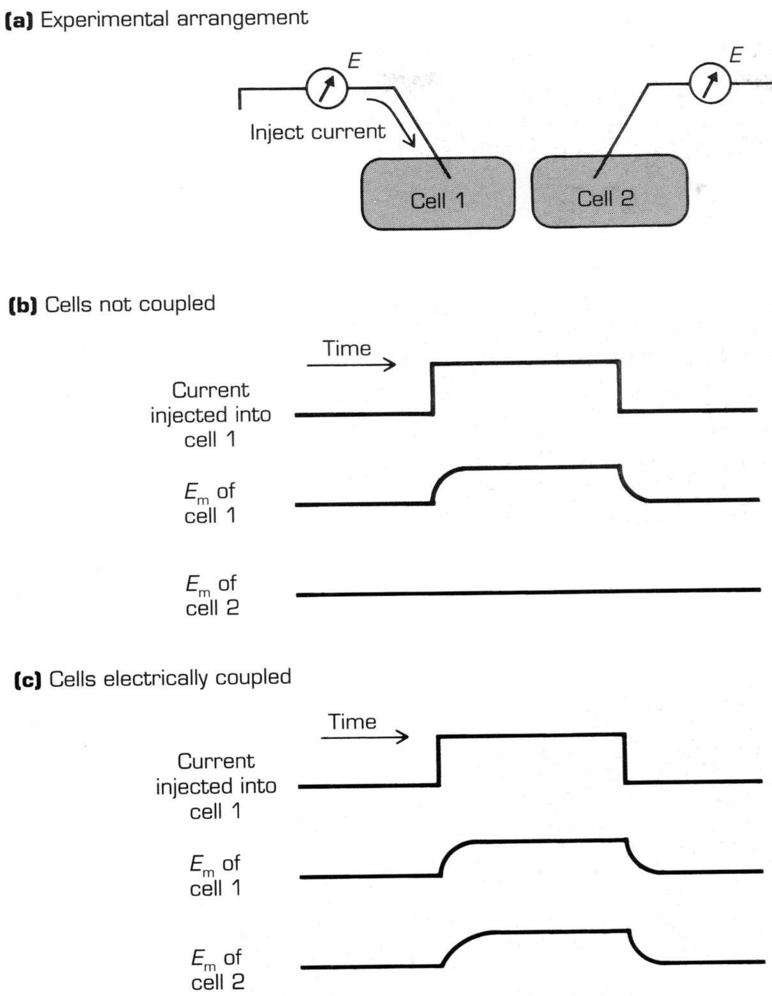

Figure 12-4 An experiment in which the membrane potentials of two cells are measured simultaneously with intracellular microelectrodes. (a) A depolarizing current is injected into cell 1. (b) If the cells are not electrically coupled, the depolarization occurs only in the cell in which the current was injected. (c) If the cells are electrically coupled via gap junctions, a depolarization occurs in cell 2, as well as in cell 1.

aligned, permitting small molecules like ions to pass directly from one cell to the other.

When electrical current can pass from one cell to another, as in cardiac muscle, those cells are said to be **electrically coupled**. Figure 12-4 illustrates an electrophysiological experiment to demonstrate this behavior. When current is injected into a cell, no response occurs in a neighboring cell if the cells are not electrically coupled. If the two cells are coupled via gap junctions, a response to the injected current occurs in both cells because the ions carrying the current inside the cell can pass directly through the gap junction. If the depolarization is large enough, an action potential will be triggered in both cells at the same time.

Generation of Rhythmic Contractions

The electrical coupling among cardiac muscle fibers can explain how contraction occurs synchronously in all the fibers of a chamber. We will now consider the control mechanisms responsible for the repetitive contractions that characterize the beating of the heart. If a heart is removed from the body and placed in an appropriate artificial environment, it will continue to contract repetitively even though it is isolated from the nervous system and the rest of the body. By contrast, a skeletal muscle isolated under similar conditions will not contract unless its nerve is activated. The rhythmic activity of the heart muscle is an inherent property of the individual muscle fibers making up the heart, and this constitutes another important difference between cardiac muscle fibers and skeletal muscle fibers. This difference can be demonstrated dramatically in experiments in which muscle tissue is dissociated into individual cells, which are placed in a dish isolated from each other and from the influence of any other cells, like nerve cells. Under these conditions, muscle cells from skeletal muscles are quiescent; they do not contract in the absence of their neural input. Cells from cardiac muscle, however, continue to contract rhythmically even in isolation. Thus, rhythmic contractions of heart muscle are due to built-in properties of the cardiac muscle cells. Before we can examine the membrane mechanism underlying this autorhythmicity, it will be necessary to look first at the action potential of cardiac muscle cells. In keeping with the different behavior of cardiac cells, this action potential has some different characteristics from the action potentials of neurons or skeletal muscle cells.

The Cardiac Action Potential

In Chapters 6 and 7, we discussed the ionic mechanisms underlying the action potential of nerve membrane. The action potential of skeletal muscle fibers is fundamentally the same as that of neurons. The cardiac action potential, however, is different from these other action potentials in several important ways. Figure 12-5 compares the characteristics of action potentials of skeletal and cardiac muscle cells. One striking difference is the difference in time-scale: cardiac action potentials can last several hundred milliseconds, while skeletal muscle action potentials are typically over in 1–2 msec. As we saw in Chapter 6, a long-lasting plateau like that of the cardiac action potential can arise from a calcium component in the action potential, resulting from the opening of voltage-dependent calcium channels. These channels open upon depolarization and allow influx of positively charged calcium ions. In addition, the plateau of the cardiac action potential is also associated with a reduction in the potassium permeability. This is due to a type of potassium channel that is open as long as the membrane potential is near its normal resting level and closes upon depolarization. This is the reverse of the behavior of the gated potassium channel we are familiar with from our discussion of nerve action potential. The reduction in potassium permeability caused by the closing of this channel

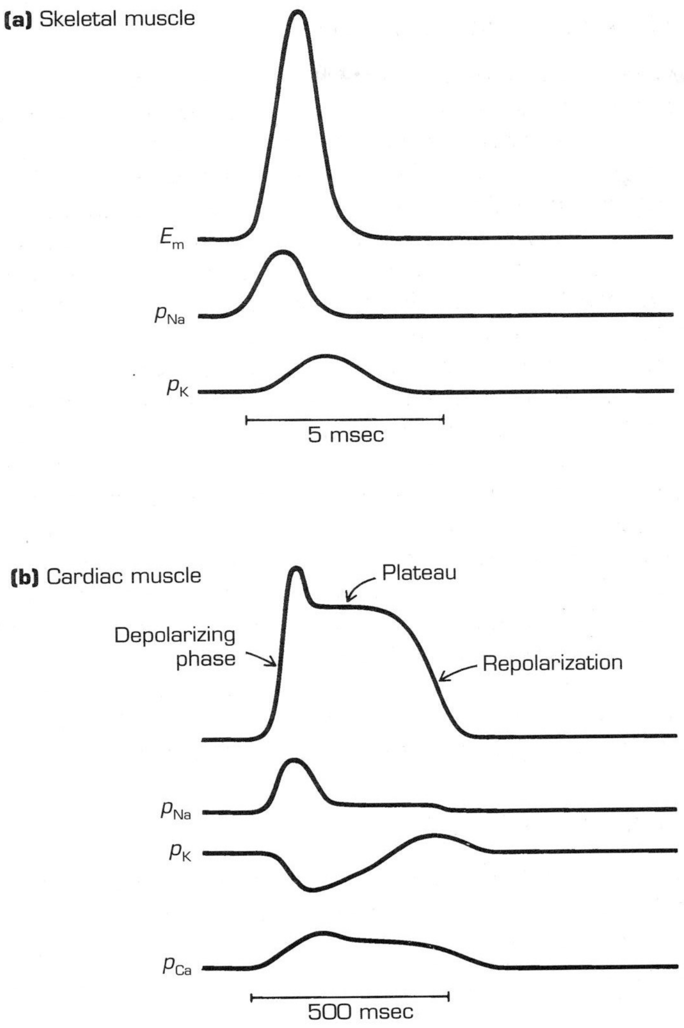

Figure 12-5 The sequence of permeability changes underlying the action potentials of (a) skeletal muscle fibers and (b) cardiac muscle fibers. Note the greatly different time-scales.

tends to depolarize the cardiac muscle fiber. Both the opening of the calcium channels and the closing of the potassium channels contribute to the plateau.

The initial rising phase of the cardiac action potential is produced by voltage-dependent sodium channels very much like those of nerve membrane. The sodium channels drive the rapid initial depolarization and are responsible for the brief initial spike of the cardiac action potential before the plateau phase sets in. Like the sodium channel of neuronal membrane, this channel rapidly closes (inactivates) with maintained depolarization. However, unlike the nerve sodium channel, this inactivation is not total; there is a small, maintained increase in sodium permeability during the plateau.

What is responsible for terminating the cardiac action potential? First, the calcium permeability of the plasma membrane slowly declines during the maintained depolarization. This decline might be a consequence of the gradual build-up of internal calcium concentration as calcium ions continue to enter the muscle fiber through the open calcium channels. Internal calcium ions are thought to have a direct or indirect action on the calcium channels, causing them to close. In addition, the potassium permeability of the plasma membrane increases, as in the nerve and skeletal muscle action potentials. This increase in potassium permeability tends to drive the membrane potential of the cardiac fiber toward the potassium equilibrium potential and thus to repolarize the fiber. There is evidence that part of this increase in potassium permeability is due to voltage-sensitive potassium channels that open in response to the depolarization during the action potential (like the *n* gates of the nerve membrane). However, the increased potassium permeability is also caused by calcium-activated potassium channels (see Chapter 6), which open in response to the rise in internal calcium concentration during the prolonged action potential.

One functional implication of the prolonged cardiac action potential is that the duration of the contraction in cardiac muscle is controlled by the duration of the action potential. The action potential and contraction of cardiac muscle fibers are compared with those of skeletal muscle fibers in Figure 12-6. In skeletal muscle, the action potential acts only as a trigger for the contractile events; the duration of the contraction is controlled by the timing of the release and

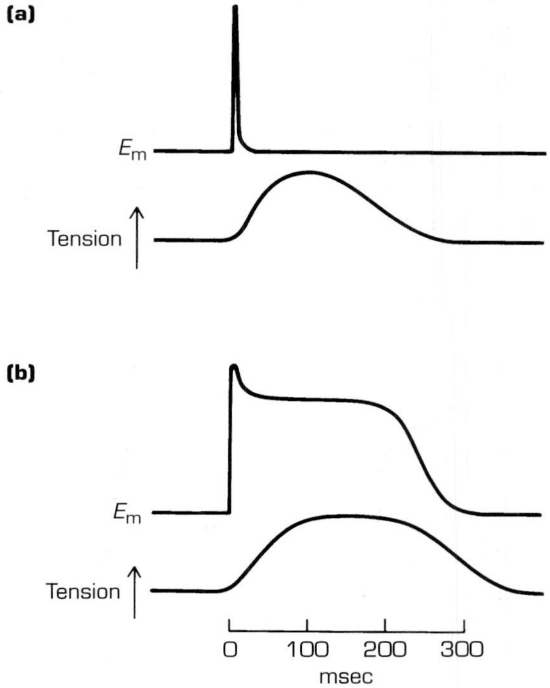

Figure 12-6 (a) In a skeletal muscle fiber, the action potential is much briefer than the resulting contraction. Thus, the action potential acts only as a trigger for the contraction, which proceeds independently of the duration of the action potential. (b) In a cardiac muscle fiber, the duration of the contraction is closely related to the duration of the action potential because of the maintained calcium influx during the plateau of the action potential. Thus, characteristics of the action potential can influence the duration and strength of the cardiac contraction.

reuptake of calcium by the sarcoplasmic reticulum, not by the duration of the action potential. In cardiac muscle fibers, however, only the initial part of the contraction is controlled by sarcoplasmic reticulum calcium; the contraction is maintained by the influx of calcium ions across the plasma membrane during the plateau phase of the cardiac action potential. For this reason, the duration of the contraction in the heart can be altered by changing the duration of the action potential in the cardiac muscle fibers. This provides an important mechanism by which the pumping action of the heart can be modulated.

The Pacemaker Potential

Although the ionic mechanism of the cardiac action potential differs in important ways from that of other action potentials, nothing in the scheme presented so far would account for the endogenous beating of isolated heart cells discussed earlier. If we recorded the electrical membrane potential of a spontaneously beating isolated heart cell, we would see a series of spontaneous action potentials, as shown in Figure 12-7. After each action potential, the potential falls to its normal negative resting value, then begins to depolarize slowly. This slow depolarization is called a **pacemaker potential**, and it is caused by spontaneous changes in the membrane ionic permeability. Voltage-clamp experiments on single isolated muscle fibers from the ventricles suggest that the pacemaker potential is due to a slow decline in the potassium permeability coupled with a slow increase in sodium and calcium permeability. When the depolarization reaches threshold, it triggers an action potential in the fiber, with a rapid upstroke caused by opening of sodium channels and a prolonged plateau produced by calcium channels. Part of the early phase of the pacemaker potential represents the normal undershoot period of an action potential (see Chapter 6), when potassium channels that were opened by depolarization during the action potential slowly close again. As these potassium channels close, the membrane potential will move in a positive direction, away from the potassium equilibrium potential.

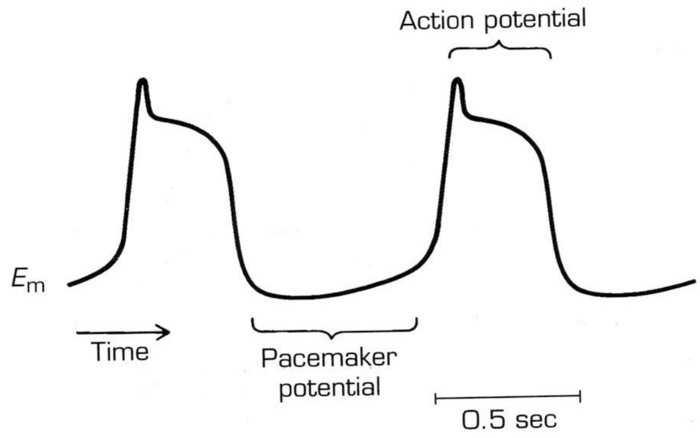

Figure 12-7 A recording of the membrane potential during repetitive, spontaneous beating in a single cardiac muscle fiber. The repolarization at the end of one action potential is followed by a slow, spontaneous depolarization called the pacemaker potential. When this depolarization reaches threshold, a new action potential is triggered.

Later phases of the pacemaker potential represent increases in sodium and calcium permeability, both of which move the membrane potential more positive, toward the sodium and calcium equilibrium potentials. The sodium permeability increases during the pacemaker potential because of the opening of nonspecific cation channels, which open at more hyperpolarized membrane potentials. As described in Chapter 8, the opening of channels with equal permeability to sodium and potassium ions (like the ACh-activated channels at the neuromuscular junction) will produce depolarization of a cell. These hyperpolarization-activated cation channels are opened in response to the membrane hyperpolarization during the undershoot of the action potential. Together with the decrease in potassium permeability, the resulting influx of sodium ions moves the membrane potential of the cardiac cell in a positive direction, toward the threshold for firing an action potential. As the membrane potential becomes more positive during the pacemaker potential, voltage-dependent calcium channels open in response to the depolarization. The resulting influx of positively charged calcium ions produces even more depolarization, ultimately triggering the next action potential in the series.

The rate of spontaneous action potentials in isolated heart cells varies from one cell to another; some cells beat rapidly and others slowly. In the intact heart, however, the electrical coupling among the fibers guarantees that all the fibers will contract together, with the overall rate being governed by the fibers with the fastest pacemaker activity. In the normal functioning of the heart, the overall rate of beating is controlled by a special set of pacemaker cells, called

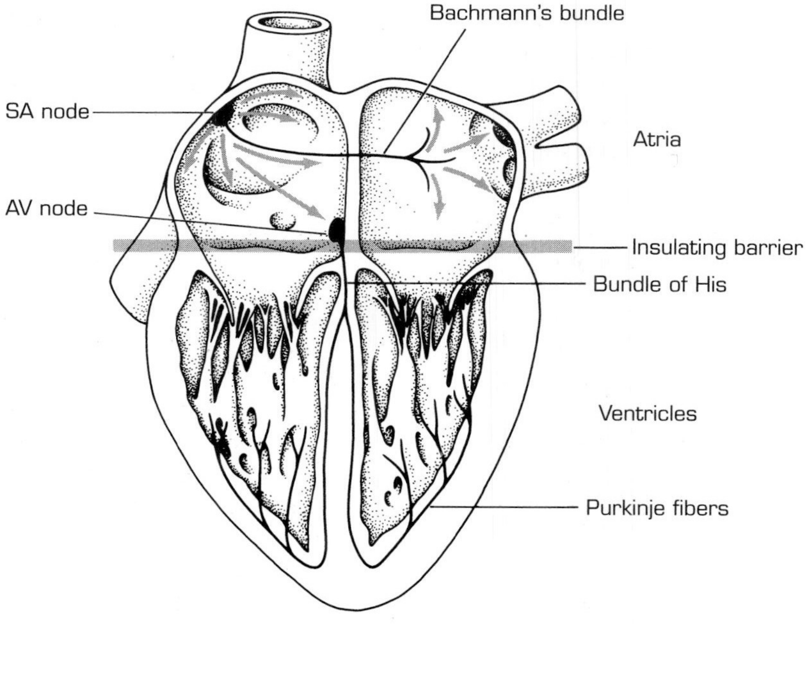

Figure 12-8 Diagram of the spread of action potentials across the heart during a single heartbeat. The excitation originates in the sinoatrial (SA) node of the right atrium and spreads throughout the atria via electrical coupling among the atrial muscle fibers. The fibers of the atria are not electrically connected to those of the ventricles. The action potential spreads to the ventricles via the atrioventricular (AV) node, which introduces a delay between the atrial and ventricular action potentials. When the wave of action potentials leaves the AV node, its spread throughout the ventricles is aided by the rapidly conducting Purkinje fibers of the bundle of His.

the **sinoatrial (SA) node**, which is located in the upper part of the right atrium. This node is indicated in the diagram of the heart in Figure 12-8. The action potential of cells in the SA node is a bit different from that of other cardiac cells. In the SA node, calcium channels play a larger role than sodium channels in triggering the action potential, as well as in sustaining the depolarization during the action potential. In the normal resting human heart, the cells of the SA node generate spontaneous action potentials at a rate of about 70 per minute. These action potentials spread through the electrical connections among fibers throughout the two atria, generating the simultaneous contraction of the atria as discussed in the first section of this chapter. This helps insure that the two atria contract together. The atrial action potentials do not spread directly to the fibers making up the two ventricles, however. This is a good thing, because we know that the contraction of the ventricles must be delayed to allow the relaxed ventricles to fill with blood pumped into them by the atrial contraction. In terms of electrical conduction, the heart behaves as two isolated units, as shown in Figure 12-8: the two atria are one unit and the two ventricles are another. The electrical connection between these two units is made via another specialized group of muscle fibers called the **atrioventricular (AV) node**. Excitation in the atria must travel through the AV node to reach the ventricles. The fibers of the AV node are small in diameter compared with other cardiac fibers. As discussed in Chapter 6, the speed of action potential conduction is slow in small-diameter fibers. Therefore, conduction through the AV node introduces a time delay sufficient to retard the contraction of the ventricles relative to the contraction of the atria. Excitation leaving the AV node does not travel directly through the muscle fibers of the ventricles. Instead, it travels along specialized muscle fibers that are designed for rapid conduction of action potentials. These fibers are called **Purkinje fibers**, and they form a fast-conducting pathway through the ventricles called the **bundle of His**. The Purkinje fibers carry the excitation rapidly to the apex at the base of the heart, where it then spreads out through the mass of ventricular muscle fibers to produce the contraction of the ventricles.

Actions of Acetylcholine and Norepinephrine on Cardiac Muscle Cells

Each muscle fiber of a skeletal muscle receives a direct synaptic input from a particular motor neuron; without this synaptic input, a skeletal fiber does not contract unless stimulated directly by artificial means. Nevertheless, we have seen that cardiac muscle fibers generate spontaneous contractions that are coordinated into a functional heartbeat by the electrical conduction mechanisms inherent in the heart itself. This does not mean, however, that the activity of the heart is not influenced by inputs from the nervous system. The heart receives two opposing neural inputs that affect the heart rate. One input comes from the cells of the parasympathetic nervous system, whose synaptic terminals in the heart release the neurotransmitter ACh. The effect of ACh is to slow the rate of depolarization during the pacemaker potential of the SA node. This has the effect of increasing the interval between successive action potentials and thus slowing the rate at which this master pacemaker region drives the heartbeat. Acetylcholine acts by increasing the potassium permeability of the muscle fiber membrane. This tends to keep the membrane potential closer to the potassium equilibrium potential and thus retard the growth of the pacemaker potential toward threshold for triggering an action potential. The second neural input to the heart comes from neurons of the sympathetic nervous system, whose synaptic terminals release the neurotransmitter norepinephrine. Activation of this input speeds the heart rate and also increases the strength of contraction. This effect is mediated via an increase in the calcium permeability that is activated upon depolarization. Thus, the parasympathetic and sympathetic divisions of the autonomic nervous system have opposite effects on the heart, just as they typically do in all other target organs as well.

Both the effect of ACh on potassium channels and the effect of norepinephrine on calcium channels are indirect effects. Recall from Chapter 9 that neurotransmitters can affect ion channels either directly (as is the case for ACh at the neuromuscular junction) or indirectly via intracellular second messengers. In the heart, the ACh released by the parasympathetic neurons binds to and activates a type of ACh receptor quite different from the nonspecific cation channel that is directly activated by ACh at the neuromuscular junction. This type of receptor is called the **muscarinic acetylcholine receptor** (because it is activated by the drug muscarine and related compounds, as well as by ACh). The ACh receptor at the neuromuscular junction is called the **nicotinic acetylcholine receptor** (because it is activated by the drug nicotine and related compounds). Muscarinic receptors are not themselves ion channels. Instead, the activated receptor binds to and stimulates GTP-binding proteins (G-proteins, see Chapter 9) that are attached to the inner surface of the membrane near the receptors. This sequence is diagrammed in Figure 12-9. In their active form, with GTP bound, the G-proteins then cause potassium channels to open, increasing the potassium permeability of the muscle cell and slowing the rate of action potentials. The effect of the G-proteins on the channel may be direct, by binding of the channel protein to one or more subunits of the active G-protein, or it may be indirect by stimulation of arachidonic acid, a second messenger produced by enzymatic cleavage of membrane lipids. The muscarinic receptor activates the G-protein by inducing the replacement of GDP by GTP at the GTP binding site. The G-protein remains active—interacting with and opening potassium channels—as long as GTP occupies the binding site on the G-protein. The activity of the G-protein is terminated by the intrinsic GTPase activity of the G-protein itself, which ultimately hydrolyzes the terminal phosphate of the GTP, converting it to the inactive GDP.

The linkage between the norepinephrine receptor of the cardiac muscle cells and the calcium channels is also mediated via G-proteins. This is summarized in Figure 12-10. The receptor on the cell surface that detects norepinephrine is a type called the **β-adrenergic receptor** (there is also a different

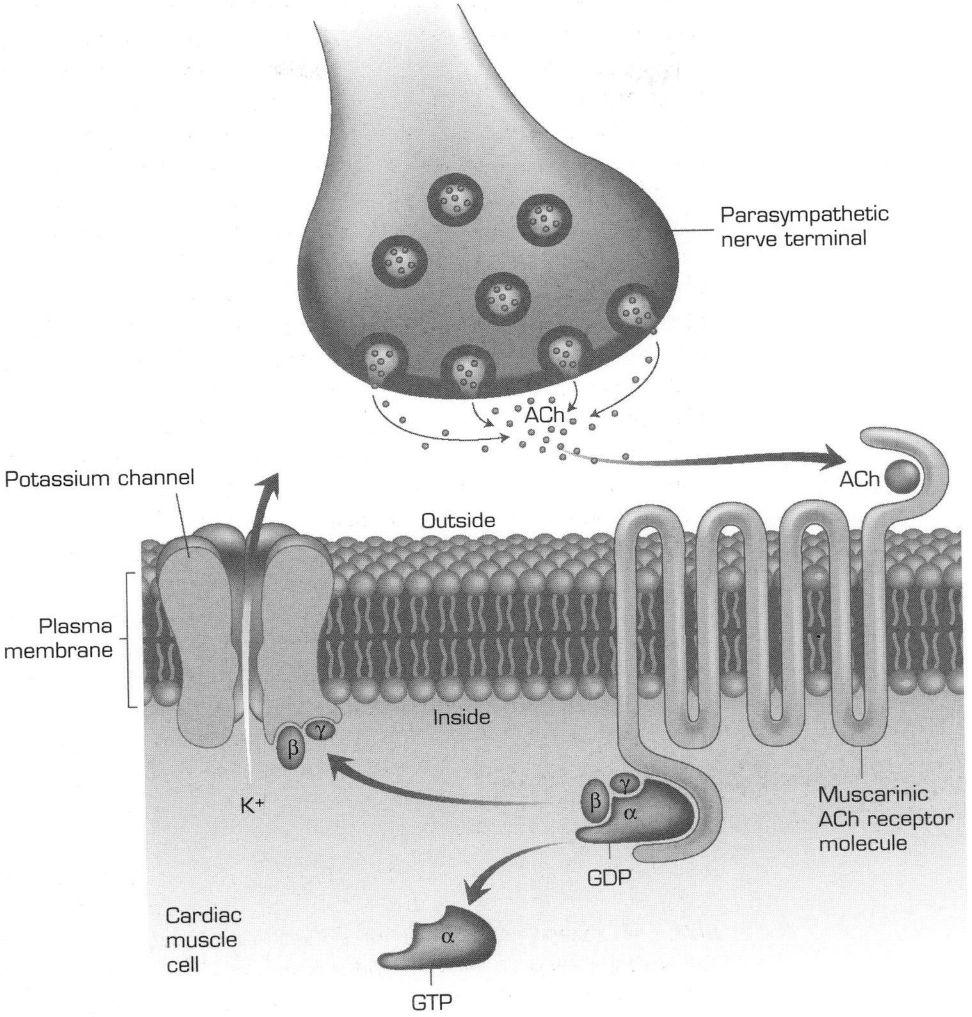

Figure 12-9 Acetylcholine indirectly opens potassium channels in cardiac muscle cells. The synaptic terminals of parasympathetic neurons release ACh, which binds to muscarinic ACh receptor molecules in the membrane of the postsynaptic muscle cell. The receptor then activates G-proteins, by catalyzing the replacement of GDP by GTP on the GTP-binding site on the α-subunit of the G-protein. The β- and γ-subunits dissociate from the α-subunit when GTP binds. The potassium channel is thought to open when the β- and γ-subunits directly interact with the channel. (Animation available at www.blackwellscience.com)

class of norepinephrine receptor called the α-adrenergic receptor, but that class is not involved in the effects of norepinephrine we are discussing here). β-Adrenergic receptors are members of the same family of receptors as the muscarinic cholinergic receptors we discussed above. Like the muscarinic

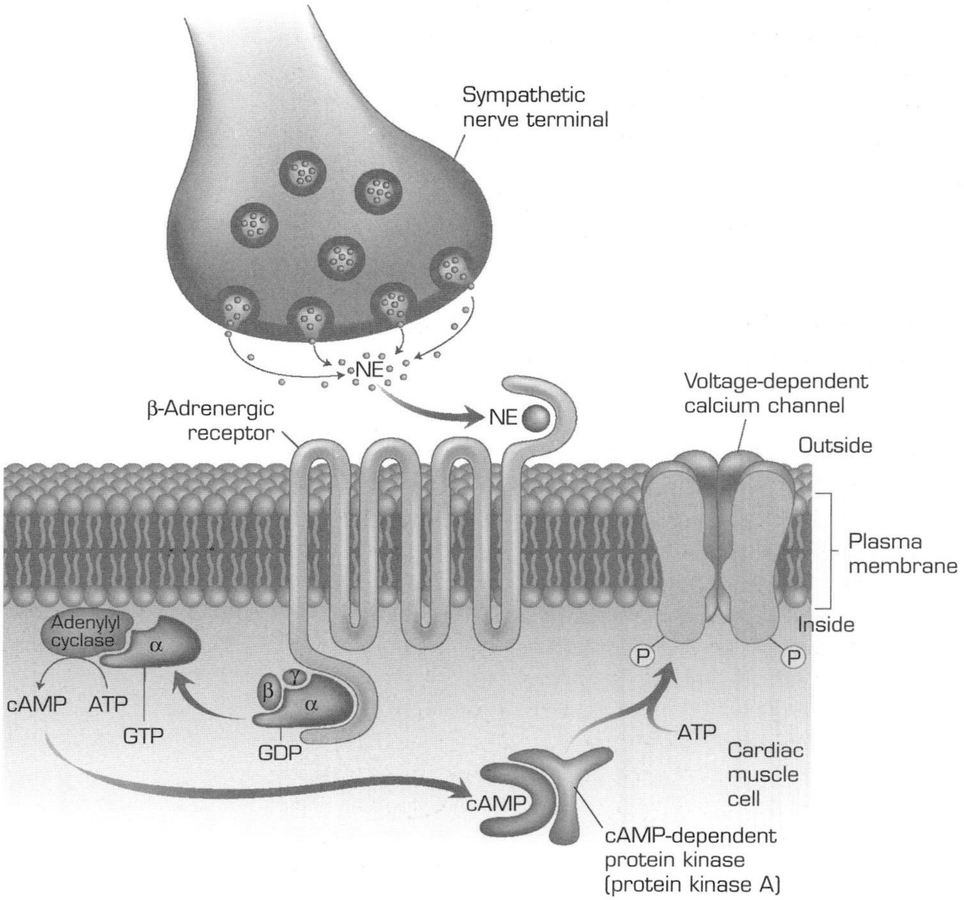

Figure 12-10 Norepinephrine promotes the activation of voltage-dependent calcium channels in cardiac muscle cells. When norepinephrine is released from the synaptic terminals of sympathetic neurons, it combines with β-adrenergic receptors in the postsynaptic membrane of the cardiac muscle cells. The activated receptor stimulates G-proteins, by catalyzing binding of GTP to the α-subunit, which then dissociates from the β- and γ-subunits. The α-subunit of the G-protein activates adenylyl cyclase, an enzyme that converts ATP into cyclic AMP. Cyclic AMP then stimulates protein kinase A, which phosphorylates calcium channel molecules. Phosphorylated calcium channels open more readily during depolarization and also remain open for a longer time. As a result, calcium influx increases during depolarization of the heart cell. (Animation available at www.blackwellscience.com)

receptor, the β-adrenergic receptor is not itself an ion channel. The receptor activates G-proteins inside the cell when norepinephrine occupies its binding site on the outside of the cell. In this case in the heart, the G-protein is a member of a class that exerts its cellular actions by changing the level of **cyclic AMP** inside the cell. The synthetic enzyme for cyclic AMP, **adenylyl cyclase**, is activated by the G-protein, causing cyclic AMP levels to rise inside the cardiac cell. Cyclic AMP binds to and stimulates **protein kinase A** (also called **cyclic-AMP-dependent protein kinase**), which in turn attaches a phosphate group to (**phosphorylates**) specific amino-acid groups of the calcium channel protein. Phosphorylation of the calcium channel is thought to be required for the channel to be able to open in response to depolarization, so an increase in cyclic AMP inside the cell translates into a greater number of openable calcium channels in the cell. In addition, each channel remains open for a longer time, on average, when it opens. Thus, phosphorylation of the channels greatly potentiates the inward calcium current that flows when the cardiac muscle cells are depolarized.

In the SA node, the triggering of the action potential depends on calcium channels. If there are more calcium channels available, the threshold potential for triggering the action potential will be lower and so the action potential will be triggered earlier during the pacemaker potential in the presence of norepinephrine. Outside of the SA node, in the muscle cells of the atria and ventricles, the role of the calcium channels is to produce the plateau phase of the action potential and to allow calcium influx, which contributes to the muscle contraction. An increase in the number of available calcium channels in these cells will increase the calcium influx during the plateau and thus increase the strength of contraction of the overall heart muscle. The combination of the increase in heart rate and the increase in strength of contraction makes the β-adrenergic receptors a powerful regulator of the amount of blood volume circulated per minute through the heart. The β-adrenergic receptors—which ultimately exert their effect by increasing the phosphorylation of voltage-activated calcium channels—are therefore targeted by many drugs that are used clinically to increase the heart output in human patients whose heart muscle has been damaged by disease.

One advantage of having the autonomic neurotransmitters exert their actions through G-protein-linked receptors, rather than by direct binding to ion channels, is that the nervous system can produce rather long-term effects on the ion channels of the heart without having to continuously provide an ongoing neural signal. Once the G-proteins are activated, they can affect channel activity for several seconds, until their activation terminates when GTP is hydrolyzed by the G-protein. Thus, ACh can be released briefly from parasympathetic nerve terminals (or norepinephrine from sympathetic nerve terminals) and still affect the heart rate for several seconds after the ACh stops being released. If instead, ACh bound to and opened a ligand-gated potassium channel in order to increase potassium permeability, the neurotransmitter would have to be continuously present, requiring the nervous system to continuously send signals from the central nervous system to the autonomic ganglia to produce a steady train of action potentials in the autonomic motor neurons. In the somatic nervous system, this is exactly what happens. Somatic motor neurons are tightly temporally coupled to the activation of their targets, the skeletal muscle fibers. This allows rapid, sub-second turn-on and turn-off of muscle activity under the control of the somatic nervous system. In general, the targets controlled by the autonomic nervous system are involved in much slower activities that are typically sustained for longer periods. Therefore, the slower and more sustained activation produced by indirect linkage between neurotransmitter receptor and ion channels seems well suited for the autonomic nervous system.

Summary

Motor systems of the nervous system can be divided into two parts, based on the motor targets that are innervated. The somatic nervous system is responsible for the control of the skeletal musculature, and thus for most of what we normally think of as the behavior of the organism. The autonomic nervous system is responsible for controlling other important organ systems, involved in maintaining the internal homeostasis of the organism. The autonomic nervous system controls the cardiovascular system, the respiratory system, the digestive system, etc. The autonomic nervous system is organized differently from the somatic nervous system. The motor neurons of the autonomic nervous system are located outside the central nervous system, in autonomic ganglia. The somatic motor neurons, by contrast, are located within the spinal cord and are thus part of the central nervous system. The autonomic nervous system is divided into the parasympathetic and the sympathetic divisions. The parasympathetic autonomic ganglia are located close to or in the target organs themselves. The sympathetic ganglia are typically located close to the central nervous system, and most of them found in two chains of ganglia, called the paravertebral ganglia, that parallel the spinal column on each side of the spinal cord.

The nerve terminals of the parasympathetic postganglionic neurons release the neurotransmitter ACh in the target organ. Acetylcholine typically acts on the target cells by activating muscarinic cholinergic receptors, which exert their postsynaptic actions by altering the level of internal second messengers —such as cyclic AMP—in the postsynaptic cell. The nerve terminals of the sympathetic postganglionic neurons release the transmitter norepinephrine, which also exerts its postsynaptic effect by altering the levels of internal second messengers. In organs that receive both sympathetic and parasympathetic innervation, the actions of ACh and norepinephrine on the target cells are usually opposite. In the heart, for example, ACh decreases heart rate and reduces cardiac output, while norepinephrine increases heart rate and cardiac output.

The muscle fibers making up the heart are specialized in a number of ways to carry out their function of efficiently pumping blood through the vessels of the circulatory system. These specializations lead to a number of differences between cardiac muscle fibers and skeletal muscle fibers, which are summarized in Table 12-1. In addition, the heart as an organ contains specific structures whose function is to coordinate the pumping activity. These structures include the SA node, the AV node, and the Purkinje fibers. The SA node is the

Table 12-1 Comparison of some properties of skeletal and cardiac muscle fibers.

Property	Skeletal muscle	Cardiac muscle
Striated	Yes	Yes
Electrically coupled	No	Yes
Spontaneously contract in absence of nerve input	No	Yes
Duration of contraction controlled by duration of action potential	No	Yes
Action potential is similar to that of neurons	Yes	No
Calcium ions make an important contribution to the action potential	No	Yes
Effect of neural input	Excite	Excite or inhibit
Division of nervous system that provides neural control	Somatic	Autonomic (parasympathetic and sympathetic)
Neurotransmitter released onto muscle fibers by neurons	ACh	ACh (parasympathetic) or Norepinephrine (sympathetic)
Effect of neurotransmitter on postsynaptic ion channels	Direct	Indirect (via G-proteins)

master pacemaker region of the heart, which controls the heart rate during normal physiological functioning of the heart. The AV node provides a path for electrical conduction between the atria and the ventricles and is responsible for the delay between atrial and ventricular contractions. The Purkinje fibers provide a rapidly conducting pathway for distributing excitation throughout the ventricles during the power stroke of a single heartbeat.

The activity of the heart is controlled by both the sympathetic and parasympathetic divisions of the autonomic nervous system. Acetylcholine released by the parasympathetic nerve terminals in the heart causes slowing of the heart rate by opening potassium channels. Norepinephrine released by the sympathetic nerve terminals increases the response of voltage-dependent calcium channels to depolarization, which increases the rate of beating and the strength of contraction. Both effects of neurotransmitters are indirect, mediated via receptors that act via GTP-binding proteins. These receptors are muscarinic receptors in the case of ACh and β-adrenergic receptors in the case of norepinephrine. The effect of the β-adrenergic receptors is to increase the levels of cyclic AMP inside the cardiac cells, which in turn promotes phosphorylation of calcium channels by protein kinase A.

appendix A Derivation of the Nernst Equation

The Nernst equation is used extensively in the discussion of resting membrane potential and action potentials in this book. The derivation presented here is necessarily mathematical and requires some knowledge of differential and integral calculus to understand thoroughly. However, I have tried to explain the meaning of each step in words; hopefully, this will allow those without the necessary background to follow the logic qualitatively.

This derivation of the Nernst equation uses equations for the movement of ions down concentration and electrical gradients to arrive at a quantitative description of the equilibrium condition. The starting point is the realization that at equilibrium there will be no net movement of the ion across the membrane. In the presence of both concentration and electrical gradients, this means that the rate of movement of the ion down the concentration gradient is equal and opposite to the rate of movement of the ion down the electrical gradient. For a charged substance (an ion), movement across the membrane constitutes a transmembrane electrical current, I. Thus, at equilibrium

$$I_C = -I_E \qquad \text{(A-1)}$$

or

$$I_C + I_E = 0 \qquad \text{(A-2)}$$

where I_C and I_E are the currents due to the concentrational and electrical gradients, respectively.

Concentrational Flux

Consider first the current due to the concentration gradient. which will be given by

$$I_C = A\Phi_C ZF \qquad \text{(A-3)}$$

In words, Equation (A-3) states that the current through the membrane of area A will be equal to the flux, Φ_C, of the ion down the concentration gradient (number of ions per second per unit area of membrane) multiplied by Z (the valence of the ion) and F (Faraday's constant; 96,500 coulombs per mole of univalent ion). The factor ZF translates the flux of ions into flux of charge and hence into an electrical current. The flux Φ_C for a given ion (call the ion Y, for example) will depend on the concentration gradient of Y across the membrane (that is, $[Y]_{in} - [Y]_{out}$) and on the membrane permeability to Y, p_Y. Quantitatively, this relation is given by

$$\Phi_C = p_Y([Y]_{in} - [Y]_{out}) \quad \text{(A-4)}$$

Note that p_Y has units of velocity (cm/sec), in order for Φ_C to have units of molecules/sec/cm^2 (remember that [Y] has units of molecules/cm^3). The permeability coefficient, p_Y, is in turn given by

$$p_Y = D_Y/a \quad \text{(A-5)}$$

where D_Y is the diffusion constant for Y within the membrane and a is the thickness of the membrane. D_Y can be expanded to yield

$$D_Y = uRT \quad \text{(A-6)}$$

where u is the mobility of the ion within the membrane and RT (the gas constant times the absolute temperature) is the thermal energy available to drive ion movement. Substituting Equation (A-6) in (A-5) and the result in (A-4) yields

$$\Phi_C a = uRT([Y]_{in} - [Y]_{out}) \quad \text{(A-7)}$$

Equation (A-7) gives us the flux through a membrane of thickness a, but we would like a more general expression that gives us the flux through any arbitrary plane in the presence of a concentration gradient. To arrive at this expression, consider the situation diagrammed in Figure A-1, which shows a segment

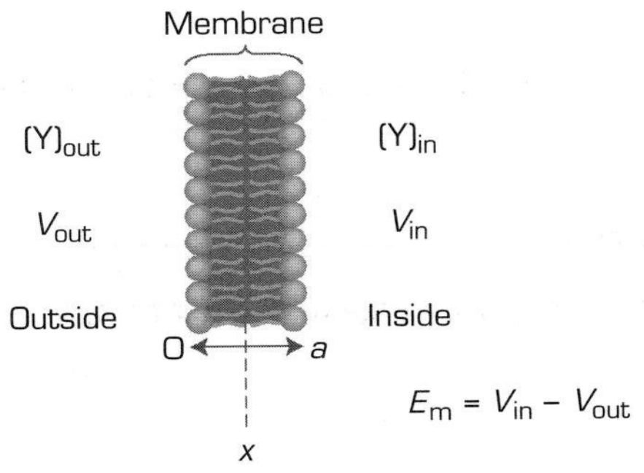

Figure A-1 Segment of membrane separating two compartments.

of membrane separating two compartments. The dimension perpendicular to the membrane is called x, and the membrane extends from 0 to a (thickness = a). In this situation, Equation (A-7) can be expressed in the form of an integral equation:

$$\Phi_C\left(\int_0^a dx\right) = uRT\left(\int_0^a dC\right) \tag{A-8}$$

Here, C stands for the concentration of the ion; therefore, in reference to Figure A-1, C_a is $[Y]_{in}$ and C_0 is $[Y]_{out}$. Differentiating both sides of Equation (A-8) yields

$$\Phi_C \, dx = uRT \, dC \tag{A-9}$$

which can be arranged to give the more general form of Equation (A-7) that we desire:

$$\Phi_C = uRT\left(\frac{dC}{dx}\right) \tag{A-10}$$

Equation (A-10) can be substituted into Equation (A-3) to give us the ionic current due to the concentration gradient.

Current Due to Electrical Gradient

Return now to the current driven by the electrical gradient, which can be expressed as

$$I_E = A\Phi_E ZF \tag{A-11}$$

The flux, Φ_E, of a charged particle through a plane at position x in the presence of a voltage gradient dV/dx will be

$$\Phi_E = uZFC\left(\frac{dV}{dx}\right) \tag{A-12}$$

Again, u is the mobility of the ion, and C is the concentration of the ion at position x. The factor ZFC is then the concentration of charge at the location of the plane; this is necessary because the voltage gradient dV/dx acts on charge and ZFC gives the "concentration" of charge at position = x. Equation (A-12) is analogous to Equation (A-10), except it is the voltage gradient rather than the concentration gradient that is of interest.

Total Current at Equilibrium

Equations (A-12), (A-11), (A-10) and (A-3) can be combined into the form of Equation (A-2) to give

$$uAZF\left(RT\frac{dC}{dx}+ZFC\frac{dV}{dx}\right)=0 \tag{A-13}$$

This requires that

$$RT\left(\frac{dC}{dx}\right)=-ZFC\left(\frac{dV}{dx}\right) \tag{A-14}$$

Equation (A-14) can be rearranged to give a differential equation that can be solved for the equilibrium voltage gradient:

$$\left(-\frac{RT}{ZF}\right)\left(\frac{dC}{C}\right)=dV \tag{A-15}$$

This can be solved for V by integrating across the membrane. Using the nomenclature of Figure A-1, the integrals are

$$-\frac{RT}{ZF}\int_{[Y]_{out}}^{[Y]_{in}}\frac{dC}{C}=\int_{V_{out}}^{V_{in}}dV \tag{A-16}$$

The solution to these definite integrals is

$$\frac{RT}{ZF}(\ln[Y]_{in}-\ln[Y]_{out})=V_{in}-V_{out} \tag{A-17}$$

or

$$\frac{RT}{ZF}\ln\left(\frac{[Y]_{out}}{[Y]_{in}}\right)=V_{in}-V_{out}=E_m \tag{A-18}$$

Equation (A-18) is the Nernst equation.

appendix B

Derivation of the Goldman Equation

The Goldman equation, or constant-field equation, is important to an understanding of the factors that govern the steady-state membrane potential. As discussed in Chapter 4, the Goldman equation describes the nonequilibrium membrane potential reached when two or more ions with unequal equilibrium potentials are free to move across the membrane. The basic strategy in this derivation is to use the flux equations derived in Appendix A to solve separately for the ionic current carried by each permeant ion and then to set the sum of all ionic currents equal to zero. The derivation is somewhat more complex than that of the Nernst equation in Appendix A, and it requires some knowledge of differential and integral calculus to follow in detail. Nevertheless, it should be possible for those without the necessary mathematics to follow the logic of the steps and thus to gain some insight into the physical mechanisms described by the equation.

When several ions are moving across the membrane simultaneously, a steady value of membrane potential will be reached when the sum of the ionic currents carried by the individual ions is zero; that is, for permeant ions A, B, and C

$$I_A + I_B + I_C = 0 \tag{B-1}$$

The first step in arriving at a value of membrane potential that satisfies this condition is to solve for the net ionic flux, 0, for each ion separately. The total flux for a particular ion will be the sum of the flux due to the concentration gradient and the flux due to the electrical gradient:

$$\Phi_T = \Phi_C + \Phi_V \tag{B-2}$$

The expressions for Φ_C and Φ_V are given by Equations (A-10) and (A-12) in Appendix A. Thus, Equation (B-2) becomes

$$\Phi_T = uRT(\mathrm{d}C/\mathrm{d}x) + uZFC(\mathrm{d}V/\mathrm{d}x) \tag{B-3}$$

If it is assumed that the electric field across the membrane is constant (this is the constant-field assumption from which the equation draws its alternative name) and that the thickness of the membrane is a, then

$$\mathrm{d}V/\mathrm{d}x = V/a \tag{B-4}$$

In that case, Equation (B-3) can be written as

$$\frac{\Phi_{\mathrm{T}}}{uRT} = \frac{\mathrm{d}C}{\mathrm{d}x} + \frac{ZFV}{RTa}C \tag{B-5}$$

This is a differential equation of the form

$$Q = \frac{\mathrm{d}C}{\mathrm{d}x} + P(x)C$$

which has a solution

$$C\ \exp\left(\int P(x)\,\mathrm{d}x\right) = \int Q\ \exp\left(\int P(x)\,\mathrm{d}x\right)\mathrm{d}x + \text{constant} \tag{B-6}$$

In this instance, $Q = \Phi_{\mathrm{T}}/(uRT)$ and $P(x) = (ZFV)/(RTa)$. Making these substitutions and integrating Equation (B-6) across the membrane of thickness a (that is, from 0 to a) gives

$$C\exp\left(\frac{ZFV}{RTa}\right)\Bigg|_0^a = \frac{\Phi_{\mathrm{T}}}{uRT}\int_0^a \exp\left(\frac{ZFVx}{RTa}\right)\mathrm{d}x \tag{B-7}$$

This becomes

$$C_a\exp\left(\frac{ZFV}{RT}\right) - C_0 = \frac{\Phi_{\mathrm{T}}}{uRT}\left[\exp\left(\frac{ZFVx}{RTa}\right)\Big/\left(\frac{ZFV}{RTa}\right)\right]\Bigg|_0^a$$

or

$$C_a\exp\left(\frac{ZFV}{RT}\right) - C_0 = \frac{\Phi_{\mathrm{T}}}{uRT}\frac{RTa}{ZFV}\left[\exp\left(\frac{ZFVa}{RTa}\right) - \exp\left(\frac{ZFV\cdot 0}{RTa}\right)\right]$$

Rearranging and combining terms yields

$$C_a\exp\left(\frac{ZFV}{RT}\right) - C_0 = \frac{\Phi_{\mathrm{T}}a}{uZFV}\left[\exp\left(\frac{ZFV}{RT}\right) - 1\right]$$

This can be solved for Φ_T to yield

$$\Phi_T = \frac{uZFV}{a}\left[\frac{C_a \exp(ZFV/RT) - C_0}{\exp(ZFV/RT) - 1}\right] \tag{B-8}$$

Now, C_a and C_0 are the concentrations of the ion just within the membrane. These concentrations are related to the concentrations in the fluids inside and outside the cell by $C_a = \beta C_{in}$ and $C_0 = \beta C_{out}$, where β is the oil–water partition coefficient for the ion in question. Substituting these in Equation (B-8) gives

$$\Phi_T = \frac{\beta uZFV}{a}\left[\frac{C_{in} \exp(ZFV/RT) - C_{out}}{\exp(ZFV/RT) - 1}\right] \tag{B-9}$$

The permeability constant, p_i, for a particular ion is given by $p_i = \beta uRT/a$, or $p_i/RT = \beta u/a$. Making this substitution in Equation (B-9) gives

$$\Phi_T = \frac{p_i ZFV}{RT}\left[\frac{C_{in} \exp(ZFV/RT) - C_{out}}{\exp(ZFV/RT) - 1}\right] \tag{B-10}$$

The flux, Φ_T, for an ion can be converted to a flow of electrical current, as required in Equation (B-1), by multiplying by ZF (the number of coulombs per mole of ion); therefore

$$I = \frac{p_i Z^2 F^2 V}{RT}\left[\frac{C_{in} \exp(ZFV/RT) - C_{out}}{\exp(ZFV/RT) - 1}\right] \tag{B-11}$$

This is the expression we need for each ion in Equation (B-1). For instance, if the three permeant ions are Na, K, and Cl with permeabilities p_{Na}, p_K, and p_{Cl}, then Equation (B-1) becomes (keeping in mind that the valence of chloride is −1)

$$\frac{F^2V}{RT}\left[\frac{p_K([K]_{in} e^{FV/RT} - [K]_{out}) + p_{Na}([Na]_{in} e^{FV/RT} - [Na]_{out})}{\exp(FV/RT) - 1}\right.$$
$$\left. + \frac{p_{Cl}([Cl]_{in} e^{-FV/RT} - [Cl]_{out})}{\exp(-FV/RT) - 1}\right] = 0$$

Multiplying through by $-\exp(FV/RT)/-\exp(FV/RT)$ and rearranging yields

$$\frac{F^2V}{RT(\exp(FV/RT) - 1)}\left[(p_K[K]_{in} + p_{Na}[Na]_{in} + p_{Cl}[Cl]_{out})\, e^{FV/RT}\right.$$
$$\left. - (p_K[K]_{out} + p_{Na}[Na]_{out} + p_{Cl}[Cl]_{in})\right] = 0$$

This requires that

$$(p_K[K]_{in} + p_{Na}[Na]_{in} + p_{Cl}[Cl]_{out})\, e^{FV/RT}$$

$$- (p_K[K]_{out} + p_{Na}[Na]_{out} + p_{Cl}[Cl]_{in})] = 0$$

or

$$e^{FV/RT} = \frac{(p_K[K]_{out} + p_{Na}[Na]_{out} + p_{Cl}[Cl]_{in})}{(p_K[K]_{in} + p_{Na}[Na]_{in} + p_{Cl}[Cl]_{out})}$$

Taking the natural logarithm of both sides and solving for V yields the usual form of the Goldman equation

$$V = \frac{RT}{F} \ln\left(\frac{p_K[K]_{out} + p_{Na}[Na]_{out} + p_{Cl}[Cl]_{in}}{p_K[K]_{in} + p_{Na}[Na]_{in} + p_{Cl}[Cl]_{out}} \right)$$

appendix C Electrical Properties of Cells

Electrical signals are fundamental to nervous system function. The electrical properties of cells are important in determining how electrical signals spread along plasma membrane. This Advanced Topic explores the electrical characteristics of cell membranes as electrical conductors and insulators. These passive electrical properties arise from the physical properties of the membrane material and from the ion channels in the membrane.

The Cell Membrane as an Electrical Capacitor

An electrical capacitor is a charge-storing device, which consists of two conducting plates separated by an insulating barrier. Because the lipid bilayer of the plasma membrane forms an insulating barrier separating the electrically conductive salt solutions of the ICF and ECF, the plasma membrane behaves as a capacitor. When a capacitor is hooked up to a battery as shown in Figure C-1, the voltage of the battery causes electrons to leave one conducting plate and to accumulate on the other plate. This charge separation continues until the resulting voltage gradient across the capacitor equals the voltage of the battery. The amount of charge, q, stored on the capacitor at that time will be given by $q = CV$, where V is the voltage across the capacitor and C is the **capacitance**

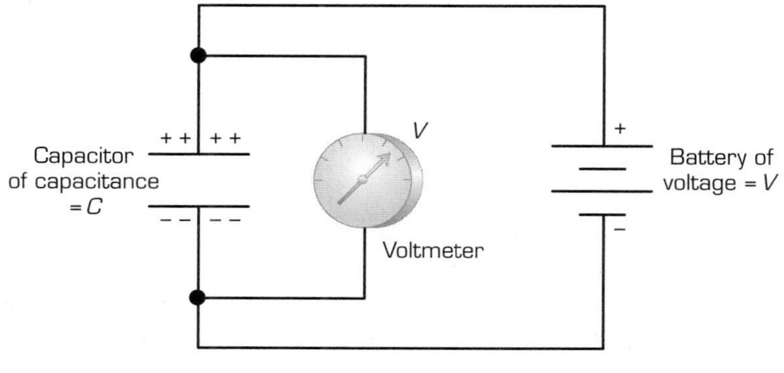

Figure C-1 When a battery is connected to a capacitor, charge accumulates on the capacitor until the voltage across the capacitor equals the voltage of the battery.

of the capacitor. Capacitance is directly proportional to the area of the plates (bigger plates can store more charge) and inversely proportional to the distance separating the two plates. Capacitance also depends on the characteristics of the insulating material between the plates, which is the lipid of the plasma membrane in cells.

The unit of capacitance is the farad (F): a 1 F capacitor can store 1 coulomb of charge when hooked up to a 1 V battery. Biological membranes have a capacitance of approximately 10^{-6} F (that is, 1 microfarad, or μF) per cm^2 of membrane area. From this value of membrane capacitance, the thickness of the insulating lipid portion of the membrane can be estimated using the following relation:

$$x = \frac{\varepsilon_0 \kappa}{C} \tag{C-1}$$

In this equation, x is the distance between the conducting plates (that is, the ICF and the ECF), C is the capacitance of the plasma membrane (1 $\mu F/cm^2$), ε_0 is the permittivity constant (8.85×10^{-8} $\mu F/cm$), and κ is the dielectric constant of the insulating material separating the two conducting plates ($\kappa = 5$ for membrane lipid). The calculated membrane thickness is approximately 4.5 nm, which is similar to the membrane thickness of approximately 7.5 nm estimated with electron microscopy. The thickness estimated from capacitance is less because it is determined by the insulating portion of the membrane, whereas the total membrane thickness, including associated proteins, is observed through the electron microscope.

Electrical Response of the Cell Membrane to Injected Current

Many electrical signals in nerve cells arise when ion channels open in the plasma membrane, allowing a flow of electrical current, carried by ions, to move across the membrane and alter the membrane potential of the cell. This situation can be mimicked experimentally by placing a microelectrode inside a cell and injecting charge into the cell through the microelectrode. Figure C-2 shows the response of a cell to injected current, considering only the capacitance of the cell membrane. If a constant current, I, is injected into the cell, then charge, q, is added to the membrane capacitor at a constant rate ($I = dq/dt$). Because $q = CV$ for a capacitor, we obtain the result:

$$I = C\frac{dV}{dt} = \text{constant} \tag{C-2}$$

In other words, dV/dt is a constant, and voltage changes linearly (that is, at a constant rate) during injection of constant current.

The response of the cell to injected current is different, however, if we take into account the presence of ion channels in the cell membrane. Ion channels

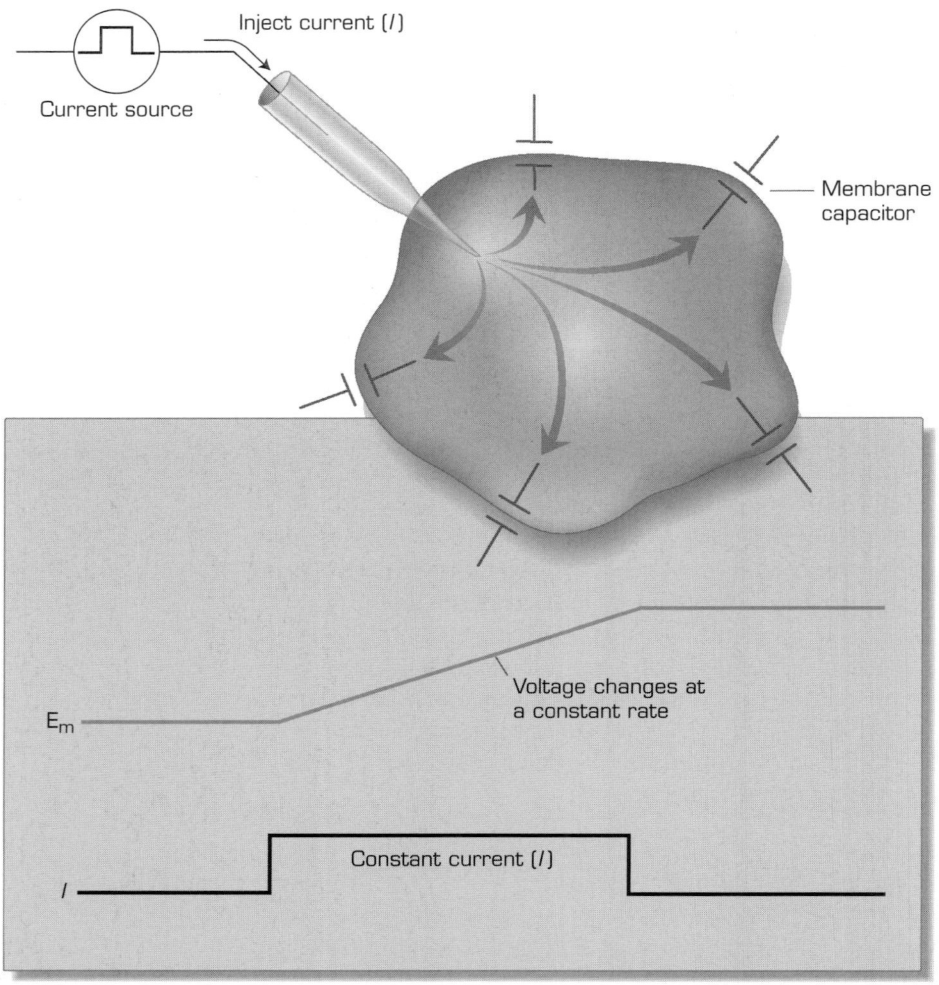

Figure C-2 The rise of voltage during injection of constant current in a cell. Only the contribution of the membrane capacitance is considered, and the effect of membrane resistance is neglected. During injection of charge at a constant rate, the resulting voltage on the membrane capacitor rises linearly.

provide a path for injected charge to move across the membrane, instead of being added to the charge on the membrane capacitor. The electrical analog of the current path provided by the ion channels is an electrical resistor. Figure C-3 illustrates the effect of adding a resistive path for current flow in the cell membrane, in parallel with the capacitance of the cell membrane. In a spherical cell, the injected current has equal access to all parts of the cell membrane at the same time. Therefore, we can combine all of the resistors and all of the capacitors for each patch of cell membrane, resulting in the analogous electrical circuit shown in Figure C-3, consisting of the combined, parallel resistance R

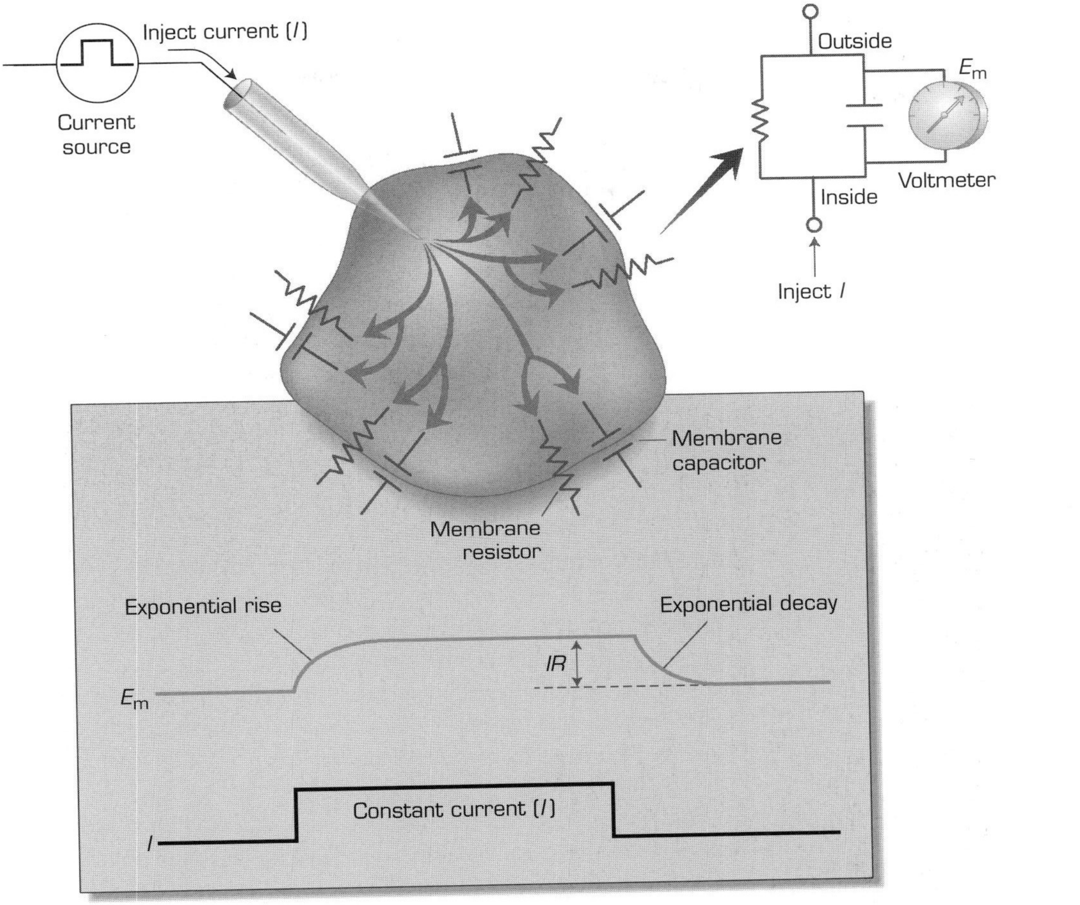

Figure C-3 The rise of voltage during injection of constant current into a spherical cell, considering both the capacitance and the resistance of the cell membrane. The membrane capacitors represent the insulating portion of the cell membrane, and the membrane resistors represent open ion channels that allow charge to move across the membrane. At the onset of the injected current, all of the injected charge initially flows onto the membrane capacitance. As the voltage on the capacitor builds up, progressively more of the current flows through the resistance. Finally, all of the current flows through the membrane resistance, and the asymptotic voltage is governed by Ohm's law ($V = IR$).

and the combined parallel capacitance C. The injected current now consists of two components: i_C, the component that flows onto the capacitor, and i_R, the component that flows through the membrane resistor, R. The capacitative current is given by Equation C-2, and the resistive current is given by Ohm's law: $i_R = V/R$. Hence, the total current is

$$I = \frac{V}{R} + C\frac{dV}{dt} \quad \text{(C-3)}$$

Solving Equation (C-3) for V yields:

$$V = IR(1 - e^{-t/RC}) \tag{C-4}$$

Thus, voltage rises exponentially during injection of a constant current, I. The product, RC, is the exponential time constant of the voltage rise, which is abbreviated τ. The asymptotic value of the voltage is IR, which is the voltage expected when all of the current is flowing through the membrane resistance. Initially, all of the injected charge flows onto the membrane capacitor, but as charge accumulates, more and more charge flows instead through the resistor, until finally all of the current flows through the resistive path. When the current injection terminates, the accumulated charge on the capacitor discharges through the parallel resistance, R. This decay of voltage is also exponential, with the same time constant, τ, given by RC.

The Response to Current Injection in a Cylindrical Cell

In a spherical cell, as in Figure C-3, the injected current flows equally to the resistors and capacitors in all parts of the membrane at the same time. However, neurons typically give rise to many long, thin neurites that extend long distances to make contact with other cells. Current injected in the cell body of the neuron, for example, must flow along the interior of a neurite to reach the portion of the cell membrane located in the neurite at a distance from the cell body. In this situation, then, current does not have equal access to all parts of the membrane.

Figure C-4 illustrates the analogous electrical circuit for a long cylindrical cell. To reach the parallel resistor and capacitor at progressively more distant portions of the cell membrane, current injected at one end of the cell must flow through the resistance provided by the interior of the cell. This resistance can be quite large for cylindrical neurites of neurons. The resistance of a cylindrical conductor is given by

$$R = \frac{r4l}{\pi d^2} \tag{C-5}$$

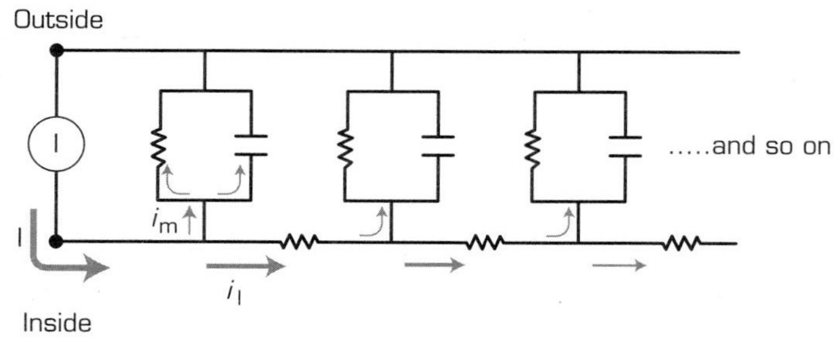

Figure C-4 The equivalent electrical circuit for a long cylindrical cell. A constant current is injected at one end of the cell. At each position along the cell, current divides into a membrane component, i_m, flowing onto the membrane resistance and capacitance at that point, and a longitudinal component, i_l, that flows through the resistance of the cell interior to more distant portions of the membrane. The amount of current remaining at each position along the cell is indicated by the thickness of the arrows.

where r is the specific resistance of the conducting material, l is the length of the cylinder, and d is the diameter of the cylinder. For the cytoplasm of a neurite, r is approximately 100 Ω cm, which is about 10^7 times worse than copper wire. Thus, a neurite 1 μm in diameter would have an internal resistance of approximately 1.3×10^6 Ω per μm of length.

The current at the site of injection divides into two components. Some of the current (designated i_m, for membrane current) flows onto the parallel membrane resistance and capacitance at the injection site. The remainder of the current (indicated by i_l, for longitudinal current) flows through the internal resistance of the neurite. At the next portion of the neurite, the current again divides into membrane and longitudinal components. Thus, the amount of current declines with distance along the neurite. In addition, current entering the parallel *RC* circuit at each position changes with time, because the voltage on the capacitance at each local position builds up as described previously for the spherical cell. As a result, the change in membrane voltage produced by current injection in the cylindrical cell varies as a function of both time and distance from the injection site.

Analysis of the electrical circuit shown in Figure C-4 leads to the following equation for membrane voltage:

$$V + \tau \partial V/\partial t = \lambda^2 \partial^2 V/\partial x^2, \text{ where } \tau = r_m c_m \text{ and } \lambda = \sqrt{r_m/r_i} \quad \text{(C-6)}$$

In this second-order, partial differential equation, r_m and c_m are the resistance and the capacitance of the amount of membrane in a 1 cm length of the cylindrical cell, and r_i is the internal resistance of a 1 cm length of the cylindrical cell. For an infinitely long cylindrical cell, the solution of Equation (C-6) is the cable equation:

$$V(x,t) = V_{\substack{x=0 \\ t=\infty}} \tfrac{1}{2} \left\{ e^{-X}\left[1 - erf\left(\frac{X}{2\sqrt{T}} - \sqrt{T}\right)\right] - e^{X}\left[1 - erf\left(\frac{X}{2\sqrt{T}} - \sqrt{T}\right)\right]\right\} \quad \text{(C-7)}$$

In this equation, $X = x/\lambda$ and $T = t/\tau$. That is, both distance and time are normalized with respect to λ and τ, which are defined in Equation (C-6). As in the exponential equation governing rise of voltage during current injection in a spherical cell, τ is the time constant of the cylindrical cell. The constant factor, λ, is called the **length constant** of the cylindrical cell.

The function *erf* in Equation (C-7) is the error function, which is defined as

$$erf(z) = 1/\sqrt{\pi} \int_{-z}^{+z} e^{-y^2}\, dy \quad \text{(C-8)}$$

The error function, *erf*(z), is the integral under a Gaussian probability distribution from $-z$ to $+z$, as illustrated graphically in Figure C-5. Note that as z increases from 0, the integral of the Gaussian function first increases rapidly,

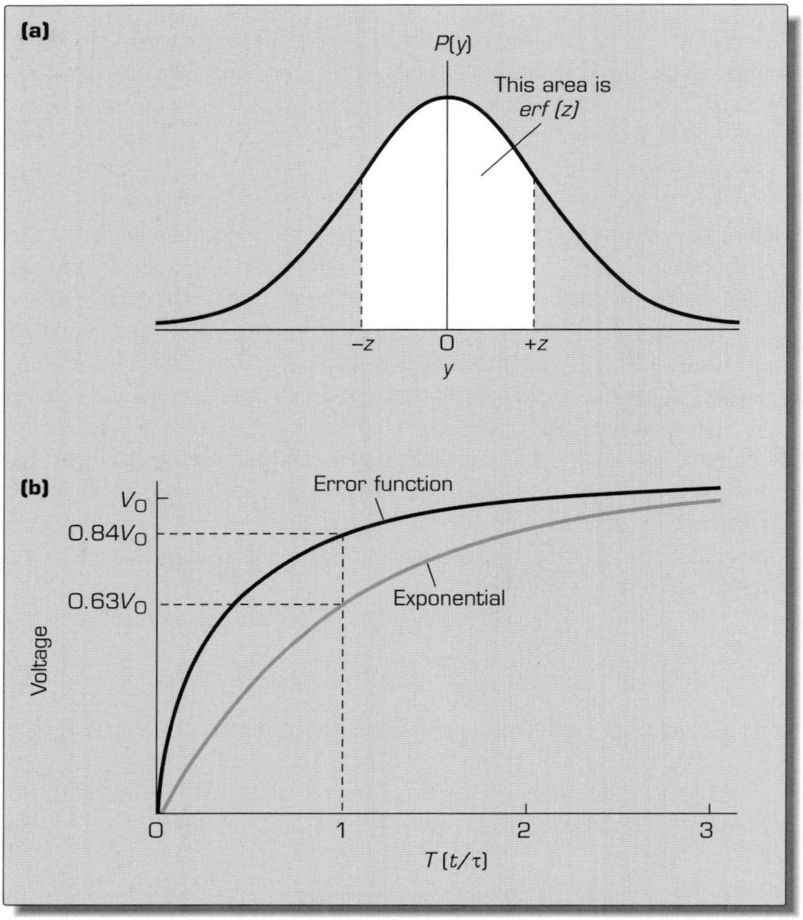

Figure C-5 The error function represents the area under a Gaussian curve. (a) The bell-shaped curve represents a Gaussian function. The error function (*erf*) of a variable, z, is the integral of the Gaussian function from $-z$ to $+z$. (b) The time-course of rise of voltage with time after onset of a constant current. The error function rises more steeply than an exponential function. Time is normalized with respect to the time constant, τ, in both cases. When $t = \tau$ (that is, $T = 1$), the error function has reached 0.84 of its final value, V_0, but the exponential function has reached 0.63 of its final value.

then progressively more slowly. The rise of $erf(T)$ with increasing T is shown in Figure C-5b, compared on the same time scale with an exponential rise. When $t = \tau$ (that is, when $T = 1$), the exponential function rises to 0.63 (that is, $1 - 1/e$) of its final, asymptotic value, whereas the error function rises to 0.84 of its asymptotic value.

Although Equation (C-7) may seem daunting, it reduces to simpler relations under certain circumstances. For example, the steady-state decay of voltage with distance from the injection site (that is, $V(x)$ at $t = \infty$) can be obtained by recognizing that dV/dt eventually becomes zero a long time after the onset of current injection. Thus, when $dV/dt = 0$, Equation (C-6) becomes

$$V = \lambda^2 d^2V/dx^2 \tag{C-9}$$

which has an exponential solution:

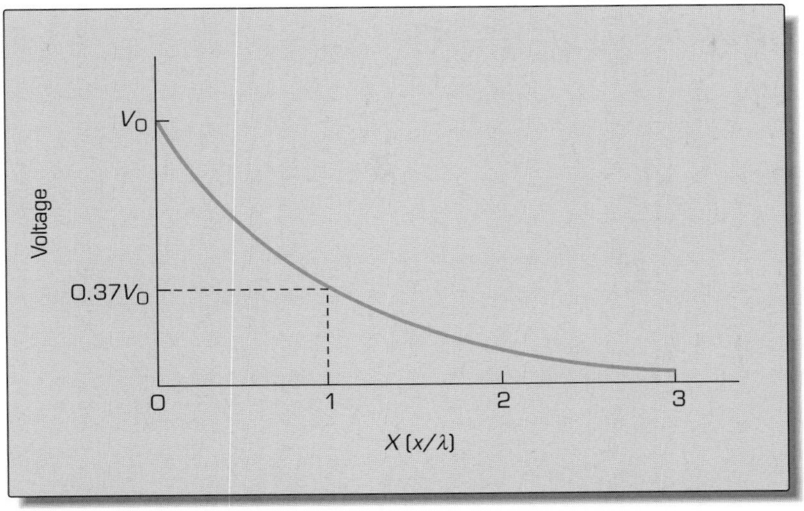

Figure C-6 The steady-state decay of voltage with distance when a constant current is injected at $X = 0$ in an infinitely long cylindrical cell. Distance is normalized with respect to the length constant, λ. At $x = \lambda$ (that is, $X = 1$), steady-state voltage is 37% of the steady-state voltage at the site of current injection, V_0.

$$V(x) = V_0 e^{-x/\lambda} \qquad \text{(C-10)}$$

In this equation, V_0 is the steady-state voltage at the injection site at $t = \infty$. Thus, in the steady state, voltage declines exponentially with distance from the injection site, and the spatial decay is governed by the length constant, λ. Figure C-6 summarizes the decline of voltage along a cylindrical neurite. At a distance λ (that is, one length constant) from the injection site, the steady-state voltage declines to $1/e$ (that is 0.37) of the voltage at the injection site.

Another special case is the rise of voltage with time at the site of current injection (that is, $V(t)$ at $x = 0$). With $x = 0$, Equation (C-7) reduces to

$$V(t) = V_0 erf(\sqrt{T}) \qquad \text{(C-11)}$$

In other words, voltage at the injection site rises with a time-course given by the error function, as shown in Figure C-7. At a distance $x = \lambda$ from the injection site, the asymptotic voltage at $t = \infty$ is $0.37V_0$, as described above, and the time-course of the rise is given by the cable equation (Equation (C-7)) with $X = 1$. This time-course is also shown in Figure C-7. Note that unlike the rapid rise at $x = 0$, the voltage at $x = \lambda$ rises with a pronounced delay, which represents the time for the injected current to begin to reach the membrane distant from the injection site. Because of the appreciable internal resistance to current flow, injected charge will flow first onto the membrane capacitance at the injection site and then in the intervening portions of membrane, before reaching more distant parts of the membrane. Thus, the rise of voltage is not only smaller but also slower at progressively greater distance from the point where current is injected into a cylindrical cell.

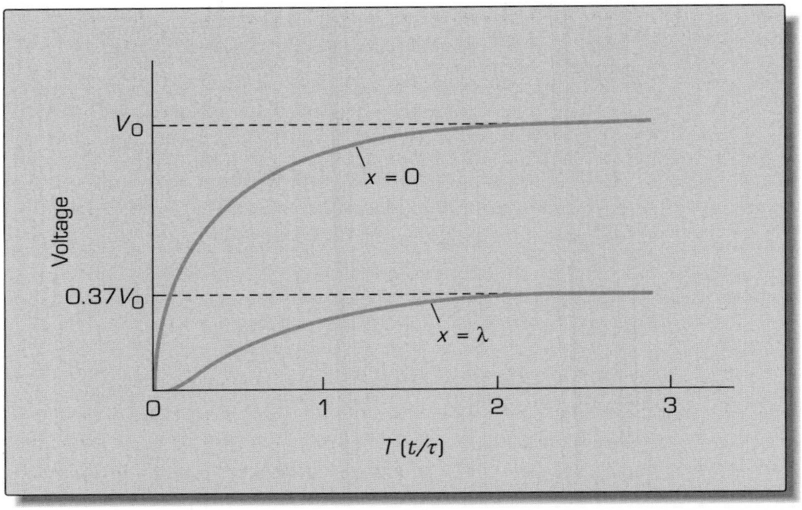

Figure C-7 The rise of voltage with time after the onset of current injection at two locations along an infinitely long cylindrical cell. Time is normalized with respect to the time constant, τ. At the site of injection ($x = 0$), the voltage rises according to the error function. At a distance of λ from the injection site ($x = \lambda$), the voltage rises with an S-shaped delay to its final value, which is 37% of the steady-state voltage at the site of current injection, V_0.

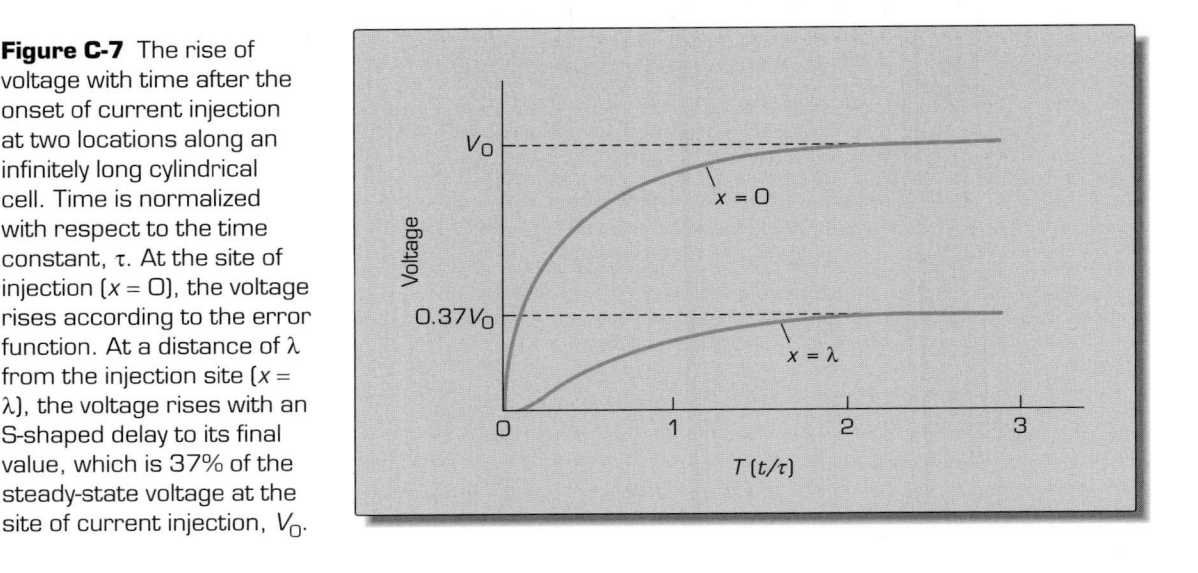

In the nervous system, the passive cable properties of neurites have functional significance for the influence exerted by a particular synaptic input on action potential firing in a postsynaptic neuron. A synaptic input located on a dendrite at a distance from the cell body of the neuron would produce a smaller, slower change in membrane potential in the cell body than a synaptic input located near the cell body. Thus, nearby synaptic inputs have greater influence on the activity of postsynaptic cells.

Suggested Readings

This section provides readings for those interested in additional information about the topics covered in this book, at both introductory and more advanced levels. General References cover the same broad topics as this book and present related material as well. References given under Specific Topics represent a range of difficulty from introductory to original papers in technical journals.

Rating system:

(*) Introductory level; for a general, nonscientist audience or beginning students

(**) Intermediate level; a general review for nonspecialists or second-level students

(***) Advanced level; for advanced students wishing greater detail

(****) Original research articles; usually intended for specialists and professional researchers

General References

Annual Reviews, Inc. publishes yearly volumes in several scientific disciplines. Articles relevant to neurobiology are commonly found in: *Annual Review of Neuroscience, Annual Review of Physiology, Annual Review of Biophysics and Biomolecular Structure, Annual Review of Biochemistry*, and *Annual Review of Cell Biology*. (***) {www.annualreviews.org}

Current Opinion in Neurobiology publishes monthly issues, each organized around a particular theme. Articles are brief and emphasize recent findings. (***) {http://reviews.bmn.com}

Hall, J.W. (ed.) *An Introduction to Molecular Neurobiology*. Sunderland, MA: Sinauer, 1992. (**)

Handbook of Physiology. Volumes published periodically by the American Physiological Society. Those on neurophysiology and cardiovascular physiology contain advanced material on topics covered in this book. Articles often require advanced knowledge of biology, chemistry, and mathematics. (***)–(****)

Hille, B. *Ionic Channels of Excitable Membranes*, 3rd ed. Sunderland, MA: Sinauer Associates, 2001. (***)

Kandel, E.R., Schwartz, J.H., and Jessell, T.M. (eds) *Principles of Neural Science*, 4th ed. New York: Elsevier, 2000. (***)

Katz, B. *Nerve, Muscle and Synapse*. New York: McGraw-Hill, 1966. (***)

Levitan, I.B., and Kaczmarek, L.K. *The Neuron, Cell and Molecular Biology*. New York: Oxford University Press, 1991. (**)

Matthews, G.G. *Introduction to Neuroscience*. Malden, MA: Blackwell Science, 2000. A study guide for students of neuroscience, with practice exam questions. (*) {http://blackwellscience.com}

Matthews, G.G. *Neurobiology: Molecules, Cells, and Systems*, 2nd ed. Malden, MA: Blackwell Science, 2001. Covers the material of this book and many other aspects of general neurobiology. (*) {http://blackwellscience.com}

Physiological Reviews is a periodical published by the American Physiological Society. Articles are usually long and comprehensive reviews of a special topics, and issues frequently include coverage of cellular and molecular neurobiology. (***) {http://physrev.physiology.org}

Scientific American publishes well-illustrated reviews written primarily for a general readership. These articles often provide a good starting point for further reading. (*) {http://www.scientificamerican.com}

Trends in Neurosciences presents brief, up-to-date reviews on very specific topics. Again, these articles are usually good starting points for more in-depth reading. Other "Trends in . . ." series (*Trends in Biochemical Sciences, Trends in Cell Biology*, and *Trends in Pharmacological Sciences*) sometimes include articles of interest to neurobiologists. (***) {http://journals.bmn.com}

Specific Topics

The Cell and its Composition

Bretscher, M.S. (1985) The molecules of the cell membrane. *Scientific American*, 253, 100–108. (*)

Burton, R.F. (1983) The composition of animal cells: solutes contributing to osmotic pressure and charge balance. *Comparative Biochemistry and Physiology [B]*, 76, 663–671. (****)

Fettiplace, R., and Haydon, D.A. (1980) Water permeability of lipid membranes *Physiological Reviews*, 60, 510. (****)

Gilles, R. *Mechanisms of Osmoregulation: Maintenance of Cell Volume*. New York: Wiley, 1979. (***)

Kwon, H.M., and Handler, J.S. (1995) Cell volume regulated transporters of compatible osmolytes. *Current Opinion in Cell Biology*, 7, 465–471. (**)

Macknight, A.D. (1988) Principles of cell volume regulation. *Renal Physiology and Biochemistry*, 11, 114–141. (****)

Orgel, L.E. (1994) The origin of life on the earth. *Scientific American*, 271, 76–83. (*)

Singer, S.J., and Nicolson, G.L. (1972) The fluid mosaic model of the structure of cell membranes. *Science*, 175, 720. (***)

Verkman, A.S. (1992) Water channels in cell membranes. *Annual Review of Physiology*, 54, 97–108. (***)

Resting Membrane Potential

Fambrough, D.M., Lemas, M.V., Hamrick, M., Emerick, M., Renaud, K.J., Inman, E.M., Hwang, B., and Takeyasu, K. (1994) Analysis of subunit assembly of the Na-K-ATPase. *American Journal of Physiology*, 266, C579–C589. (****)

Glynn, I.M. (1988) How does the sodium pump pump? *Society of General Physiologists Series*, 43, 1–17. (***)

Hodgkin, A.L., and Horowicz, P. (1958) The influence of potassium and chloride ions on the membrane potential of single muscle fibers. *Journal of Physiology*, 148, 127. (****)

Hodgkin, A.L., and Katz, B. (1949) The effects of sodium ions on the electrical activity of the giant axon of the squid. *Journal of Physiology*, 108, 37. (****)

Kaplan, J.H. (1985) Ion movements through the sodium pump. *Annual Review of Physiology*, 47, 535–544. (***)

Neher, E., and Sakmann, B. (eds) *Single Channel Recording*, 2nd ed. New York: Plenum Press, 1995. (***)

Sather, W.A., Yang, J., and Tsien, R.W. (1994) Structural basis of ion channel permeation and selectivity. *Current Opinion in Neurobiology*, 4, 313–323. (**)

Action Potential

Armstrong, C.M. (1992) Voltage-dependent ion channels and their gating. *Physiological Reviews*, 72, S5–S13. (***)

Bezanilla, F., and Stefani, E. (1994) Voltage-dependent gating of ionic channels. *Annual Review of Biophysics and Biomolecular Structure*, 23, 819–846. (****)

Catterall, W.A. (1992) Cellular and molecular biology of voltage-gated sodium channels. *Physiological Reviews*, 72, S15–S48. (***)

Catterall, W.A. (1995) Structure and function of voltage-gated ion channels. *Annual Review of Biochemistry*, 64, 493–531. (***)

Hodgkin, A.L., and Huxley, A.F. (1952) Quantitative description of membrane current and its application to conduction and excitation in nerve. *Journal of Physiology*, 117, 500. (****)

Hodgkin, A.L., Huxley, A.F., and Katz, B. (1952) Measurement of current voltage relations in the membrane of the giant axon of *Loligo*. *Journal of Physiology*, 116, 424. (****)

Kallen, R.G., Cohen, S.A., and Barchi, R.L. (1993) Structure, function, and expression of voltage-dependent sodium channels. *Molecular Neurobiology*, 7, 383–428. (****)

Neher, E., and Sakmann, B. (1992) The patch clamp technique. *Scientific American*, 266:3, 28–35. (*)

Pallotta, B.S., and Wagoner, P.K. (1992) Voltage-dependent potassium channels since Hodgkin and Huxley. *Physiological Reviews*, 72, S49–S67. (***)

Synaptic Transmission

Augustine, G.J., Burns, M.E., DeBello, W.M., Pettit, D.L., and Schweizer, F.E. (1996) Exocytosis—proteins and perturbations. *Annual Review of Pharmacology and Toxicology*, 36, 659–701. (***)

Augustine, G.J., Charlton, M.P., and Smith, S.J. (1987) Calcium action in synaptic transmitter release. *Annual Review of Neuroscience*, 10, 633–693. (***)

Bennett, M.K., and Scheller, R.H. (1994) A molecular description of synaptic vesicle membrane trafficking. *Annual Review of Biochemistry*, 63, 63–100. (***)

Bertolino, M., and Llinas, R.R. (1992) The central role of voltage-activated and receptor-activated calcium channels in neuronal cells. *Annual Review of Phamacology and Toxicology*, 32, 399–421. (***)

Brown, D.A. (1990) G-proteins and potassium currents in neurons. *Annual Review of Physiology*, 52, 215–242. (***)

Calakos, N., and Scheller, R.H. (1996) Synaptic vesicle biogenesis, docking, and fusion—a molecular description. *Physiological Reviews*, 76, 1–29. (***)

Catterall, W.A. (1999) Interactions of presynaptic Ca^{2+} channels and snare proteins in neurotransmitter release. *Annals of the New York Academy of Science*, 868, 144–159. (**)

Clapham, D.E. (1994) Direct G protein activation of ion channels? *Annual Review of Neuroscience*, 17, 441–464. (***)

Del Castillo, J., and Katz, B. (1954) Quantal components of the end-plate potential. *Journal of Physiology*, 124, 560. (****)

Dunlap, K., Luebke, J.I., and Turner, T.J. (1995) Exocytotic Ca^{2+} channels in mammalian central neurons. *Trends in Neurosciences*, 18, 89–98. (**)

Fatt, P., and Katz, B. (1952) Spontaneous subthreshold activity at motor nerve endings. *Journal of Physiology*, 117, 109. (****)

Geppert, M., and Südhof, T.C. (1998) RAB3 and synaptotagmin, the yin and yang of synaptic membrane fusion. *Annual Review of Neuroscience*, 21, 75–95. (**)

Gilman, A.G. (1987) G proteins: transducers of receptor-generated signals. *Annual Review of Biochemistry*, 56, 615–649. (**)

Hamm, H.E., and Gilchrist, A. (1996) Heterotrimeric G proteins. *Current Opinion in Cell Biology*, 8, 189–196. (**)

Heuser, J.E. (1989) Review of electron microscopic evidence favoring vesicle exocytosis as the structural basis for quantal release during synaptic transmission. *Quarterly Journal of Experimental Physiology*, 74, 1051–1069. (***)

Heuser, J.E., and Reese, T.S. (1981) Structural changes after transmitter release at the frog neuromuscular junction. *Journal of Cell Biology*, 88, 564–580. (****)

Jahn, R., and Südhof, T.C. (1999) Membrane fusion and exocytosis. *Annual Review of Biochemistry*, 68, 863–911. (**)

Jahn, R., and Südhof, T.C. (1994) Synaptic vesicles and exocytosis. *Annual Review of Neuroscience*, 17, 219–246. (**)

Katz, B., and Miledi, R. (1967) The timing of calcium action during neuromuscular transmission. *Journal of Physiology*, 189, 535. (****)

Linder, M.E., and Gilman, A.G. (1992) G proteins. *Scientific American*, 267:1, 56–61. (*)

Matthews, G. (1996) Neurotransmitter release. *Annual Review of Neuroscience*, 19, 219–233. (***)

Rothman, J.E. (1996) The protein machinery of vesicle budding and fusion. *Protein Science*, 5, 185–194. (***)

Tsien, R.W., Lipscombe, D., Madison, D., Bley, K., and Fox, A. (1995) Reflections on Ca^{2+}-channel diversity, 1988–1994. *Trends in Neurosciences*, 18, 52–54. (**)

Van der Kloot, W., and Molgo, J. (1994) Quantal acetylcholine release at the vertebrate neuromuscular junction. *Physiological Reviews*, 74, 899–991. (***)

von Gersdorff, H., and Matthews, G. (1999) Electrophysiology of synaptic vesicle cycling. *Annual Review of Physiology*, 61, 725–752. (***)

Wickman, K., and Clapham, D.E. (1995) Ion channel regulation by G proteins. *Physiological Reviews*, 75, 865–885. (***)

Skeletal Muscle

Ashley, C.C., and Ridgeway, E.B. (1968) Simultaneous recording of membrane potential, calcium transient and tension in single muscle fibres. *Nature*, 219, 1168. (****)

Bourne, G.H. *The Structure and Function of Muscle*, 2nd ed. New York: Academic Press, 1972 (Vol. I), 1973 (Vols. II and III), 1974 (Vol. IV). (***)

Buchtal, E., and Schmalbruch, H. (1980) Motor unit of mammalian muscle. *Physiological Reviews*, 60, 90. (***)

Freund, H.-J. (1983) Motor unit and muscle activity in voluntary motor control. *Physiological Reviews*, 63, 387. (***)

Hoyle, G. (1983) *Muscles and Their Neural Control*. New York: Wiley. (**)

Huxley, H.E. (1973) Muscular contraction and cell motility. *Nature*, 243, 445. (**)

Huxley, H.E. (1996) A personal view of muscle and motility mechanisms. *Annual Review of Physiology*, 58, 1–19. (**)

Schneider, M.F. (1994) Control of calcium release in functioning skeletal muscle fibers. *Annual Review of Physiology*, 56, 463–484. (***)

Heart

Brown, H.F. (1982) Electrophysiology of the sinoatrial node. *Physiological Reviews*, 62, 505. (***)

Campbell, D.L., Rasmusson, R.L., and Strauss, H.C. (1992) Ionic current mechanisms generating vertebrate primary cardiac pacemaker activity at the single cell level: an integrative view. *Annual Review of Physiology*, 54, 279–302. (***)

Clapham, D.E. (1994) Direct G protein activation of ion channels? *Annual Review of Neuroscience*, 17, 441–464. (***)

Deal, K.K., England, S.K., and Tamkun, M.M. (1996) Molecular physiology of cardiac potassium channels. *Physiological Reviews*, 76, 49–67. (***)

Hartzell, H.C. (1988) Regulation of cardiac ion channels by catecholamines, acetylcholine and second messenger systems. *Progress in Biophysics and Molecular Biology*, 52, 165–247. (***)

Hartzell, H.C., Méry, P.-F., Fischmeister, R., and Szabo, G. (1991) Sympathetic regulation of cardiac calcium current is due exclusively to cAMP-dependent phosphorylation. *Nature*, 351, 573–576. (****)

Irisawa, H., Brown, H.F., and Giles, W. (1993) Cardiac pacemaking in the sinoatrial node. *Physiological Reviews*, 73, 197–227. (***)

Kobilka, B. (1992) Adrenergic receptors as models for G protein-coupled receptors. *Annual Review of Neuroscience*, 15, 87–114. (***)

Linder, M.E., and Gilman, A.G. (1992) G proteins. *Scientific American*, 267:1, 56–61. (*)

Szabo, G., and Otero, A.S. (1990) G protein mediated regulation of K^+ channels in heart. *Annual Review of Physiology*, 52, 293–305. (***)

Wickman, K., and Clapham, D.E. (1995) Ion channel regulation by G proteins. *Physiological Reviews*, 75, 865–885. (***)

Index

Note: Page numbers followed by "f" refer to illustrations.

W9-CTN-200

Be My Disciples

Peter M. Esposito
President

Jo Rotunno, MA
Publisher

Susan Smith
Director of Project Development

Program Advisors
Michael P. Horan, PhD
Elizabeth Nagel, SSD

GRADE ONE

PARISH EDITION

The Subcommittee on the Catechism, United States Conference of Catholic Bishops, has found this catechetical series, copyright 2013, to be in conformity with the Catechism of the Catholic Church.

NIHIL OBSTAT
Rev. Msgr. Robert Coerver
Censor Librorum

IMPRIMATUR
† Most Reverend Kevin J. Farrell DD
Bishop of Dallas
August 22, 2011

The *Nihil Obstat and Imprimatur* are official declarations that the material reviewed is free of doctrinal or moral error. No implication is contained therein that those granting the *Nihil Obstat and Imprimatur* agree with the contents, opinions, or statements expressed.

Acknowledgements

Excerpts are taken and adapted from the *New American Bible* with Revised New Testament and Revised Psalms ©1991, 1986, 1970, Confraternity of Christian Doctrine, Washington, D.C., and are used by permission. All Rights Reserved. No part of the *New American Bible* may be reproduced in any form without permission in writing from the copyright owner.

Excerpts are taken or adapted from the English translation of the Rite of Baptism, ©1969; Rite of Confirmation (Second Edition), ©1975, International Committee on English in the Liturgy, Inc. (ICEL). All rights reserved.

Excerpts are taken and adapted from the English translation of the *Roman Missal*, ©2010, International Commission on English in the Liturgy, Inc. (ICEL) All rights reserved.

Copyright © 2013 RCL Publishing LLC
All rights reserved. Be My Disciples is a trademark of RCL Publishing LLC. This publication, or parts thereof, may not be reproduced in any form by photographic, electronic, mechanical, or any other method, for any use, including information storage and retrieval, without written permission from the publisher. Send all inquiries to RCL Benziger, 8805 Governor's Hill Drive, Suite 400, Cincinnati, OH 45249.

Toll Free 877-275-4725
Fax 800-688-8356

Visit us at www.RCLBenziger.com
and www.ByMyDisciples.com

20701 ISBN 978-0-7829-1570-9 (Student Edition)
20711 ISBN 978-0-7829-1576-1 (Catechist Edition)

1st printing
Manufactured for RCL Benziger in Cincinnati, OH, USA. December, 2011

Contents

Welcome to *Be My* Disciples

Jesus wants you to be his **disciple**! He wants you to know about him and follow him. This year you will learn many new things about Jesus. You will learn how to be a good disciple.

All About Me

My name is

____

I am a child of God.

Unit 1: We Believe, Part One

You will learn about God's Son, Jesus.

Look on page 38. Find out the name of Jesus' mother. Trace her name on the line.

Mary

Unit 2: We Believe, Part Two

You will learn about the Holy Trinity.

Look on page 66. Find out the name of the helper Jesus promised to send. Trace the helper's name on the line.

Holy Spirit

Unit 3: We Worship, Part One

You will learn that each of the Church's seasons tell us something about Jesus.

Look on page 87. Find out when the Church celebrates that Jesus was raised from the dead. Trace the word on the line.

Easter

Unit 4: We Worship, Part Two

You will learn how our Church celebrates and prays.

Look on page 137. Learn the name of the most important celebration of the Church. Trace the word on the line.

Mass

Unit 5: We Live, Part One

You will learn how to live the Ten Commandments.

Look on page 167. Learn which commandment teaches us to worship only God. Trace the word on the line.

First

Unit 6: We Live, Part Two

You will learn to live as a child of God.

Look on page 218. Learn who gave us the Our Father. Trace the name under the picture.

Jesus

Listen to God's Word

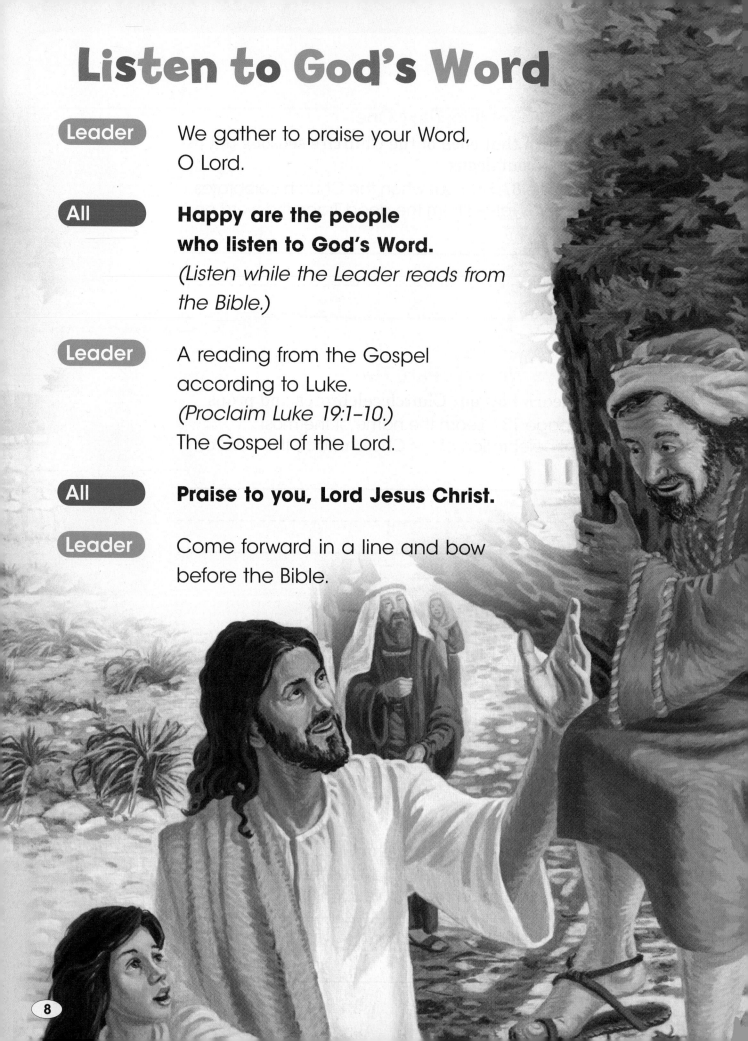

Leader We gather to praise your Word,
O Lord.

All **Happy are the people
who listen to God's Word.**
(Listen while the Leader reads from the Bible.)

Leader A reading from the Gospel
according to Luke.
(Proclaim Luke 19:1–10.)
The Gospel of the Lord.

All **Praise to you, Lord Jesus Christ.**

Leader Come forward in a line and bow
before the Bible.

We Believe
Part One

UNIT 1

Time for Children

The day was getting late. Jesus was tired. His friends wanted him to rest. But moms and dads started bringing their children to see Jesus. Jesus' friends said to them, "Go away. Jesus is tired. He has no time for children now."

"Wait!" Jesus said to his friends. "I always have time for children. Let the children come to me."

The children rushed to Jesus. Jesus welcomed and blessed them all. Jesus said with a big smile, "Look, this is what heaven is like."

BASED ON MARK 10:13–16

What I Have Learned

What is something you already know about these faith words?

Creation

Jesus

Faith Words to Know

Put an **X** next to the faith words you know.
Put a **?** next to the faith words you need to learn more about.

Faith Words

____ Bible	____ Creator	____ Son of God
____ faith	____ wonder	____ kindness

A Question I Have

What question would you like to ask about the Bible?

The Bible

What is a favorite book that someone reads to you?
Why is it your favorite?

When we listen to a story from the Bible, we hear God's Word to us. In the Bible we read:

Listen to God's word and keep it.
Then you will be blessed. Based on Luke 11:28

What do these words from the Bible ask you to do?

Disciple Power

Faithful

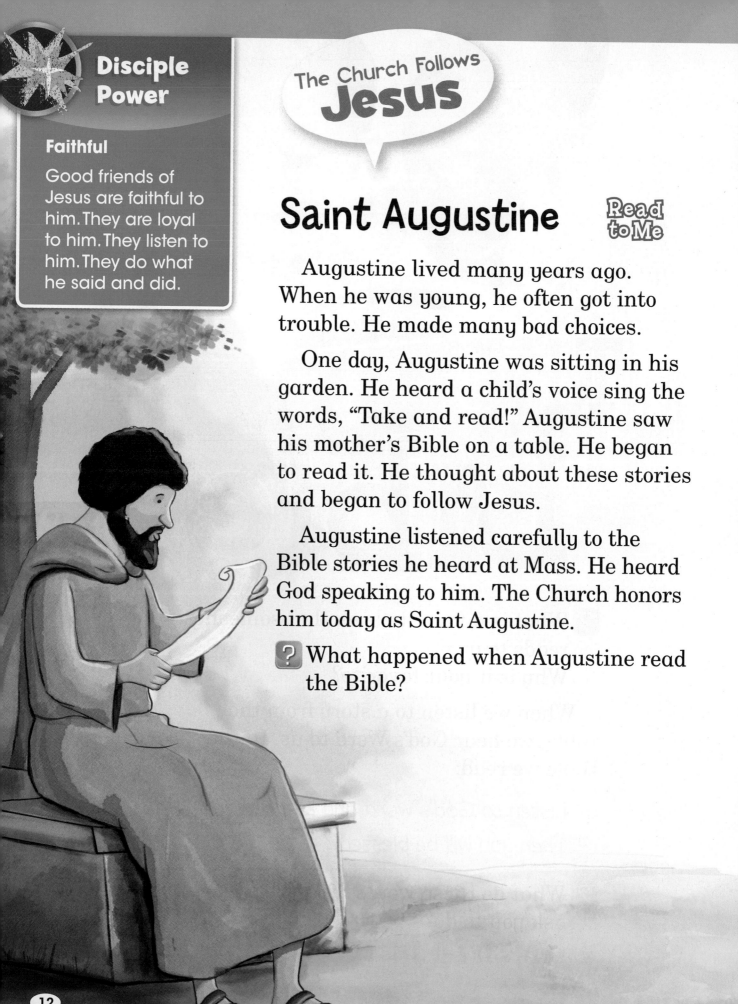

Good friends of Jesus are faithful to him. They are loyal to him. They listen to him. They do what he said and did.

The Church Follows **Jesus**

Saint Augustine

Read to Me

Augustine lived many years ago. When he was young, he often got into trouble. He made many bad choices.

One day, Augustine was sitting in his garden. He heard a child's voice sing the words, "Take and read!" Augustine saw his mother's Bible on a table. He began to read it. He thought about these stories and began to follow Jesus.

Augustine listened carefully to the Bible stories he heard at Mass. He heard God speaking to him. The Church honors him today as Saint Augustine.

? What happened when Augustine read the Bible?

About the Bible

God chose people to help write the **Bible**. The Bible is the written Word of God. It is a holy book because it is God's very own Word to us. The Bible also tells us about God's love for us.

Faith Focus
What does the Bible tell us about God?

Faith Word
Bible
The Bible is the written Word of God. It is God's very own Word to us.

Activity

Draw or write about your favorite story from the Bible. Share your story with a partner.

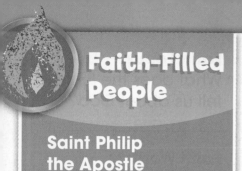

Faith-Filled People

Saint Philip the Apostle

Philip was one of the first twelve Apostles. The Apostles were the first leaders of the Church. Philip was a curious man. He wanted to know everything he could about Jesus and his teachings.

A Man Learns about God

One day a man was reading the Bible. Philip was a follower, or disciple, of Jesus. He saw the man and ran up to him.

Philip asked, "Do you understand what you are reading?"

The man said. "No. I need help."

Philip told the man about God's love.

Philip told the man about Jesus.

The man became a follower of Jesus.

He became a member of the Church.

BASED ON ACTS OF THE APOSTLES 8:26–40

? What did Philip tell the man?

Learning about God

Philip helped the man to understand a story in the Bible. He helped the man become a follower of Jesus.

At Mass, we listen to readings from the Bible. The priest or deacon helps us understand what we heard. This helps us learn how to follow Jesus.

Catholics Believe

Readings at Mass

God speaks to us through the Bible. The readings at Mass teach us about God's love and help us learn how to follow Jesus.

Activity

When the Bible is read to you, what do you do? Trace the dotted lines to find out.

listen

I Follow Jesus

The Bible is God's Word to you. When you listen to the Bible at Mass, God is speaking to you. When you and your family read the Bible at home, God is speaking to you. When you do these things, you are a faithful and loyal follower of Jesus.

Activity

Reading God's Word

Draw you and your family reading the Bible at home.

My Faith Choice

Check (√) how you will listen to God speaking to you in the Bible.

This week I will

- ☐ listen to the readings at Mass.
- ☐ ask someone to read a Bible story to me.

Pray, "Thank you, Holy Spirit, for helping me listen to the Word of God and follow Jesus. Amen."

Chapter Review

Draw lines to finish the sentences.

1. The Bible tells us about	Word of God.
2. We hear the Word of God	God's love for us.
3. The Bible is the written	follow Jesus.
4. Augustine read the Bible and began to	at Mass.

TO HELP YOU REMEMBER

1. The Bible is God's Word to us.
2. Stories in the Bible teach us about God's love.
3. We listen to the Bible at Mass.

A Listening Prayer

Leader O God, open our ears.
Help us listen to your Word.

All **Help us listen to your Word.**

Leader Listen to the Word of God.
Then think about what you hear.

Reader Act as children of God.
Obey your parents. Love others,
just as Jesus did.

BASED ON EPHESIANS 5:1, 6:1

Reader *Hold up the Bible and say:*
The Word of the Lord.

All **Thanks be to God.**

With My Family

This Week . . .

In chapter 1, "The Bible," your child learned:

- God is the real author of the Bible.
- The Bible is the inspired, written Word of God.
- The Holy Spirit inspired the human writers of the Bible to assure that God's Word would be accurately communicated.
- A faithful follower of Jesus reads the Bible and follows the teachings of the Church.

For more about related teachings of the Church, see the *Catechism of the Catholic Church*, 101–133, and the *United States Catholic Catechism for Adults*, pages 11–15.

Sharing God's Word

Read together Acts of the Apostles 8:26–40 about Philip the Apostle. Or read the adaptation of the story on page 14. Talk about why it is important to read the Bible every day.

We Live as Disciples

The Christian home and family is a school of discipleship. Choose one or more of the following activities to do as a family or design a similar activity of your own.

- Throughout the week choose a time to read the Bible as a family. Talk about ways the Bible passage or story you read helps your family live as a Catholic family.
- Help your child develop good habits that help him or her become a faithful follower of Jesus. Build on the things your child is already doing; for example, praying each day, helping out at home with chores, or treating others kindly.

Our Spiritual Journey

In this section, you will learn some of the major spiritual disciplines of the Church. These disciplines help us form the good habits of living as faithful followers of Jesus. Daily prayer is one of those disciplines. In this chapter, your child prayed and listened to Scripture. Read and pray together the prayer on page 17. This type of prayer is called *lectio divina*.

For more ideas on ways your family can live as disciples of Jesus, visit **www.BeMyDisciples.com**

CHAPTER 2

God Loves Us

Name the people who know and love you. How do they show you that they love you?

The Bible tells us that God loves us. Listen to these words from the Bible about God's love for you.

Lord, you see me and know me.
You know when I sit and when I stand.
You know what I think and where I go.
You know everything I do. BASED ON PSALM 139:1–6

What do these words say about God?

Disciple Power

Generosity

Followers of Jesus are generous. We share our things with others. We pray for them. We show generosity to them.

The Church Follows **Jesus**

Saint Rose of Lima

Read to Me

Rose knew and loved God. She knew that God loved her. She helped others know about God's love.

Rose lived with her family in Lima in the country of Peru. Rose helped take care of the family garden. She grew flowers and food. She sold flowers and gave the money to the poor and the sick. This made them feel better.

Rose was kind and generous. She helped people learn how much God loved them.

Saint Rose of Lima shows us how to love God and help others. You can pray to Saint Rose. Ask her to help you share God's love too.

Activity

Dear Saint Rose,

Help me to show my love for

____

____

We Know God Loves Us

God knows us and loves us all the time.

God wants us to know and love him too.

The Bible has many stories of people who had **faith** in God. They listened to God. They came to know and **believe** in him.

Here is a Bible story about faith. Abraham and Sarah lived a long time before Jesus. God chose Abraham to be a great leader. God made him a promise. God said,

You will be the father of many nations.
I will bless you and your wife Sarah.
You will soon become the parents of a son.

BASED ON GENESIS 17:4, 15–16

Abraham and Sarah listened to God and did what he asked. They had faith in God and believed in his promises.

? What did God promise Abraham?

Faith Focus
Who helps us to know God and believe in him?

Faith Words

faith
Faith is a gift from God. It helps us to know God and to believe in him.

believe
To believe means to have faith in God. It means to give yourself to God with all your heart.

Faith-Filled People

Isaac

Isaac is the son whom God promised Abraham and Sarah. The name Isaac means "he laughs." Isaac brought much joy and happiness to his parents.

Jesus Helps Us to Know God

Many years after Abraham and Sarah died, God sent his Son Jesus to us. Jesus is the Son of God.

Jesus helps us best to know God and his love. Jesus helps us to believe in God and have faith in him.

Jesus taught over and over again how much God loves us. He taught us that God is love.

Activity

Color the spaces. Make the **X**s one color and the **O**s another color. Find out who teaches us the most about God.

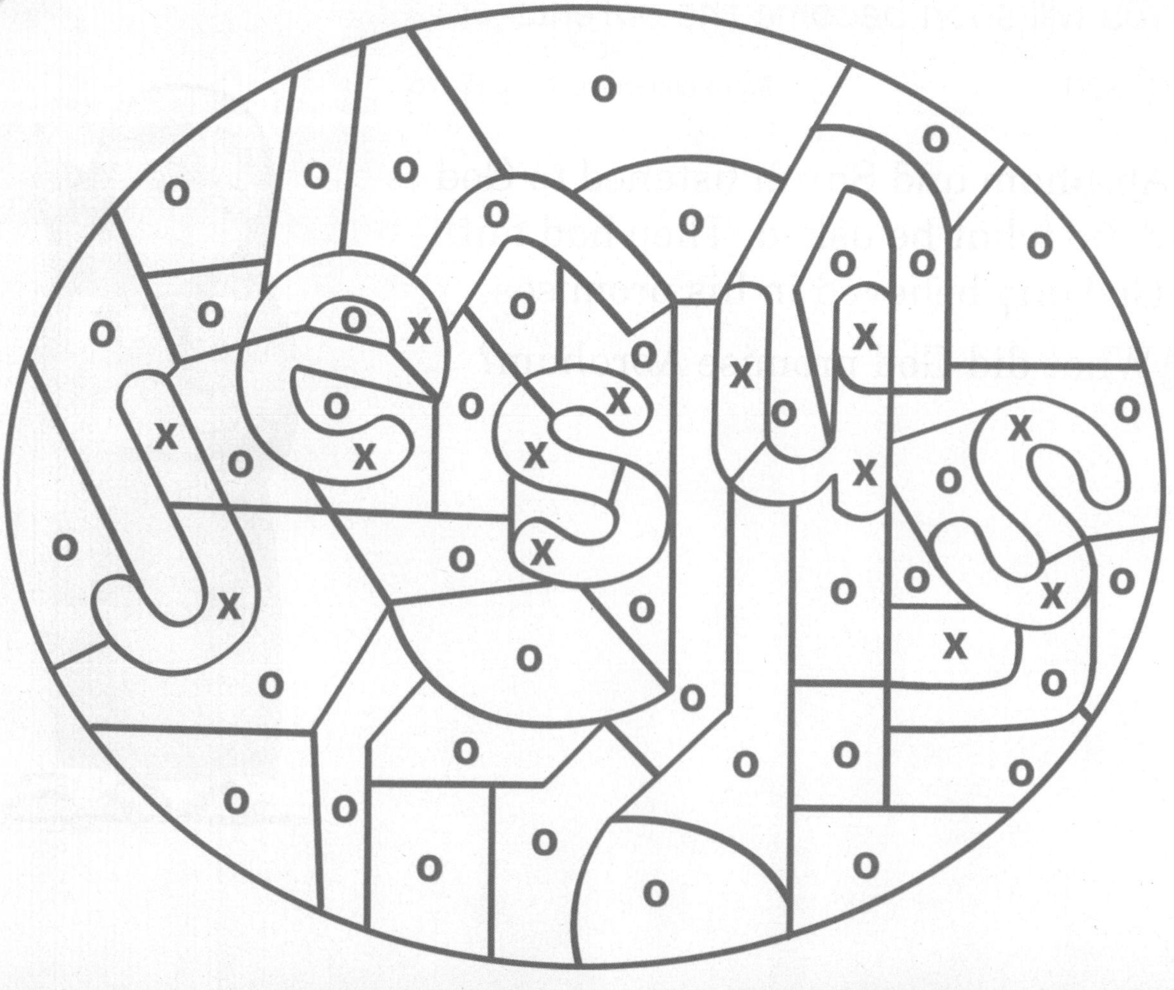

Our Family Helps Us Know God

God gave us the gift of a family. Our families help us grow in our faith. They help us know God and believe in him. Our families help us give ourselves to God with all our hearts.

Catholics Believe

Sign of the Cross

Catholics pray the Sign of the Cross. This shows we have faith in God. We pray, "In the name of the Father, and of the Son, and of the Holy Spirit. Amen."

Activity

Trace the words. Discover one important thing about God.

God

loves us.

I Follow Jesus

Your family and the Church help you to learn how much God loves you. You can help your family and friends learn how much God loves them. You can treat them the way Jesus asked. You can be kind and generous to them.

Activity

Sharing God's Love

In one heart, draw people helping you learn about God. In the second heart, draw yourself sharing God's love.

My Faith Choice

Check (√) what will you do. This week I will help others know how much God loves them. I will

- ☐ tell others about God.
- ☐ show my family I love them.
- ☐ thank God for his love.

Pray, "Thank you, Holy Spirit, for helping me to show my love for God. Amen."

Chapter Review

Complete the sentences. Color the ☐ next to the best choice.

1. To __________ means to have faith in God.

☐ believe

☐ hope

2. Faith is a gift from __________.

☐ our friends

☐ God

TO HELP YOU REMEMBER

1. God's gift of faith helps us come to know him and believe in him.
2. Jesus is the Son of God. He helps us to know how much God loves us and to have faith in God.
3. Our family and our Church help us to know, love, and serve God.

Sign of the Cross

We pray the Sign of the Cross to begin our prayers. Pray the Sign of the Cross with your class.

In the name of the Father,

and of the Son,

and of the Holy Spirit.

Amen.

With My Family

This Week . . .

In chapter 2, "God Loves Us," your child learned:

- God has revealed himself and invites us to believe in him and his love for us.
- Jesus Christ reveals the most about God and his love for us.
- Jesus is the Son of God. He is the fullness of God's Revelation.
- Our family and our Church help us grow in faith in God and in love for him.

For more about related teachings of the Church, see the *Catechism of the Catholic Church*, 80–95 and 142–175, and the *United States Catholic Catechism for Adults*, pages 35–47.

Sharing God's Word

Read together John13:31-35 from your family Bible or from a children's version of the Bible. Emphasize that when we treat one another as Jesus told his disciples to do, we show our love for God and for one another. We also show others how much God loves them.

We Live as Disciples

The Christian home and family is a school of discipleship. It is the first place where children learn to live as disciples. Choose one or more of the following activities to do as a family or design a similar activity of your own.

- Compile a list of the names of people who have helped or who are helping your family grow in faith and in your love for God. Pray for these people at a family meal.
- Name the ways your family is generous to each other and to other people. Remind your children that when they are generous they are living as Jesus taught.

Our Spiritual Journey

Generosity is a habit of being a disciple of Jesus. Generously sharing our spiritual and material blessings with others, especially people in need, is one of the foundational spiritual disciplines, or practices, of the Christian life. This discipline is known as almsgiving. Make almsgiving one of the hallmarks of your family's life. Pray together: Dear Jesus, give me a generous heart.

For more ideas on ways your family can live as disciples of Jesus, visit **www.BeMyDisciples.com**

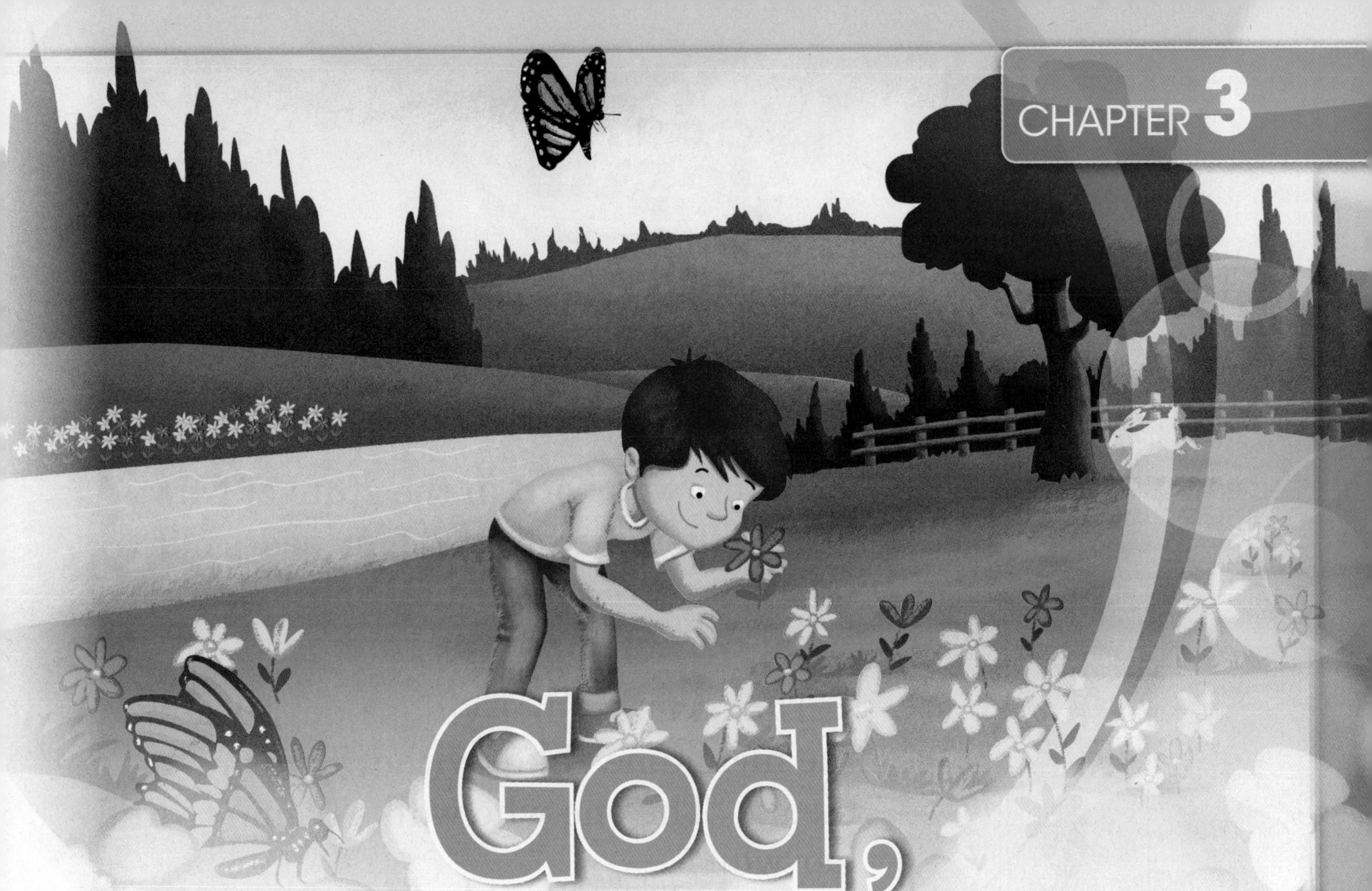

CHAPTER 3

God, Our Father and Creator

What is your favorite part of creation?

Close your eyes and see all the beautiful things in the world. Listen to what the Bible tells us about the world.

O God, everything you made is wonderful! — BASED ON PSALM 136:4

What do these words from the Bible tell you about the world God made?

Disciple Power

Wonder

The word "wonderful" comes from the word "wonder." Wonder is a special gift from God. God gives us this gift to help us come to know how good he is.

The Church Follows Jesus

Kateri Tekakwitha

Read to Me

Kateri was born in the state of New York. When Kateri was four years old, her eyes were harmed by an illness. She could hardly see in the sunlight.

The people of her village gave Kateri the nickname Tekakwitha. This name means "The one who walks trying to find her way."

Kateri loved the outdoors. She helped grow corn, beans, and squash. She picked roots of plants in the forest to make medicines. The beauty of the forest reminded her of God. It filled her with wonder. Kateri went there to talk with God and listen to him.

The Catholic Church honors her as Blessed Kateri Tekakwitha. The things she did and said show us how to live as disciples of Jesus.

? Why was being in the forest important to Kateri?

God Made Everything

God is the **Creator** of the world.
God alone made Heaven and Earth.
He made everything out of love.
The Bible tells us,

God looked at everything he made.
He saw that it was very good.

BASED ON GENESIS 1:31

Faith Focus
What do the things we see and hear in creation tell us about God?

Faith Words

Creator
God is the Creator. He made everything out of love and without any help.

image of God
We are created in the image of God. We are children of God.

Activity

Think of your favorite part of God's creation. Draw a picture of it. Share what it tells you about God.

Faith-Filled People

Catherine of Siena

Catherine enjoyed looking at creation. Things in nature reminded Catherine how much God loves us. This helped her to grow in her love for God. The Church celebrates the feast day of Saint Catherine of Siena on April 29.

God Creates People

God is the Creator of all people. He creates every person to be an **image of God.** In the Bible we read,

> *God made people in his image. He blessed them and told them to take care of everything he made. God said everything he made was very good.*
>
> BASED ON GENESIS 1:26–31

God loves every person. We are very special to God. He created us to be happy with him now on Earth and forever in Heaven.

? Why are you special to God?

God Is Our Loving Father

Every person is created by God. God the Creator is our loving Father.

This is why the Bible tells us we are children of God.

> *Jesus helped us to know and believe that God is our Father. He taught us to pray, "Our Father, who art in heaven . . ."*
>
> BASED ON LUKE 11:2

God the Father loves us and knows each of us by name. Jesus told us that God the Father cares for all his creation. He cares for all people.

We show we love God our Father when we take care of ourselves. We show our love for God when we take care of creation.

Catholics Believe

The Our Father

The Church prays the Our Father every day. We pray the Our Father at Mass. We tell God we know he loves and cares for us. We love God.

Activity

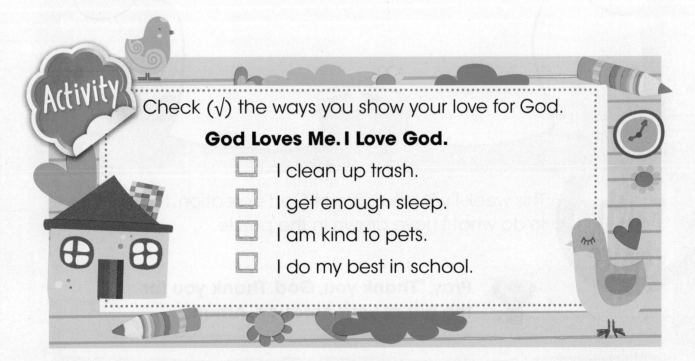

Check (√) the ways you show your love for God.

God Loves Me. I Love God.

- [ ] I clean up trash.
- [ ] I get enough sleep.
- [ ] I am kind to pets.
- [ ] I do my best in school.

I Follow Jesus

God is wonderful. He is so very good to us. The world shows us how wonderful God is. The world is God's gift to everybody. One way you can say thank you to God is to help take care of the things in the world.

Activity

Caring for God's Creation

In the puzzle piece, draw a picture of yourself taking care of something in God's creation.

My Faith Choice

This week I will take care of God's creation. I will try to do what I have drawn in the puzzle.

Pray, "Thank you, God. Thank you for the gift of your creation. Amen."

Chapter Review

Draw lines to complete the sentences.

Column A	**Column B**
1. Jesus	made everything out of love.
2. People	taught us to call God our Father.
3. God	are made in the image of God.

TO HELP YOU REMEMBER

1. God is the Creator. He made the whole world out of love.
2. God created people in the image of God.
3. Jesus taught us that God is our Father.

Thank You, God!

A rebus prayer uses pictures to help us pray. Use a word for each picture. Pray the prayer together.

ALL **Thank you, God, for your**  **.**

Reader 1 You made the  and .

ALL **Thank you, God, for your**  **.**

Reader 2 You made  and .

ALL **Thank you, God, for your**  **.**

Reader 3 You made the .

ALL **Thank you, God, for your**  **.**
You made ME! Based on Psalm 148

With My Family

This Week . . .

In chapter 3, "God, Our Father and Creator," your child learned:

- God is the Creator. All God's creation is good. Everything good exists because God created it out of love.
- God created every person in his image. He created people with all their differences.
- God is our Father. There is no limit to his love for his children.
- We respond to God the Creator's love by helping to take care of creation.

For more about related teachings of the Church, see the *Catechism of the Catholic Church*, 232–248 and 268–314, and the *United States Catholic Catechism for Adults*, pages 53–56, 67–68.

Sharing God's Word

Read together the Bible story in Genesis 1:26–31 about the creation of people. Or read the adaptation of the story on page 30. Emphasize that every person is an image of God.

We Live as Disciples

The Christian home and family is a school of discipleship. It is the first place where children learn to live as disciples. Choose one or more of the following activities to do as a family or design a similar activity of your own.

- God created each person out of love. Take turns sharing what you like about each person.
- Invite your child to take part in keeping your home clean. Explain how this is one way of thanking God for his many gifts to your family.

Our Spiritual Journey

Prayer is one of the main spiritual disciplines of the Christian life. Giving thanks to God is one of the five main forms of prayer. Invite everyone to close their eyes and see their favorite part of creation. Think of how much God loves us and silently pray, "God you are so wonderful."

For more ideas on ways your family can live as disciples of Jesus, visit **www.BeMyDisciples.com**

CHAPTER 4

Jesus, the Son of God

How do you celebrate your birthday?

Birthdays are wonderful days. Saint Luke tells us about the birthday of Jesus. He tells us:

> *Mary and Joseph came to Bethlehem. They had to stay in a stable with animals. During the night, Jesus was born. Mary wrapped him in cloth. She laid him in a manger.* BASED ON LUKE 2:1–7

What else do you know about the birth of Jesus?

Disciple Power

Kindness

We live the virtue of kindness by treating others as we want to be treated.

The Church Follows Jesus

We Have Room!

Read to Me

Daniella and everyone in San Carlos was excited. It was almost time for Christmas. It was time to celebrate Las Posadas.

For nine nights, the people walked together in the streets. Two people were chosen to be Mary and Joseph. Everyone walked behind them. They carried lighted candles.

Mary and Joseph knocked on many doors. Joseph said, "My wife will soon have a baby. Do you have room for us in your home?" All answered, "We have no room." Finally, one family said to Joseph, "We have room! Come in."

Daniella was very excited. Her family was the one who answered, "We have room! Come in."

? How did Daniella and her family show kindness to Mary and Joseph?

The Son of God

At Christmas each year we remember and celebrate the birth of Jesus. Jesus is the only son of Mary and the **Son of God**. Jesus is truly God and truly man.

The Bible tells us that angels told shepherds about the birth of Jesus. We read,

The shepherds rushed and found Mary, Joseph, and Jesus. They told everyone about Jesus and praised God for all they heard and saw.

BASED ON LUKE 2:15–17, 20

Faith Focus
Who is Jesus?

Faith Words

Son of God
Jesus is the Son of God. Jesus is truly God and truly man.

Holy Family
The Holy Family is the family of Jesus, Mary, and Joseph.

Activity

Draw a picture of you telling others about Jesus. Do what the shepherds did.

Faith-Filled People

Anne and Joachim

Saint Anne and Saint Joachim were the parents of Mary. They were the grandparents of Jesus. They helped Mary to love and trust God. The Church celebrates their feast day on July 26.

The Holy Family

Mary is the mother of Jesus, the Son of God. Joseph is the foster father of Jesus. We call Jesus, Mary, and Joseph the **Holy Family**. The Holy Family lived in a town called Nazareth.

Mary and Joseph showed their love for Jesus. They took very good care of Jesus as he was growing up. Jesus grew in his love of God and of people.

Activity

Write the names *Jesus*, *Mary*, and *Joseph* under their pictures.

Jesus Shares God's Love

When Jesus grew up, he taught others about God. He shared God's love with everyone. Jesus showed us how to treat people. Jesus treated everyone with kindness and respect.

Respect means to treat every person as a child of God. We are to treat everyone with kindness and respect. We are to share God's love with people.

Catholics Believe

Lord, Have Mercy

At Mass we pray, "Lord, have mercy." Mercy is another word for kindness. Jesus told us that people who are kind and show mercy are blessed by God.

Activity

Color the ♡s in the photos of people showing kindness and respect.

I Follow Jesus

God is always kind to people. Jesus shared God's kindness with people. You are a disciple of Jesus. You are kind to people. You treat them with respect. When you do these things, you are a sign of God's love.

Activity

In the kite draw yourself being kind to someone.

My Faith Choice

This week I will do what I drew in the kite. I will

____

____

Pray, "Thank you, Holy Spirit, for helping me to be kind to others as Jesus taught. Amen."

Chapter Review

Circle the word that best completes each sentence.

1. Jesus is the __________ of God.

   (Son) Angel

2. __________ is the mother of Jesus.

   Anne Mary

3. __________ is the foster father of Jesus.

   Joachim Joseph

TO HELP YOU REMEMBER

1. Jesus is the only son of Mary and the Son of God.
2. The family of Jesus, Mary, and Joseph is the Holy Family.
3. Jesus shared God's love with everyone.

Jesus, I Love You

We show that we love Jesus by treating people as he did. Learn to sign this prayer.

"Jesus, I love you."

Pray this prayer in the morning and at night. Teach your family to sign the prayer. Ask them to pray it with you.

Jesus

I love you.

With My Family

This Week . . .

In chapter 4, "Jesus, the Son of God," your child learned:

- Jesus is the only son of Mary and the Son of God.
- Gabriel announced to Mary that she would become the mother of the Savior, the Son of God who she was to name Jesus.
- The Son of God became truly human without giving up being God. This mystery of faith is called the Incarnation. Jesus is truly God and truly man.
- We call Jesus, Mary, and Joseph the Holy Family. Jesus' life in the Holy Family prepared him for the work the Father sent him to do.

For more about related teachings of the Church, see the *Catechism of the Catholic Church*, 456–478 and 512–560, and the *United States Catholic Catechism for Adults*, pages 77–87, 143–149.

Sharing God's Word

Read together Luke 2:1–20 about the shepherds who rushed to see the newly born Jesus. Or read the adaptation of the story on page 37. Emphasize that Jesus is truly God and truly man. He is the only son of Mary and the Son of God.

We Live as Disciples

The Christian home and family is a school of discipleship. Choose one or more of the following activities to do as a family or design a similar activity of your own.

- Talk together about the ways that family members are kind to each other. Explain how acts and words of kindness show a person's love for God.
- Choose to do a family activity that shows kindness to people who are not members of your family. For example, as a family visit someone who is lonely or help an elderly neighbor.

Our Spiritual Journey

The Great Commandment is the guiding precept of the Christian life. It is the summary or foundational principle of human as well as Christian living. In this chapter your child signed an act of love using American Sign Language. Encourage your child to teach you to sign the prayer on page 41. Pray it often together.

For more ideas on ways your family can live as disciples of Jesus, visit **www.BeMyDisciples.com**

Unit 1 **Review**

Name ______________________

A. Choose the Best Word

Complete the sentences. Color the circle next to the best choice.

1. The Bible is __________ own Word to us.

○ the Church's ○ God's

2. Faith is a gift from God that helps us to know God and to __________ in him.

○ love ○ believe

3. Jesus treated __________ people with respect.

○ all ○ some

4. Jesus is the __________ of God.

○ Man ○ Son

5. __________ is the Mother of Jesus.

○ Mary ○ Anne

B. Show What You Know

Circle the numbers next to the words that tell about Jesus.

1. Son of God

2. Holy Spirit

3. Loving Father

4. taught others about God

5. shared God's love with everyone

C. Connect with Scripture

What was your favorite story about Jesus in this unit? Draw something that happened in the story. Tell your class about it.

D. Be a Disciple

1. *What saint or holy person did you enjoy hearing about in this unit? Write the name here. Tell your class what this person did to follow Jesus.*

2. *What can you do to be a good disciple of Jesus?*

We Believe
Part Two

UNIT 2

The Last Supper

On the night before he died, Jesus ate a special meal with his Apostles. Here is what Jesus said and did.

Jesus took some bread. He gave thanks to God. He broke the bread. He shared the bread with his friends and said, "Eat this bread. It is my body."

Then Jesus took a cup filled with wine. He gave the cup to his friends and said, "Take this and drink. This is the cup of my blood. When you eat this bread and drink this wine, you remember me."

BASED ON 1 CORINTHIANS 11:23–26

What I Have Learned

What is something you already know about these faith words?

Mary

____

The Holy Trinity

____

Faith Words to Know

Put an **X** next to the faith words you know. Put a **?** next to the faith words you need to learn more about.

Faith Words

____ angels	____ hope	____ Church
____ courage	____ Holy Spirit	____ Catholic

A Question I Have

What question would you like to ask about the Church?

____

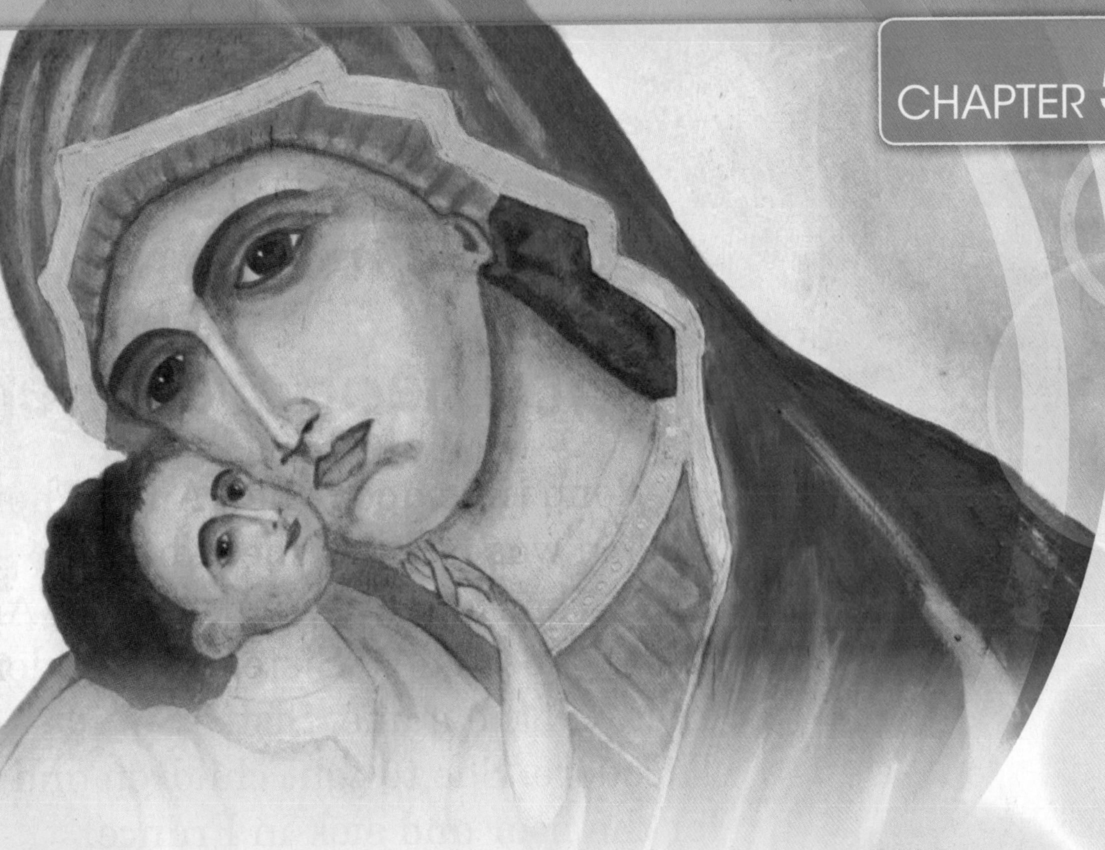

CHAPTER 5

Mary, the Mother of Jesus

What is your favorite family story?

In the Bible, we hear stories about Mary. The angel Gabriel said to Mary:

Hail Mary, God is with you.

Based on Luke 1:42

What do these words from the Bible tell you about God?

Disciple Power

Courage

The virtue of courage helps us to trust in God and live our faith.

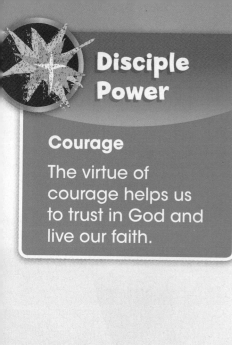

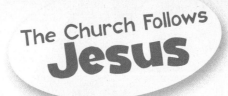

The Church Follows Jesus

Read to Me

Saint Théodore Guérin

A long time ago, when Anne-Thérèse Guérin was a child, she wanted to serve God. When she was 25 years-old, Anne-Thérèse became a Sister of Providence and took the name, Sister Saint Théodore. She taught children and cared for the poor and sick in France.

After 16 years, a bishop in the United States needed some help. So Mother Théodore went with five nuns to America to build schools and orphanages.

At first, Mother Théodore and the nuns were cold and lonely. Over the years, people learned to trust the sisters. Like Mary, Mother Théodore lived a life of courage. She always trusted God.

The Church honors Mother Théodore Guérin as a saint and celebrates her feast day on October 3.

? How do you show that you trust God?

God Loves Mary

Faith Focus
What does the Bible tell us about Mary?

Faith Word
angels
Angels give honor and glory to God. They are God's messengers and helpers.

Angels are messengers of God. God sent the angel Gabriel to a young woman named Mary. The angel gave Mary this message from God. Gabriel said,

> "You are blessed, Mary. The Holy Spirit will come to you. You will have a baby. The baby's name will be Jesus. He will be called the Son of God."
>
> Mary listened carefully to the angel Gabriel. Then she said to Gabriel, "Yes, I will do what God wants me to do."
>
> BASED ON LUKE 1:28, 31, 35, 38

Activity

Find a partner. Act out what happened when the angel Gabriel gave Mary the message from God. One of you will take the part the angel. One of you will be Mary.

Faith-Filled People

Saint Juan Diego

Juan Diego walked many miles to Mass everyday. One day, Mary appeared to him on Tepeyac Hill. She told Juan to build a church on this site and sent him to the bishop. Soon a church was built. People from all over the world visit Mary's church.

Say Yes to God

Mary said yes to God. Mary had faith in God and trusted him. Mary loved God with her whole heart.

God asks us to have faith in him too. God asks us to trust him and to love him with our whole heart.

We have faith and trust that God will always be with us. We trust that God always loves us. We show we love God when we say yes to him as Mary did.

Activity

Check (√) ways you can say yes to God. I say yes to God when I

___ Pray every day.

___ Act mean.

___ Help at home.

___ Share my toys.

___ Play fairly.

___ Say "Thank you."

God Chose Mary

God chose Mary to be the mother of Jesus. Mary cared for Jesus. We call Mary the Mother of God. Mary is very special.

Jesus wants Mary to love and care for us, too. He gave her to us as our special mother. Mary prays for us.

We celebrate the feast of Mary, the Mother of God on January 1. She prays to her son, Jesus, for us.

? Why is Mary our special mother?

Catholics Believe

Feast Days

The Church honors and shows our love for Mary on special days each year. These are called feast days. Each year on January 1 we celebrate the feast of Mary, the Mother of God. This is a holy day of obligation. We have the responsibility to take part in the celebration of Mass.

I Follow Jesus

Mary showed her faith and love for God. Courage can help you show your faith in God. You show your faith and love for God by what you say and what you do.

Activity

I Trust in God

Choose one way you can show your courage as a follower of Jesus. Draw or write about it in this space.

My Faith Choice

This week I will show my faith and love for God. I will

____

____

Pray, "Thank you, God, for helping me to show my faith and love for you."

Chapter Review

Complete the sentences. Color the ○ next to the best choice .

1. Mary said, "____" to God.

● Yes ○ No

2. Courage helps us to ____ God.

○ trust ○ know

3. Saint Juan Diego walked to ____ every day.

○ school ○ Mass

TO HELP YOU REMEMBER

1. God chose Mary to be the mother of Jesus.
2. The Bible tells us about Mary's faith in God.
3. Mary love and trusts God.

Psalm Prayer

Psalms are prayers in the Bible. We pray a psalm during Mass. Pray together:

Leader We listen to God's Word, like Mary. Happy are the people who listen to God's Word. BASED ON PSALM 1:1–2

All **Happy are the people who listen to God's Word.**

Leader We say yes to God, like Mary.

All **Happy are the people who listen to God's Word.**

With My Family

This Week . . .

In chapter 5, "Mary, the Mother of Jesus," your child learned:

- The Gospel account of the Annunciation tells us about the angel Gabriel announcing to Mary that God had chosen her to be the mother of Jesus.
- Mary is the mother of Jesus, the Son of God. Mary is the Mother of God.
- The Gospel account of the Annunciation shares with us Mary's faith and trust in God and her love for him.
- Courage helps us trust in God and live our faith, even in difficult times.

For more about related teachings of the Church, see the *Catechism of the Catholic Church*, 484–507, and the *United States Catholic Catechism for Adults*, pages 141–149.

Sharing God's Word

Read together Luke 1:26–38, the Gospel account of the Annunciation. Or read the adaptation of the story on page 48. Emphasize Mary's faith and trust in God and her love for him.

We Live as Disciples

The Christian home and family is a school of discipleship. Choose one or more of the following activities to do as a family or design a similar activity of your own.

- Teach your child the Mass responses "Thanks be to God" and "Praise to you, Lord Jesus Christ." Guide your child to use these responses properly when your family takes part in the celebration of the Mass.
- Courage is the virtue that helps us trust God and live our faith. Help your child to recognize the ways your family is living this virtue. Remind them that when they say yes to God, they are living as disciples of Jesus.

Our Spiritual Journey

The Psalms are a confession of faith in song. From the times of David until the present, the praying of the Psalms has nourished the faith of the People of God. Such prayer is both personal and communal. Memorize psalm verses such as the one on page 53 and integrate praying them spontaneously to respond to the various circumstances of your life.

For more ideas on ways your family can live as disciples of Jesus, visit **www.BeMyDisciples.com**

CHAPTER 6

Jesus Shares God's Love

? How do family members show love for one another?

Jesus always shares God's love with people. He said,

"Let the children come to me. If you want to enter God's Kingdom, become like a child." Then Jesus took the children in his arms and blessed them.

BASED ON MARK 10:14–16

? How do you show others that God loves them?

Disciple Power

Hope

The virtue of hope helps us to remember that one day we may live in happiness with God forever in Heaven.

The Church Follows Jesus

Saint Gianna

Read to Me

Saint Gianna Beretta Molla was a wife, a mother, and a doctor. She cared for many people in her life. They all remembered her smile and her care for others.

Gianna believed that caring for the sick showed God's love. If her patients did not have money to pay her, she let them give her food. Sometimes she paid for their medicine herself.

Gianna was a doctor who cared for children. She helped mothers learn how to take care of themselves and their children.

In 1955 Gianna married Pietro Molla and soon they had three children. She helped them and all people have hope in God.

? Who are some of the people who share God's love with you?

Jesus Loves Us

Jesus always shared God's love with people. He helped people in many ways. Jesus forgave the people who hurt him.

Some people did not want Jesus to teach and help others. They had Jesus killed on a **cross**. This is called the Crucifixion.

Because he loved us, Jesus died on a cross for all of us. He forgave the people who put him on the cross. He forgives us when we sin. Jesus died so that we could live with him forever in Heaven.

Faith Focus
How did Jesus show his love for us?

Faith Words

cross
The cross is a sign of God's love. It reminds us that Jesus died on a cross so that we could live forever in Heaven.

Resurrection
God's raising Jesus from the dead to new life.

Activity

Trace the words. Pray the prayer with a partner to thank Jesus for his love.

Thank you,
Jesus.

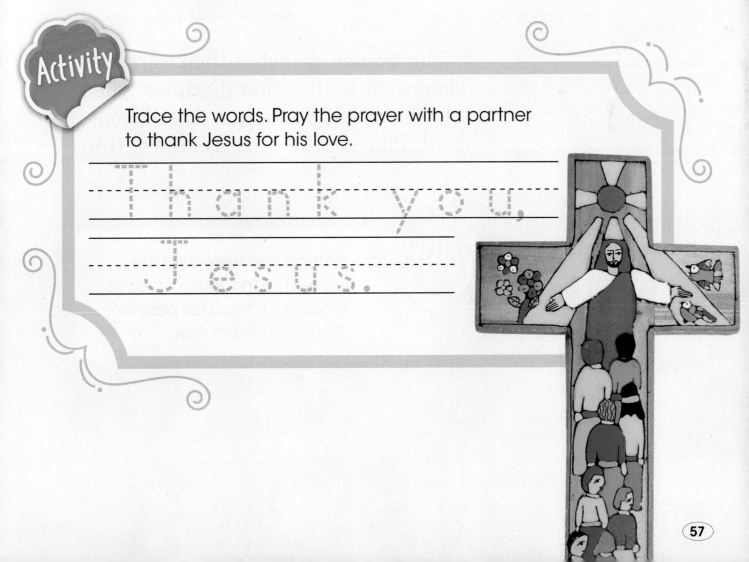

Faith-Filled People

Saint Mary Magdalene

Mary Magdalene was a disciple of Jesus. She was one of the women who went to the tomb. The women were the first ones to know that Jesus was raised from the dead. The Church celebrates her feast day on July 22.

Jesus Is Alive

After Jesus died on the cross, his friends buried his body in a tomb. Three days later some women who were disciples, or followers, of Jesus went to the place where Jesus was buried. The women were surprised at what they saw and heard. The Bible tells us,

> *When the women came to the tomb, they saw men in white robes. "Jesus is not here," the men said. "He has been raised from the dead. Go and tell the other disciples of Jesus."*
>
> BASED ON LUKE 24:1–4, 6; MATTHEW 28:7

The women did what they were told. They went to the other disciples. They told them that Jesus was raised from the dead. This is called the **Resurrection**.

Activity

Pretend you were with the women at the tomb of Jesus. What would you tell people? Share with your class.

Jesus Returned to His Father

After Jesus was raised from the dead, he stayed with his disciples for forty days. The Risen Jesus told his disciples to tell everyone in the world about him. Jesus told the disciples to invite everyone to believe in him and to be baptized.

Then Jesus returned to his Father in Heaven. We call this the Ascension. After we die, we hope that we too will return to God the Father in Heaven.

Catholics Believe

Candlemas Day

Each year the Church blesses candles on February 2. This day is called Candlemas Day. We use these candles in our churches and in our homes. They remind us of the Risen Jesus, the Light of the world. We, too, are to be lights in the world.

Activity

Listen to the story of Jesus' Ascension into Heaven. Act it out with your friends.

I Follow Jesus

The virtue of hope helps us to trust in God's love. When you tell others about Jesus, you are sharing God's love with people. You are a light in the world.

Activity

Jesus Is Alive!

Make a poster that tells people about Jesus. Use your poster as a reminder to act as a follower of Jesus.

My Faith Choice

This week I will share my poster. I will tell someone about Jesus. I will say:

____

____

Pray, "Thank you, Jesus, for teaching me how to be a light in the world."

Chapter Review

Draw lines from the words in Column A to the sentences that they complete in Column B.

Column A	Column B
1. forgives	**a.** Jesus was ______ from the dead.
2. raised	**b.** Jesus died on a ______ for all of us.
3. cross	**c.** Jesus ______ us when we sin.

TO HELP YOU REMEMBER

1. Jesus loved us so much that he gave his life for us.
2. Jesus' rising from the dead is called his Resurrection.
3. Jesus returned to his Father in Heaven.

An Act of Hope

The Church gives us a special prayer called the Act of Hope. In this prayer, we tell God we always trust in his love for us. His word to us is always true. Pray this prayer together.

O my God,
you always love us.
You are always good to us.
You word to us is always true.
With your help, we hope that
we will live with you in Heaven.
Amen.

With My Family

This Week . . .

In chapter 6, "Jesus Shares God's Love," your child learned:

- ▶ Jesus showed his great love for us by dying on the cross.
- ▶ Three days after his death, Jesus was raised from the dead. Forty days later, Jesus ascended, or returned, to his Father in Heaven.
- ▶ Before he ascended to Heaven, Jesus commanded the disciples to evangelize the world. This means they were to tell all people about Jesus and his teaching. They were to make disciples of all people and to baptize them.
- ▶ Hope is the virtue that helps us remember and trust in God's love. We hope that one day we will live in happiness with God forever in Heaven.

For more about related teachings of the Church, see the *Catechism of the Catholic Church*, 561, 620–621, 629, 656–665, and the *United States Catholic Catechism for Adults*, pages 77–87.

Sharing God's Word

Read Luke 24:1–12, the account of the Resurrection. Or read the adaptation of the story on page 58. Emphasize that as the first disciples did, we are to tell people about Jesus.

We Live as Disciples

The Christian home and family is a school of discipleship. Choose one or more of the following activities to do as a family or design a similar activity of your own.

- ▶ Jesus tells us that we are to be lights in the world. Each night at dinner, light a candle as part of your mealtime prayer. Take turns telling about how each family member was a light in the world that day.
- ▶ It is difficult to know everyone in your parish. Each month make an effort to introduce yourselves as a family to one new family in your parish.

Our Spiritual Journey

Our spiritual pilgrimage is a journey of hope. It is with confidence that we trust that God's promise of eternal life will come true. Learn and help your child learn an act of hope. Pray it regularly.

For more ideas on ways your family can live as disciples of Jesus, visit **www.BeMyDisciples.com**

CHAPTER 7

The Holy Spirit, Our Helper

? Who are some of the people who help you to learn new things?

Everyone needs teachers and helpers. The Holy Spirit is the special teacher and helper whom Jesus sent to us.

Jesus told his disciples, "God, my Father, will send you the Holy Spirit. The Holy Spirit will be your helper". BASED ON JOHN 14:26

? What do you know about the Holy Spirit?

Disciple Power

Counsel

Counsel is another word for the help that a good teacher gives us. Counsel is a gift of the Holy Spirit. This gift helps us choose to live as followers of Jesus.

Read to Me

Signs of the Holy Spirit

We can learn about God in many different ways, through words and pictures. Some churches have stained-glass windows. They may show Jesus, Mary, the saints, or symbols of the Holy Spirit. Flames of fire and a white dove are two symbols of the Holy Spirit.

The light shining through stained-glass windows reminds us of God's love for us. The Holy Spirit helps us to share that love with others.

Activity

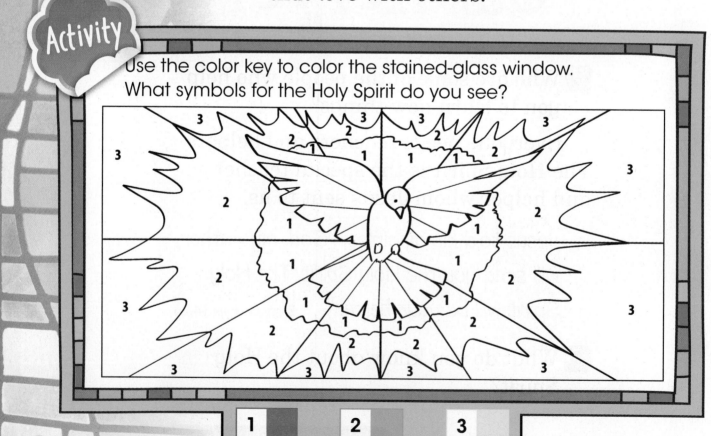

Use the color key to color the stained-glass window. What symbols for the Holy Spirit do you see?

The Holy Spirit Is with Us

Jesus taught us that there is only one God. Jesus is the Son of God. He taught us about God the Father and God the **Holy Spirit**.

Jesus taught us that there is one God in Three Divine Persons. He taught that there is one God, who is God the Father, God the Son, and God the Holy Spirit. This is called the **Holy Trinity**.

Faith Focus
How does the Holy Spirit help and teach us?

Faith Words

Holy Spirit
The Holy Spirit is the Third Person of the Holy Trinity. The Holy Spirit is always with us to be our helper.

Holy Trinity
The Holy Trinity is one God in Three Divine Persons—God the Father, God the Son, and God the Holy Spirit.

Activity

A shamrock helps us remember that there are Three Divine Persons in one God. Trace the name of these Persons in the three Leaves.

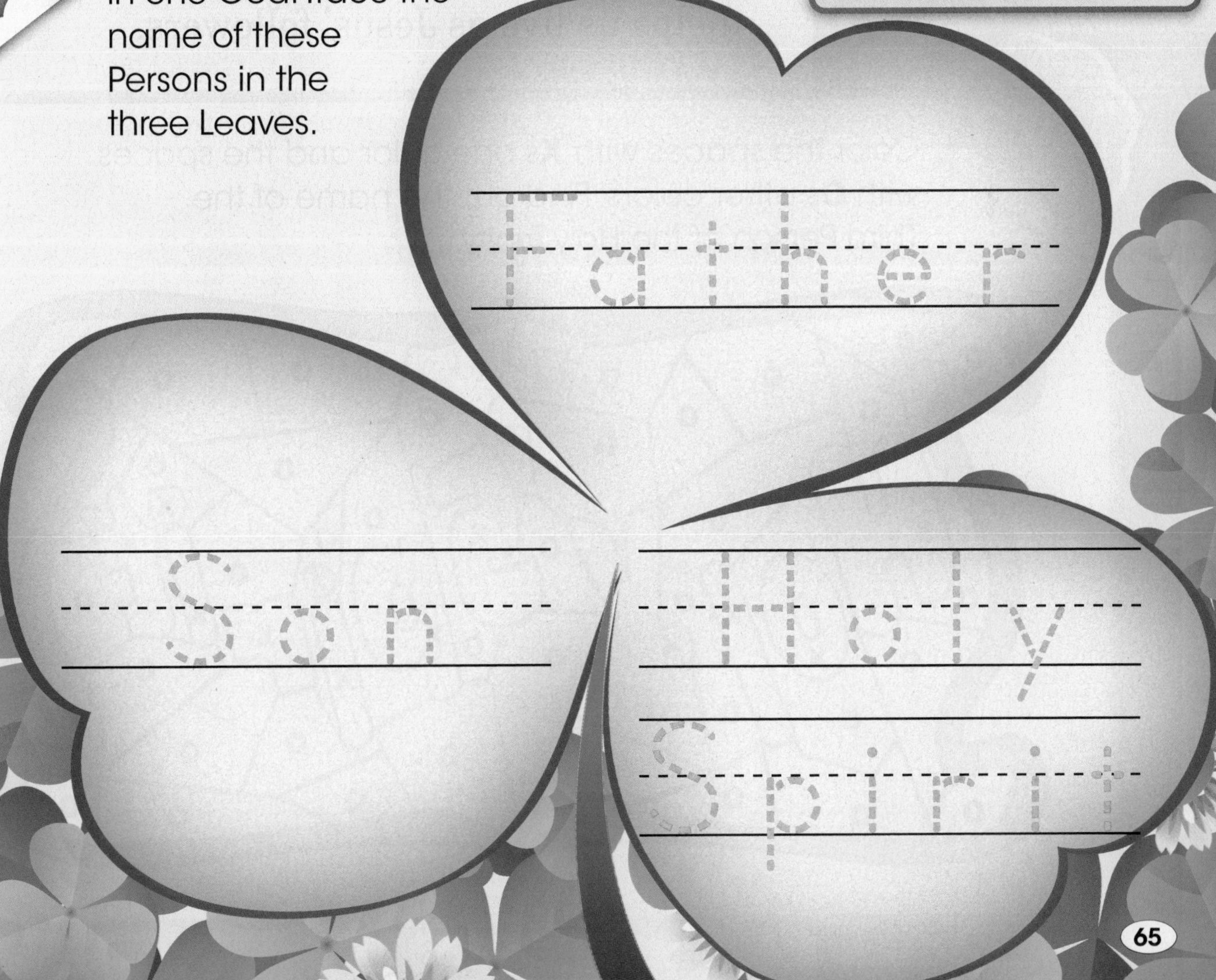

Faith-Filled People

Saint Patrick

Saint Patrick was a bishop. He taught people in Ireland about the Holy Trinity. The Church celebrates the feast day of Saint Patrick on March 17.

Jesus' Promise

Jesus made a promise to his friends. He promised that God the Father would send them a helper. Jesus said,

> *The Father will give you a helper who will always be with you.*
>
> Based on John 14:16

God the Holy Spirit is the helper whom the Father would send.

Jesus told his friends that the Holy Spirit would be their teacher and helper.

The Holy Spirit helps us understand what Jesus said and did. The Holy Spirit helps us live as Jesus' followers.

Activity

Color the spaces with **X**s one color and the spaces with **O**s other colors. Find out the name of the Third Person of the Holy Trinity.

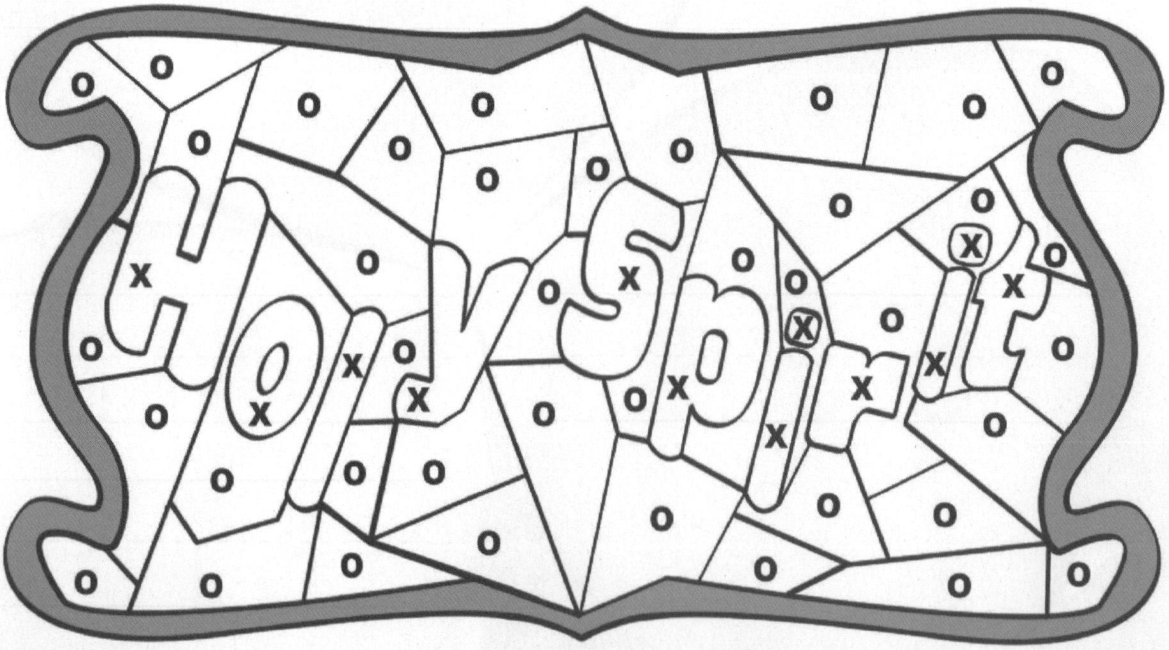

The Gift of the Holy Spirit

The Holy Spirit is the Third Person of the Holy Trinity. We first receive the gift of the Holy Spirit at Baptism. The Holy Spirit is always with us.

The Holy Spirit teaches us to pray. The Holy Spirit helps us learn what Jesus taught.

The Holy Spirit helps and teaches us to follow Jesus. Jesus told his followers,

Love one another as I love you.

Based on John 13:34

The Holy Spirit helps and teaches us to love God and one another.

Catholics Believe

Signs and Symbols

Signs and symbols help us to understand the meaning of what God has told us. The Church uses a beautiful white dove as a sign of the Holy Spirit.

Activity

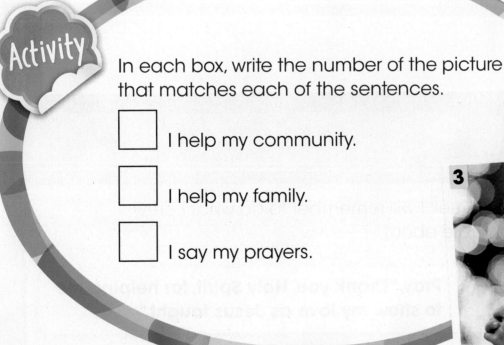

In each box, write the number of the picture that matches each of the sentences.

☐ I help my community.

☐ I help my family.

☐ I say my prayers.

I Follow Jesus

God the Holy Spirit is always with you. The Holy Spirit is your helper and teacher. The Holy Spirit helps you to make good choices as a follower of Jesus.

Activity

Teach Me to Love

Draw about the Holy Spirit helping you show love.

My Faith Choice

This week I will remember to do what I drew or wrote about.

Pray, "Thank you, Holy Spirit, for helping me to show my love as Jesus taught."

Chapter Review

Circle the names in the puzzle. Share what each name tells about God.

Father **Son** **Holy Spirit**

Q	F	A	T	H	E	R
W	S	O	N	E	O	P
H	O	L	Y	C	M	S
L	S	P	I	R	I	T

TO HELP YOU REMEMBER

1. The Holy Spirit helps and teaches us to pray.
2. The Holy Spirit helps us to know what Jesus taught.
3. The Holy Spirit helps and teaches us to do what Jesus asked us to do.

Come, Holy Spirit

Learn this prayer to the Holy Spirit. Pray it together. Use gestures to pray.

Come,
Holy Spirit,
fill our hearts with
the fire of
your love.
Amen.

With My Family

This Week . . .

In chapter 7, "The Holy Spirit, Our Helper," your child learned:

- The Holy Trinity is the mystery of one God in Three Divine Persons. Before Jesus died, he promised the disciples that he would not leave them alone, and that the Father would send them the Advocate.
- The Holy Spirit is the Advocate whom the Father sent and who is always with us. The Holy Spirit helps us to know, believe, and live what Jesus taught.
- Counsel is a gift of the Holy Spirit that helps us to make good decisions, as Jesus taught.

For more about related teachings of the Church, see the *Catechism of the Catholic Church*, 232–248 and 683–741, and the *United States Catholic Catechism for Adults*, pages 101–110.

Sharing God's Word

Read together the Gospel story in John 14:15–19. Emphasize that the Holy Spirit, the Advocate, is always with us to teach and help us to live as Jesus taught.

We Live as Disciples

The Christian home and family is a school of discipleship. Choose one or more of the following activities to do as a family, or design a similar activity of your own.

- Make prayer cards, using the Prayer to the Holy Spirit on page 69. Decorate the cards with signs and symbols of the Holy Spirit. Keep the cards to remind you that the Holy Spirit is always with your family as teacher and helper.
- This week your child learned about the Holy Trinity. Now is a good time to review the Sign of the Cross with your child. Talk about how the Sign of the Cross names all three Persons of the Holy Trinity.

Our Spiritual Journey

To give counsel is one of the Spiritual Works of Mercy. Make the Holy Spirit the center of your decision-making process and teach your child to do the same. Teach your child to respect the counsel of trusted adults, such as parents, teaches, and older family members. In this chapter, your child learned a prayer to the Holy Spirit. Read and pray together the prayer on page 69.

For more ideas on ways your family can live as disciples of Jesus, visit **www.BeMyDisciples.com**

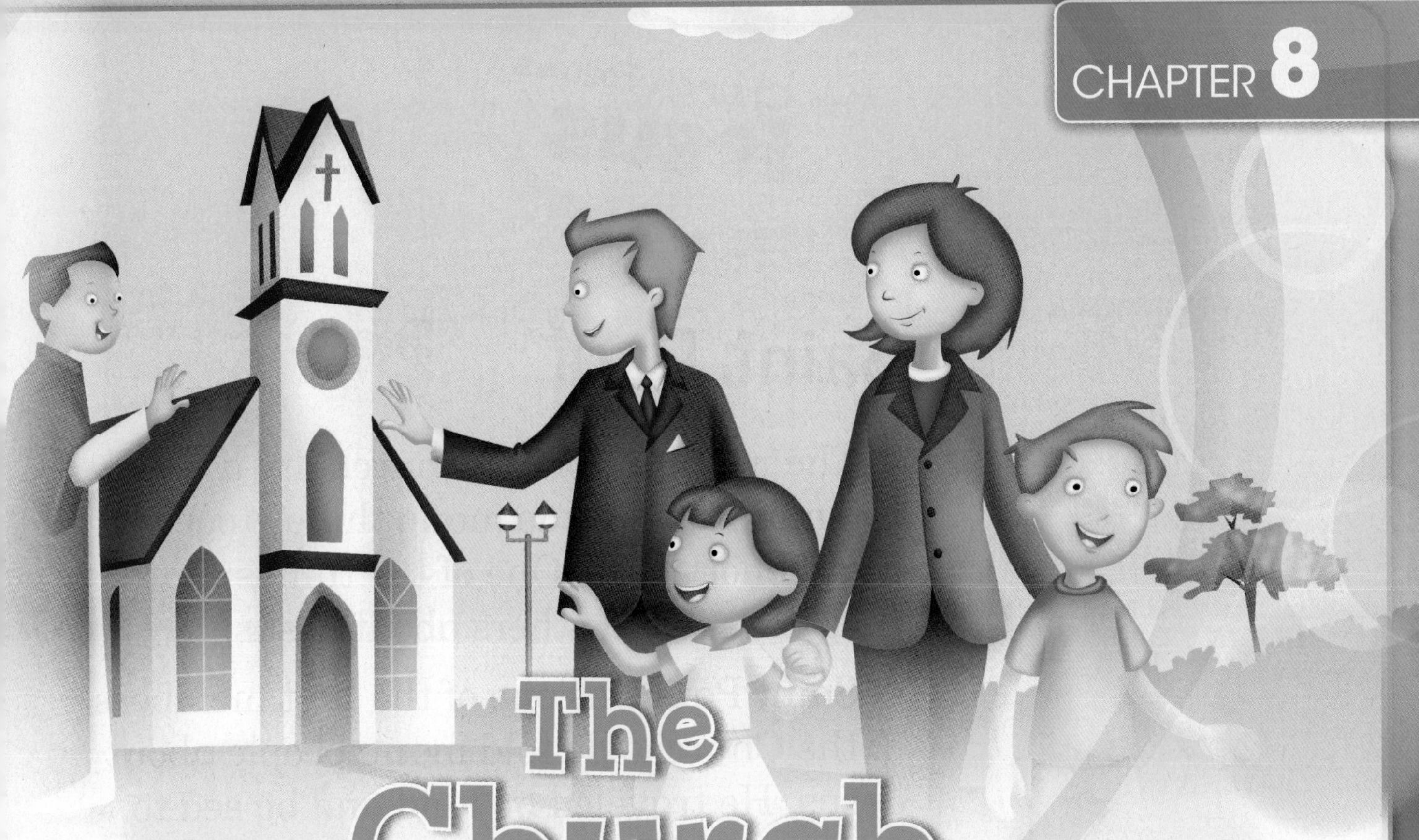

CHAPTER 8

The Church

What do you do together as a family?

Each of us belongs to our family. We also belong to the family of the Church. The Bible tells us:

The first members of the Church spent time together. They remembered Jesus. They shared all they had with one another. They prayed together. They broke and shared bread together. Together they praised God.

BASED ON ACTS OF THE APOSTLES 2:42

What does your family do at church?

Disciple Power

Reverence

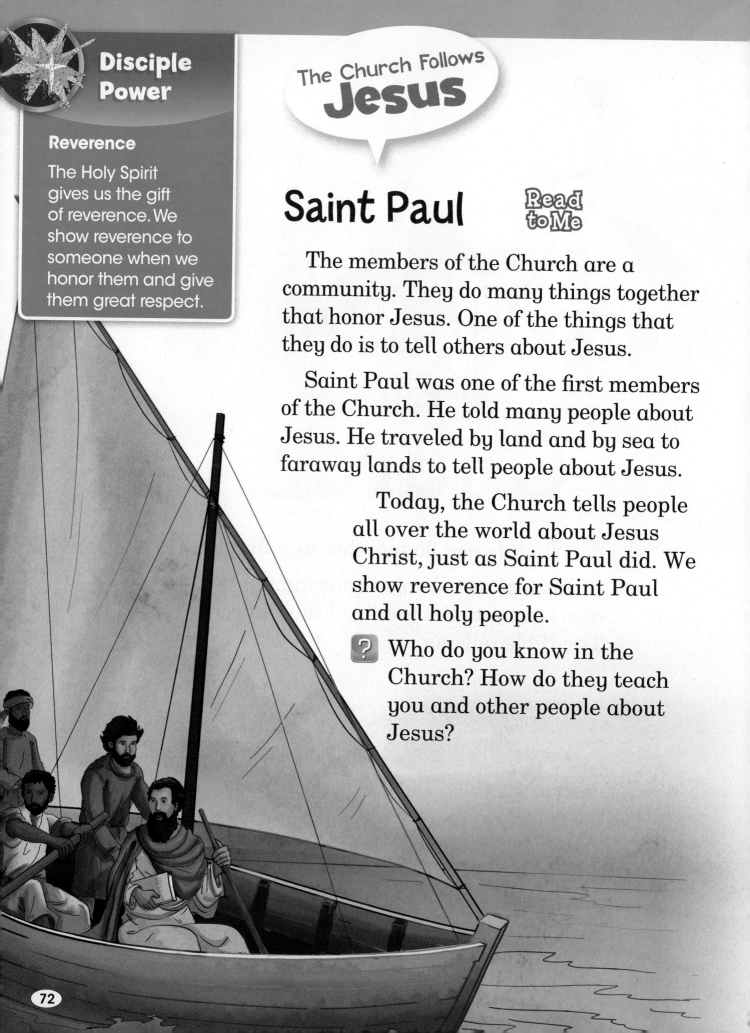

The Holy Spirit gives us the gift of reverence. We show reverence to someone when we honor them and give them great respect.

The Church Follows **Jesus**

Saint Paul

Read to Me

The members of the Church are a community. They do many things together that honor Jesus. One of the things that they do is to tell others about Jesus.

Saint Paul was one of the first members of the Church. He told many people about Jesus. He traveled by land and by sea to faraway lands to tell people about Jesus.

Today, the Church tells people all over the world about Jesus Christ, just as Saint Paul did. We show reverence for Saint Paul and all holy people.

? Who do you know in the Church? How do they teach you and other people about Jesus?

Our Church Family

The People of God who believe in Jesus and live as his followers are called the **Church**.

After the Risen Jesus returned to his Father in Heaven, the disciples went to the city of Jerusalem. The Holy Spirit came to the them as Jesus had promised.

Read what happened:

The disciples were together in a room. The power of the Holy Spirit filled them. BASED ON ACTS OF THE APOSTLES 2:1–4

The Holy Spirit helped the disciples tell people about Jesus. They invited people to be baptized. The work of the Church began.

Faith Focus
Who helps us to live as followers of Jesus?

Faith Words

Church
The Church is the People of God who believe in Jesus and live as his followers.

Catholics
Catholics are followers of Jesus and members of the Catholic Church.

Activity

Write or draw something Jesus did that you can share with others. Tell a partner. The Holy Spirit will help you.

Faith-Filled People

Saint Peter the Apostle

Saint Peter the Apostle was one of the first disciples of Jesus. Jesus chose Peter to be the first leader of the whole Church. The Church celebrates the feast day of Saints Peter and Paul on June 29.

We Are Catholics

The Catholic Church goes all the way back to Jesus and the Apostles. We belong to the Catholic Church. We join the Catholic Church when we are baptized.

Catholics are followers of Jesus Christ. We do what Jesus taught us. We learn about God and his love for us. We teach others about Jesus. We work together to help others.

We pray together and share our love for Jesus.

Activity

Find the letter that goes with each number.
Write the letter on the line above the number.
Find out three things that Catholics do.

A	B	C	D	E	F	G	H	I	J	K	L	M
1	2	3	4	5	6	7	8	9	10	11	12	13
N	**O**	**P**	**Q**	**R**	**S**	**T**	**U**	**V**	**W**	**X**	**Y**	**Z**
14	15	16	17	18	19	20	21	22	23	24	25	26

L	E	A	R	N
12	5	1	18	14

16	18	1	25

8	5	12	16

The Saints

Members of the Church show us how to live as followers of Jesus. Some of these people are called saints. Saints are grown-ups and children from all over the world. They now live with God in Heaven. The Church has named many saints.

Mary, Mother of Jesus, is the greatest saint of all. We can pray to Mary and the other saints. All of the saints help us to l ive as children of God. They want us to live as followers of Jesus. They want us to be happy with God on Earth and in Heaven.

? Who shows you how to live as a child of God? How do they show you?

Catholics Believe

The Pope

The pope is the leader of the whole Catholic Church. The pope helps us to live as followers of Jesus.

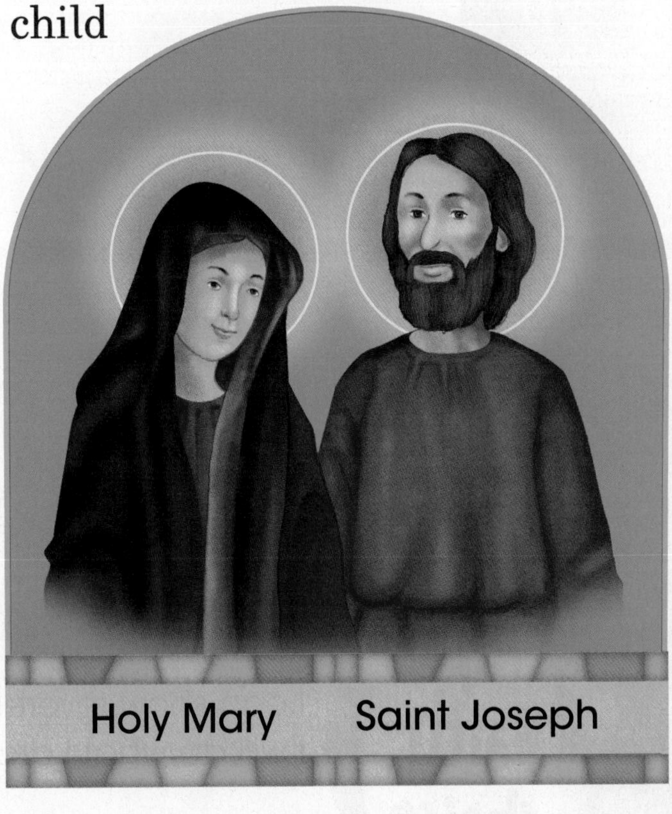

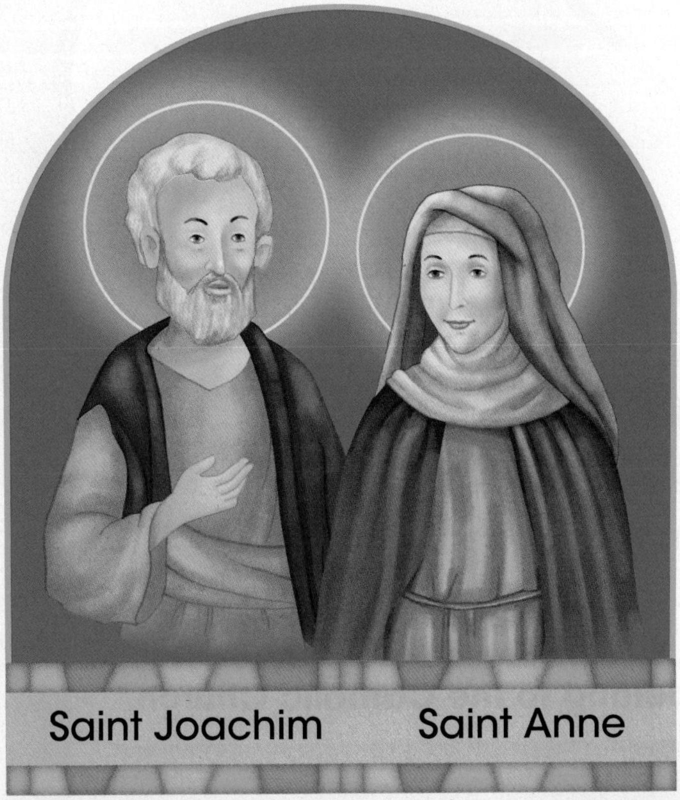

I Follow Jesus

The Holy Spirit gives you the gift of reverence. This gift helps you to honor God. You honor God when you serve him and others as Jesus taught. You show that you are a good Catholic.

Activity

Honoring God and Others

In one footstep, show how you will honor God. In the other footstep, show how you will honor others.

My Faith Choice

I can show reverence to God and others. This week I will do what I drew in the footsteps above.

Pray, "Thank you, God, for helping me to show that I belong to the Catholic Church."

Chapter Review

Color the circle next to the word that best complete each sentence.

1. The ______ came to Jesus' followers on Pentecost.

   ○ Saints ● Holy Spirit

2. ______ is the greatest saint.

   ○ Peter ○ Mary

3. Paul told many people about ______.

   ○ Jesus ○ Mary

TO HELP YOU REMEMBER

1. The Holy Spirit helps all members of the Church.
2. The Church helps us to do what Jesus taught us.
3. The saints help us to live as followers of Jesus.

Litany of the Saints

We praise and thank God for the saints in a litany prayer. Pray together.

Leader: Holy Mary, Mother of God,

All: **pray for us.**

Leader: Saint Paul

All: **pray for us.**

Leader: Saint Anne, Mother to Mary,

All: **pray for us.**

Leader: All holy men and women,

All: **pray for us.**

With My Family

This Week . . .

In chapter 8, "The Church," your child learned:

- The Church began on Pentecost. On Pentecost, the Holy Spirit came upon the disciples, and they received the power to go out and preach to others about Jesus. The work that Jesus gave to the Church began.
- God has called us together in Christ to be his Church, the new People of God. Christ is the Head of the Church, the Body of Christ. We are members of the Church. We believe in Jesus Christ and in everything he revealed to us.
- We work together as the Body of Christ to share our love for Jesus with others. The saints provide us with examples of how to live as disciples of Jesus Christ in the world today.
- The Holy Spirit gives us the gift of reverence. This gift inspires us to honor God by serving him and others.

For more about related teachings of the Church, see the *Catechism of the Catholic Church*, 737–741 and 748–801, and the *United States Catholic Catechism for Adults*, pages 111–123.

Sharing God's Word

Read together the Bible story in Acts 2:1–41 about Pentecost or read the adaptation of the story on page 73. Emphasize that on Pentecost the Holy Spirit came to the disciples, and the disciples began the work of the Church.

We Live as Disciples

The Christian home and family is a school of discipleship. Choose one or more of the following activities to do as a family, or design a similar activity of your own.

- Identify and name ways that you live as members of the Catholic Church. For example, we take part in Mass, we help the poor and hungry, we help a neighbor in need, or we visit the sick.
- The saints show us how to live as followers of Jesus. If your parish is named after a saint, take time this week to find out more about the saint. Talk about how this saint or another saint, if your parish is not named after a saint, helps you live as a Christian family.

Our Spiritual Journey

The Church is the Communion of Saints. When we die, our life is changed but not ended. The saints of the Church continue to be our companions on our earthly journey. In this chapter, your child prayed part of the Litany of the Saints. Read and pray together the prayer on page 77.

For more ideas on ways your family can live as disciples of Jesus, visit **www.BeMyDisciples.com**

Unit 2 Review

Name ____________________

A. Choose the Best Word

Complete the sentences. Color the circle next to the best choice.

1. God chose __________ to be the mother of Jesus.

○ Gabriel ○ Mary

2. Jesus __________ us so much that he gave his life for us.

○ loved ○ missed

3. Jesus returned to his Father in __________.

○ Heaven ○ Nazareth

4. The __________ helps and teaches us to pray.

○ Creator ○ Holy Spirit

5. The__________ of the Church help us to live as followers of Jesus.

○ saints ○ angels

B. Show What You Know

Circle the numbers next to the words that tell about the Holy Trinity.

1. God the Father

2. Mary, the Mother of God

3. Jesus the Son

4. the Holy Spirit

5. the People of God

6. the saints

C. Connect with Scripture

What was your favorite story about Jesus in this unit? Draw something that happened in the story. Tell your class about it.

D. Be a Disciple

1. *What saint or holy person did you enjoy hearing about in this unit? Write the name here. Tell your class what this person did to follow Jesus.*

2. *What can you do to be a good disciple of Jesus?*

We Worship
Part One

UNIT **3**

Come, Follow Me

Jesus looked out on the water and saw Simon and Andrew fishing. He called to them, "Come, follow me. I will teach you how to catch people, instead of fish."

The two brothers said, "Yes!" Off they went to follow Jesus.

Soon, Jesus spied two more fishermen named James and John. They were fixing their fishing nets.

"Come, follow me," Jesus said. The brothers said, "Yes!" Off they went to follow Jesus.

BASED ON MARK 1:16–20

What I Have Learned

What is something you already know about these faith words?

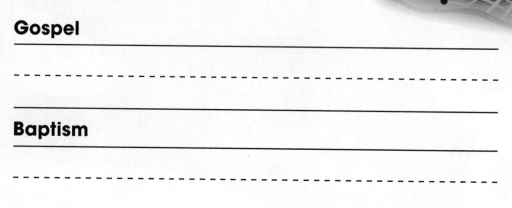

Gospel

Baptism

Faith Words to Know

Put an **X** next to the faith words you know.
Put a **?** next to the faith words you need to learn more about.

Faith Words

___ Easter
___ modesty
___ Marriage
___ Sacraments
___ goodness

A Question I Have

What question would you like to ask about the Sacraments?

CHAPTER 9

The Church Celebrates Jesus

? What is your favorite season, or time, of the year?

The Church has seasons too. Let us listen to what the Bible tells us about the seasons of the year. In the Bible, God tells us:

There is a season for everything. There is a time of the year for everything.

BASED ON ECCLESIASTES 3:1

? What is your favorite time of the year that you celebrate with the Church?

Disciple Power

Prudence

Prudence helps us ask advice from others when making important decisions. A prudent person makes good choices. Our Church family helps us to make good choices.

The Church Follows **Jesus**

Celebrating Sunday

Read to Me

Sunday is the Lord's Day. Maya Lopez and her family keep Sunday holy in many ways.

Maya and her family gather with their Church family to worship God at Mass. Every Sunday they remember that Jesus was raised from the dead.

Sunday is a special family day too. Maya's family spends time together. Sometimes they visit relatives. Sometimes they gather for a special dinner. They celebrate that their family is part of God's family.

? How does your family celebrate Sunday as the Lord's Day?

The Seasons of the Church's Year

The different times of the **Church's year** are called its seasons. Each season of the Church's year tells us something about Jesus. All year long we remember God's love for us.

Advent, Christmas, Lent, Easter, and Ordinary Time are the seasons and time of the Church's year. Each season of the Church's year has its own color. This helps us to remember the season of the Church's year we are celebrating.

Faith Focus
What is the Church's year?

Faith Words

Church's year
The Church's year is made up of four seasons. They are Advent, Christmas, Lent, and Easter.

Easter
Easter is a season of the Church's year. It is the time of the year when we celebrate that Jesus was raised from the dead.

Advent

Advent is the first season of the Church's year. The Advent season is four weeks long. The Advent wreath reminds us to prepare for Christmas. We get our hearts ready for Jesus. The color for Advent is purple.

Activity

Color three candles purple and one candle pink in the Advent wreath. Tell how your parish or your family uses an Advent wreath to celebrate Advent.

Faith-Filled People

Saint Joseph

Saint Joseph was the husband of Mary and the foster father of Jesus. An angel told Joseph that Mary was going to have a baby. An angel told Mary and Joseph to give the baby the name Jesus. The Church celebrates the feast day of Saint Joseph on March 19.

Christmas

Christmas comes after Advent. During the Christmas season we remember the birth of Jesus. He is God's Son who came to live on Earth with us. Jesus is God's greatest gift to us.

The Church's celebration of Christmas is not just one day. The season of Christmas lasts two or three weeks. We use the color white to celebrate Christmas.

Activity

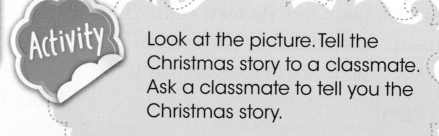

Look at the picture. Tell the Christmas story to a classmate. Ask a classmate to tell you the Christmas story.

Lent

During Lent we remember that Jesus died for us on the cross. We also get ready for Easter. The season of Lent begins on Ash Wednesday and lasts forty days. The color for Lent is purple.

Easter

During **Easter** we celebrate that Jesus was raised from the dead. This is the most important time of the Church's year. The season of Easter lasts about seven weeks. The Easter candle is lighted to remind us that Jesus is risen. The color for Easter is white.

Ordinary Time

During Ordinary Time we listen to Bible stories about what Jesus said and did. We learn to be followers of Jesus. The color for Ordinary Time is green.

Catholics Believe

Holy Days of Obligation

In addition to Sunday, Catholics have the responsibility to take part in Mass on other days. These days are called holy days of obligation.

Activity

Color the symbols for Lent, Easter, and Ordinary Time. Use the colors of the seasons.

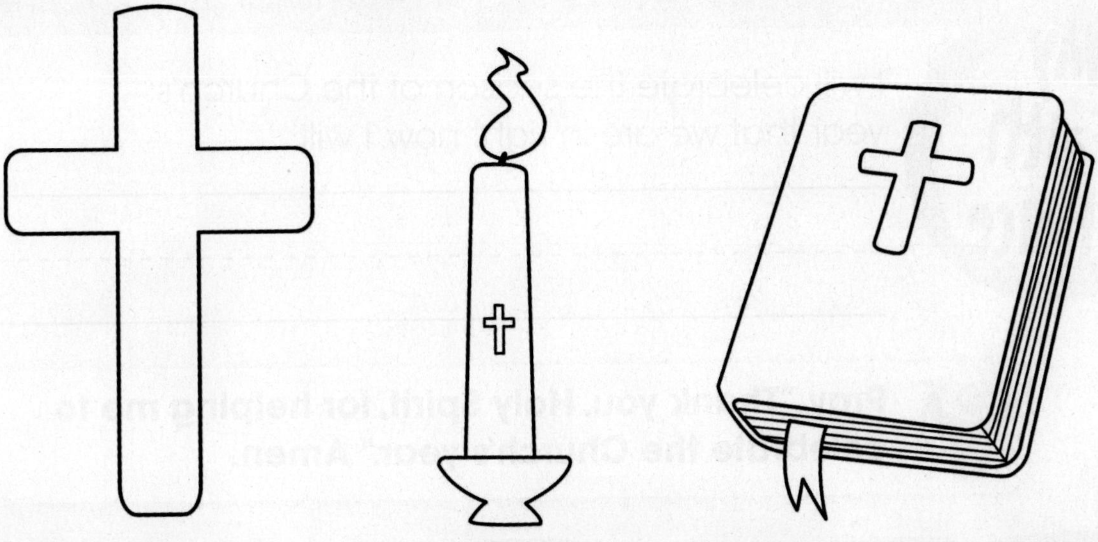

I Follow Jesus

When you celebrate the Church's seasons, you are making a good choice. Prudence helps you to make that good choice and others too.

Activity

Celebrating Jesus All Year

Look at the picture. Which season of the Church does it show?

My Faith Choice

I will celebrate the season of the Church's year that we are in right now. I will

Pray, "Thank you, Holy Spirit, for helping me to celebrate the Church's year." Amen.

Chapter Review

Draw lines to match the Church seasons with what we celebrate.

Season	What We Celebrate
1. Easter	We get ready for Easter.
2. Christmas	We celebrate that Jesus was raised from the dead.
3. Lent	We get ready for Christmas.
4. Advent	We remember the birth of Jesus.

TO HELP YOU REMEMBER

1. The Church has special times and seasons of the year.
2. The Church's year is made up of Advent, Christmas, Lent, Easter, and Ordinary Time.
3. Sunday is the Lord's Day.

Lord, We Praise You

When we worship God, we tell him that only he is God. Pray this prayer of praise together.

Jesus taught us to praise God.

Lord, we praise you.

In the morning and the night,

Lord, we praise you.

In the summer and the fall,

Lord, we praise you.

In the winter and the spring

Lord, we praise you.

With My Family

This Week . . .

In chapter 9, "The Church Celebrates Jesus," your child learned.

- The Church's year has special seasons just as the calendar year has. The seasons and time of the Church's year are Advent, Christmas, Lent, Easter, and Ordinary Time.
- Sunday is the Lord's Day.
- During the Church's year we join with Christ all year long and share in his work of Salvation. All year long we give thanks and praise to God.
- The virtue of prudence helps us to consistently make good choices. This includes taking part in Mass on Sunday.

For more about related teachings of the Church, see the *Catechism of the Catholic Church,* 1163–1173, and the *United States Catholic Catechism for Adults,* pages 173, 175, 178.

Sharing God's Word

Read together Psalm 150. Emphasize that throughout the liturgical year, the Church gives praise and thanksgiving to God. Talk about the ways in which your family is already giving thanks and praise to God.

We Live as Disciples

The Christian home and family is a school of discipleship. Choose one or more of the following activities to do as a family, or design a similar activity of your own.

- When you take part in Mass this week, look around and listen for all the signs that tell you what season of the Church year the Church is now celebrating. Point them out to your child and talk about them with her or him.
- Choose an activity that helps you celebrate the current liturgical season as a family at home. For example, during Advent you can use an Advent calendar to help anticipate and prepare for Christmas.

Our Spiritual Journey

Praising God is one of the five main forms of prayer that are part of the Church's tradition. In this chapter, your child prayed a prayer of praise on page 89. Pray this version of a prayer of praise as a family.

For more ideas on ways your family can live as disciples of Jesus, visit **www.BeMyDisciples.com**

CHAPTER 10

Signs of God's Love

? What special days and times does your family celebrate?

One time Jesus took part in a special celebration. John the Baptist baptized Jesus in the Jordan River.

> *As Jesus came up out of the water, he saw the clouds disappear. The Holy Spirit, like a dove, came down upon him.*
>
> *A voice from the sky said, "You are my Son, the One I love."* BASED ON MARK 1:10–11

? Why does God the Father love Jesus so much?

Disciple Power

Hospitality

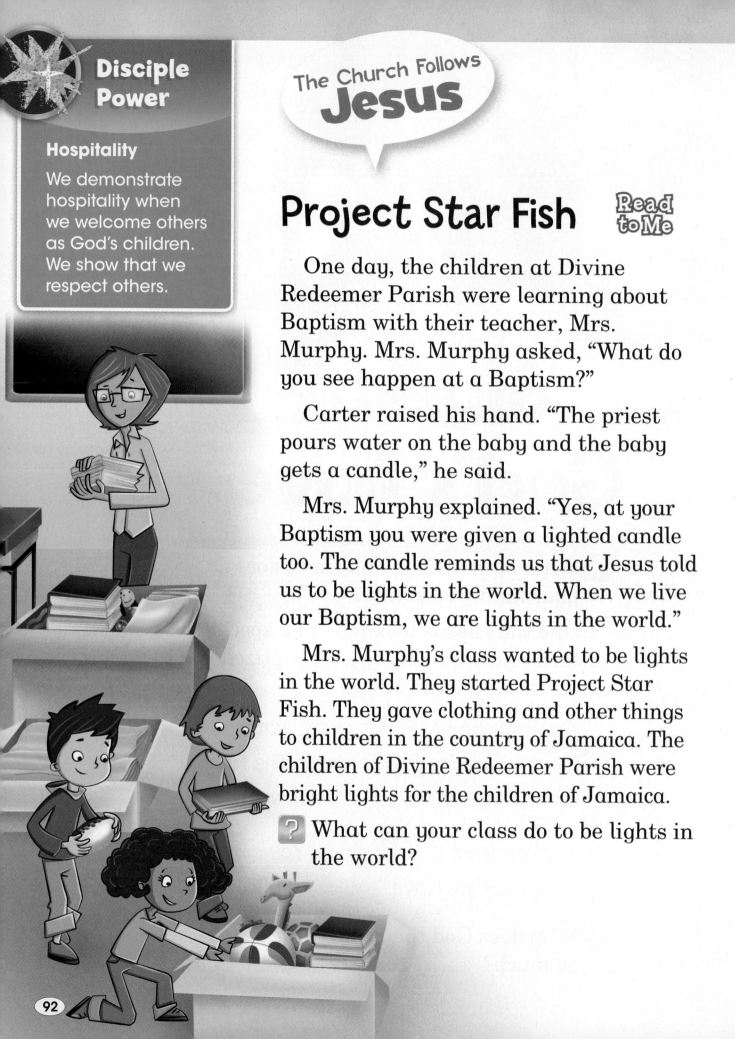

We demonstrate hospitality when we welcome others as God's children. We show that we respect others.

The Church Follows

Jesus

Project Star Fish

Read to Me

One day, the children at Divine Redeemer Parish were learning about Baptism with their teacher, Mrs. Murphy. Mrs. Murphy asked, "What do you see happen at a Baptism?"

Carter raised his hand. "The priest pours water on the baby and the baby gets a candle," he said.

Mrs. Murphy explained. "Yes, at your Baptism you were given a lighted candle too. The candle reminds us that Jesus told us to be lights in the world. When we live our Baptism, we are lights in the world."

Mrs. Murphy's class wanted to be lights in the world. They started Project Star Fish. They gave clothing and other things to children in the country of Jamaica. The children of Divine Redeemer Parish were bright lights for the children of Jamaica.

? What can your class do to be lights in the world?

God Is with Us

Jesus gave the Church seven special signs and celebrations of God's love. We call these celebrations **sacraments**. The Seven Sacraments celebrate that God is with us. They are:

Baptism

Confirmation

Eucharist

Reconciliation

Anointing of the Sick

Holy Orders

Matrimony

In the Seven Sacraments, God shares his love and life with us. Each of the sacraments helps us to grow closer to God.

Faith Focus

What do we celebrate at Baptism and Confirmation?

Faith Words

sacraments
The sacraments are the seven signs and celebrations of God's love that Jesus gave the Church.

Baptism
Baptism is the first sacrament that we celebrate. In Baptism, we receive the gift of God's life and become members of the Church.

Activity

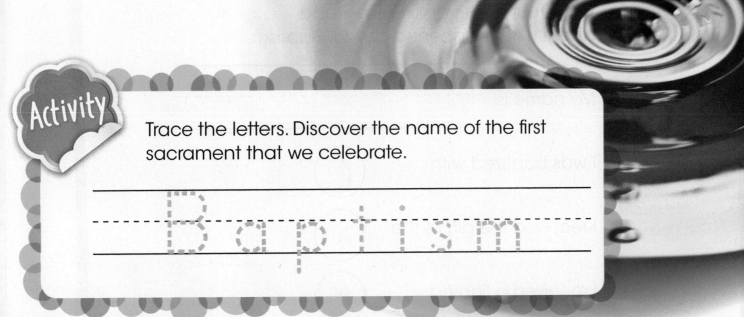

Trace the letters. Discover the name of the first sacrament that we celebrate.

Baptism

Faith-Filled People

Godparents

Godparents help us to grow in faith. They show us how to love God and other people as Jesus taught.

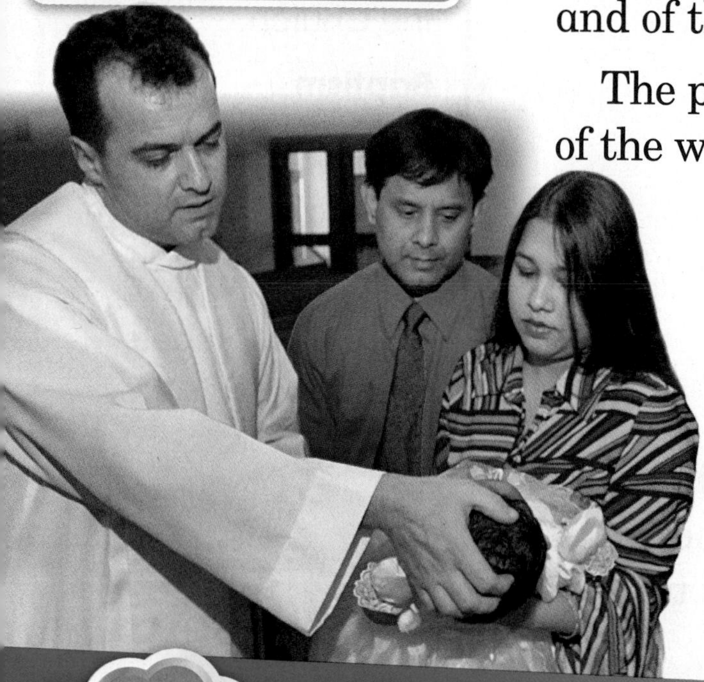

We Celebrate Baptism

Baptism is the first sacrament that we celebrate. We become members of the Church.

The priest or deacon pours water on our heads or puts us in the water three times. As he does this, he says, "I baptize you in the name of the Father, and of the Son, and of the Holy Spirit. Amen."

The pouring of water and the saying of the words tell us we receive the gift of God's life in Baptism. We receive the gift of the Holy Spirit.

Original Sin and any other sins that we have committed are forgiven. Original Sin is the first sin committed by Adam and Eve. We are born with this sin.

Activity

Write your name on the line. Read about what happened at your Baptism.

My name is ______________________________.

I was baptized with  .

I received the gift of the  .

I received a lighted  .

We Celebrate Confirmation

We celebrate Confirmation after we are baptized. Sometimes, we celebrate Confirmation right after Baptism, on the same day. If we celebrate Baptism when we are infants, we usually celebrate Confirmation when we are older.

At Confirmation, the bishop, or the priest named by the bishop, leads the celebration of Confirmation. During the celebration, he rubs special oil on the front of our heads. The oil is called Sacred Chrism. As he rubs the Chrism, he says, "Be sealed with the gift of the Holy Spirit."

The bishop or priest then says, "Peace be with you." We respond, "And also with you." The Holy Spirit teaches and helps us to live our Baptism. He helps us live as followers of Jesus. He teaches and helps us to be lights in the world.

Catholics Believe

Baptism Candle

At Baptism, we receive a lighted candle. This reminds us that we are to live as followers of Jesus, who is the Light of the world. We are to be lights in the world.

Activity

Finish this prayer to the Holy Spirit.

Holy Spirit, help me to

____

____

______________________________. Amen.

I Follow Jesus

At your Baptism, you became a member of the Church. At Confirmation, the Holy Spirit will give you special help to be a light in the world and welcome others.

Activity

Lights in the World

Draw one way you can be God's light in the world at home, at school, or in your neighborhood.

My Faith Choice

I want to be a light in the world. This week, I will:

____

____.

Pray, "Thank you, Holy Spirit, for helping me to live as a follower of Jesus. Amen."

Chapter Review

Complete the sentences. Color the O next to the best choice.

1. There are _____ Sacraments.

   O three O seven

2. _____ gave the Church the Sacraments.

   O Jesus O The saints

3. _____ is the first sacrament that we receive.

   O Eucharist O Baptism

TO HELP YOU REMEMBER

1. In Baptism, God shares his love and life with us.
2. In Baptism, we receive the gift of the Holy Spirit.
3. In Confirmation, we are sealed with the gift of the Holy Spirit to help us to live our Baptism.

Thank You, Lord

Pray this thank you prayer.

Leader Let us thank God for the gift of water.

All **Thank you, Lord.**

Leader In Baptism, water is a sign that we receive the gift of God's life.

All **Thank you, Lord.**.

Leader Come and dip your fingers in the water. Make the Sign of the Cross.

All **Amen!**

With My Family

This Week . . .

In chapter 10, "Signs of God's Love," your child learned:

- Baptism is the first sacrament we receive.
- Through Baptism God makes us sharers in his life and love. We are reborn as children of God and receive the gift of the Holy Spirit. Original Sin and personal sins are forgiven. We become members of the Church, the Body of Christ.
- Confirmation strengthens the graces of Baptism.
- Hospitality is welcoming others as children of God. We show that we respect others.

For more about related teachings of the Church, see the *Catechism of the Catholic Church*, 1113–1130, and 1210–1274, and the *United States Catholic Catechism for Adults*, pages 183–197 and 203–209.

Sharing God's Word

Read together Matthew 5:14–16. Emphasize that at Baptism we are joined to Jesus, the Light of the world. Talk about how your family members are living their Baptism and are lights in the world.

We Live as Disciples

The Christian home and family is a school of discipleship. Choose one or more of the following activities to do as a family, or design a similar activity of your own.

- Make thank-you cards for godparents. Thank your godparents for helping you grow in faith.
- Sign your child on her or his forehead with a small sign of the cross before your child leaves for school. Remind your children that they are to be lights in the world.

Our Spiritual Journey

Baptism is the doorway to the Christian life. The ritual of blessing ourselves with holy water reminds us of our Baptism. Integrate the use of this ritual into your daily life. Perhaps, begin your family prayers by inviting everyone to bless themselves with holy water while praying the Sign of the Cross.

For more ideas on ways your family can live as disciples of Jesus, visit **www.BeMyDisciples.com**

CHAPTER 11

We Follow Jesus

? What good news have you heard this week?

Followers of Jesus have the best good news to share. Listen to what Jesus tells about sharing that good news:

> *Jesus would soon return to his Father in Heaven. He told his disciples "Go into the whole world. Tell everyone the good news I shared with you."*
>
> Based on Mark 16:15

? Who has shared the Good News of Jesus with you?

Disciple Power

Goodness

Goodness is a sign that we are living our Baptism. When we are good to people, we show them that they are children of God. When we are good to people, we honor God.

The Church Follows **Jesus**

Read to Me

Saint Francis of Assisi

God's love filled the heart of Francis. So he sang about the Good News of Jesus.

Everywhere Francis went he told everyone about Jesus. He shared with everyone how much God loved them. He told people that God loves us so much that he gave us Jesus.

Everything good that Francis saw reminded him of how much God loves us. Today we honor Francis as a saint.

? What good thing in God's creation reminds you of God's love?

The Good News of Jesus

Jesus told everyone the Good News of God's love. Jesus chose followers to help him share this Good News with all people. Disciples share the Good News of Jesus.

Jesus chose Matthew to be one of his first disciples. Matthew was one of the Apostles. Matthew wrote about the Good News of Jesus. He wrote the Good News about Jesus in his **Gospel**. The word gospel means "good news."

Faith Focus
Why do we share the Gospel with others?

Faith Word
Gospel
The Gospel is the Good News that Jesus told us about God's love.

Activity

Color the ♡s next to the ways you can share the Good News of God's love.

- ♡ Tell people about Jesus.
- ♡ Say "Thank you" to someone who is kind to me.
- ♡ Make a get-well card for a friend who is sick.
- ♡ Be rude to someone who is not kind to me.

Faith-Filled People

Blessed Teresa of Calcutta

Mother Teresa was born in the country of Albania. When she was in high school, she knew God was calling her to serve the poorest of the poor. She founded the Missionaries of Charity to help her do that work. Today, Missionaries of Charity share the Good News of God's love with poor, sick, and dying people around the world.

Tell the Good News

The last story in Matthew's Gospel tells about Jesus returning to his Father in Heaven. In this story, we hear the important work that Jesus gave to his disciples:

Jesus told his disciples, "Go to every land you can. Invite all people to be my disciples. Baptize them in the name of the Father, and of the Son, and of the Holy Spirit. Teach them what I have taught you."

BASED ON MATTHEW 28:19–20

? How do Christians today tell others about the Good News?

Followers of Jesus Christ

The disciples of Jesus traveled to small villages and to large cities. They walked. They rode donkeys. They traveled in ships.

They did the work that Jesus had given them to do. They told everyone the Good News of Jesus Christ. They baptized people. Many people became followers of Jesus Christ. People called the followers of Jesus Christians.

When we hear the Gospel, we come to know Jesus better. We grow in faith. We grow in our love for God, for ourselves, and for other people.

Catholics Believe

The Four Gospels

Saints Matthew, Mark, Luke, and John each wrote a gospel about Jesus. The four gospels are part of the New Testament.

Activity

Trace the way to Jesus. Find and circle the things that followers of Jesus share with others.

I Follow Jesus

You can be a disciple of Jesus. There are many good ways to tell others about Jesus. The Holy Spirit helps you to tell people about the Good News of Jesus.

Activity

Tell the Good News

Imagine that you are one of the children in the picture. Write what you want to tell other people about Jesus.

____

____

My Faith Choice

Check (√) what you will do this week. I will share what I have written about Jesus with

- ☐ my parents
- ☐ my grandparents
- ☐ a friend
- ☐ someone at church

Pray, Tell the Holy Spirit what you have decided to do. In your own words, ask him to help you.

Chapter Review

Remember the name for Jesus' Good News. Color the spaces with ***X****s red. Color the spaces with* ***Y****s another color.*

TO HELP YOU REMEMBER

1. The Gospel is the Good News that Jesus told about God's love.
2. Jesus told his disciples to tell everyone the Good News that he shared with them.
3. When we listen to the Gospel, we come to know Jesus better and to grow in faith.

Lord, Help Us to Listen

At Mass, we pray silently before we listen to the Gospel. We trace a small cross on our foreheads, on our lips, and over our hearts. Learn to pray in this new way.

Jesus, be in my thoughts,

on my lips,

and in my heart. Amen.

With My Family

This Week . . .

In chapter 11, "We Follow Jesus," your child learned:

- The Gospel is the Good News about Jesus. Matthew, Mark, Luke, and John are the four gospel writers.
- The last story in Matthew's Gospel is about Jesus telling his disciples to preach the Gospel and to baptize people.
- Christians are to treat people with goodness. Goodness is a fruit of the Holy Spirit. It is a sign we know that every person is a child of God. In doing so, we are cooperating with the graces we received at Baptism.

For more about related teachings of the Church, see the *Catechism of the Catholic Church*, 124–133 and 849–856, and the *United States Catholic Catechism for Adults*, pages 79–85.

Sharing God's Word

Read together Matthew 28:19–20, the story about the commissioning Jesus' disciples. Or read the adaptation of the story on page 102. Emphasize that Jesus told the disciples to invite all people to be his disciples. Talk about how your family shares the Gospel with others.

We Live as Disciples

The Christian home and family is a school of discipleship. Choose one or more of the following activities to do as a family, or design a similar activity of your own.

- Saint Francis of Assisi sang about the Good News of Jesus. Invite each family member to share their favorite song or hymn that tells about Jesus. Be sure that everyone explains why the song or hymn is their favorite.
- Invite family members to share one thing that they would like everyone in the family to know about Jesus.

Our Spiritual Journey

When we are good to people, we show them respect as children of God. We are a sign to them of how much God loves them. We grow in respect for ourselves. We are all children of God. Practice the prayer form on page 105 with your child. Remind them that we say these words silently when the priest introduces the Gospel at Mass.

For more ideas on ways your family can live as disciples of Jesus, visit **www.BeMyDisciples.com**

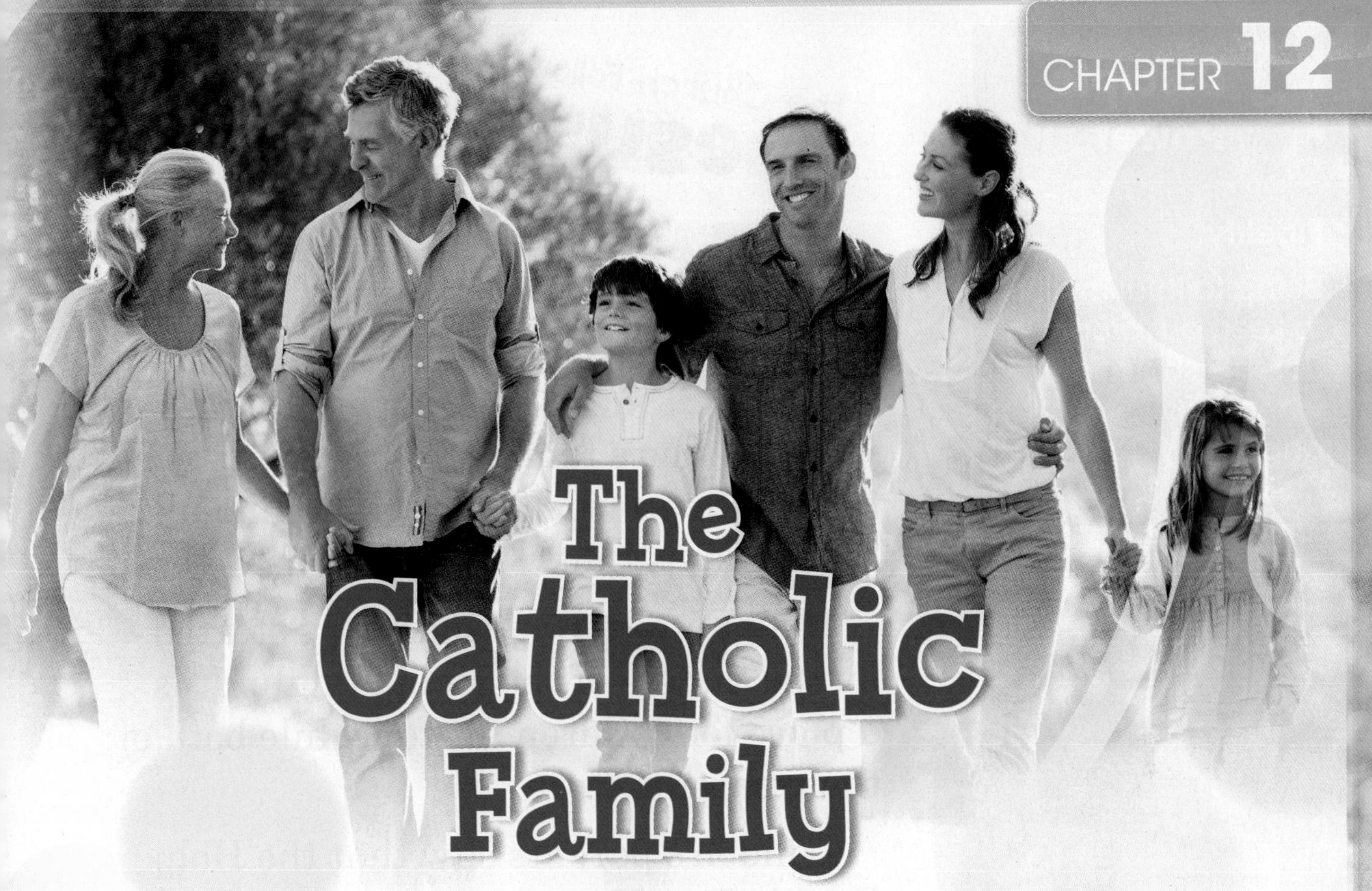

CHAPTER 12

The Catholic Family

Who belongs to your family?

Like you, Jesus grew up in a family. Listen to what the Bible tells us about Jesus' family:

When Jesus was a baby, Mary and Joseph presented him to the Lord in the Temple in Jerusalem. When they returned to the family home in Nazareth, Jesus grew up there. He came to know what God wanted him to do.

BASED ON LUKE 2:22, 39–40

What do these words from the Bible tell you about Jesus' family?

Disciple Power

Fidelity

Being faithful means to keep our promises. Parents show fidelity when they love and care for their children.

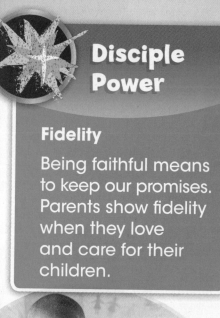

The Church Follows Jesus

Helping Families

Read to Me

In Landon's family, both parents need to work to provide food, clothes, and many other things that a family needs.

Holy Family Day Home was started by the Sisters of the Holy Family. It is a place for children to stay while both of their parents work.

Landon's parents saw that the Holy Family Day Home was a safe place for Landon to be after school when they were at work. The children learn to respect themselves and others. Landon and the other children play and learn after school.

Holy Family Day Home helps parents care for their children.

? What are some things you enjoy doing after school?

The Gift of Marriage

When a man and a woman love each other very much, they marry each other. They make a promise to love each other and live as a family their whole lives.

The Gospel of John tells us about a **marriage**:

Jesus and Mary, his mother, and the disciples went to a wedding. During the party, the married couple ran out of wine. Jesus blessed six large jars of water, and the water became wine. Jesus gave the wine to the married couple.

BASED IN JOHN 2:1–11

Faith Focus
How do our families help us to grow as children of God?

Faith Words

marriage
A marriage is the lifelong promise of love made by a man and a woman to live as a family.

Matrimony
Matrimony is the sacrament that Catholics celebrate when they marry.

Activity

Look at the pictures of the families on this page. Draw a picture of your family.

Faith-Filled People

Saint Elizabeth Ann Seton

Elizabeth Ann Seton was the first person born in the United States who was named a saint. Elizabeth and her husband, William, were the parents of five children. They showed God's love to each other and to their children.

Families Are Signs of God's Love

When Catholics get married, they celebrate the Sacrament of **Matrimony**. A husband and a wife sometimes receive the wonderful gift of children from God. They become parents.

There are many different kinds of families. All families are called to love God and one another. Families are to be signs of God's love in their homes and in the world.

A family is a blessing from God. Members of a family share their love with God and with one another. They pray together. They respect one another. They say and do kind things for one another. They take care of one another. They honor and respect each other as children of God.

Activity

Write your family name. Tell your class how your family is a sign of God's love.

____

____

____

The Family of God

At Baptism, we become part of God's family, the Church. Our Church family helps us to grow as Catholics.

The Church teaches us about the Holy Family. Mary, Joseph, and Jesus are the Holy Family.

Our family is the church of the home. Our family helps us to grow in faith. It helps us to live our faith. It teaches us to pray and to care for others as Jesus did.

Catholics Believe

The Family Church

Our families help us to know and love Jesus. They help us to live as disciples of Jesus. That is why we call our family "the family Church" or "the church of the home."

Activity

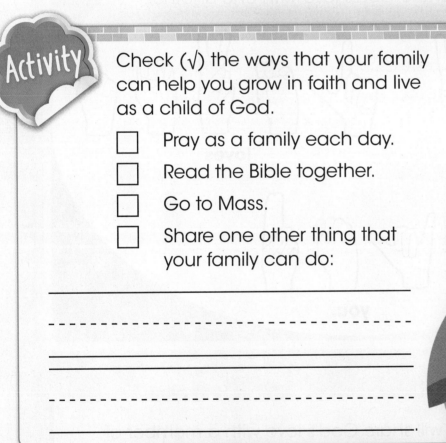

Check (√) the ways that your family can help you grow in faith and live as a child of God.

- ☐ Pray as a family each day.
- ☐ Read the Bible together.
- ☐ Go to Mass.
- ☐ Share one other thing that your family can do:

____

____

____.

I Follow Jesus

You are part of your family. You are part of God's family, the Church. God the Holy Spirit helps you to love your family. When you do this, you honor and respect your parents and the other members of your family.

Activity

Sharing Family Love

Learn to sign these words. Teach the signs to your family. Share God's love with one another.

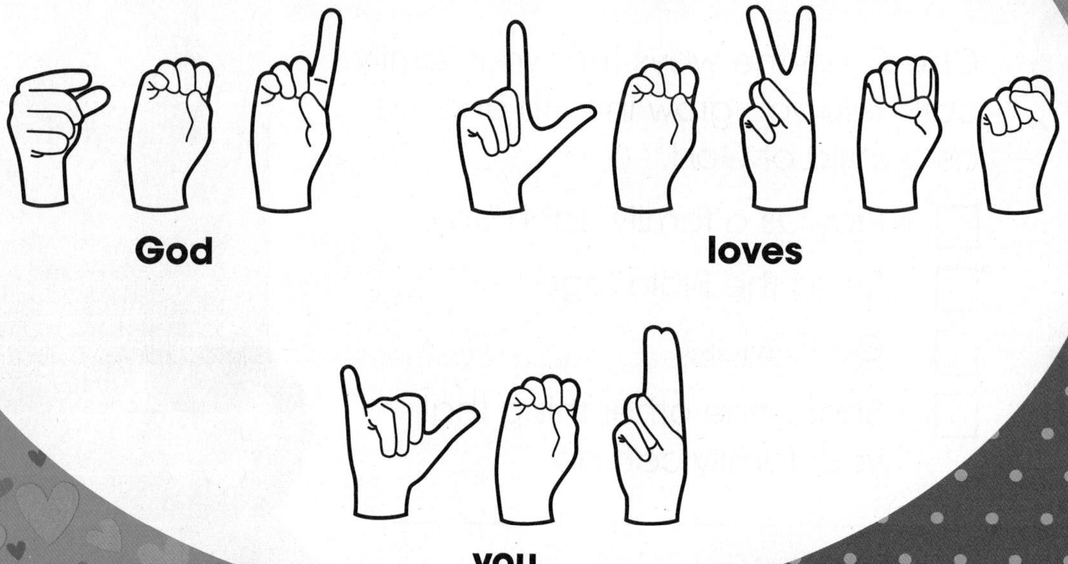

My Faith Choice

This week I will share God's love with a member of my family. I will sign the message, "God loves you," for them.

Pray, "Thank you, Holy Spirit, for helping me to share God's love with my family. Amen."

Chapter Review

Read this poem. Fill in the blanks with rhyming words.

1. Families, families, everywhere ______________

show each other love and ______________.

2. They tell us of God's love, you see. ______________

God loves each of us, you and ______________.

TO HELP YOU REMEMBER

1. Christian families are signs of Jesus' love in the world.
2. Members of a family share their love for God with one another.
3. Our family helps us to live our faith.

A Family Blessing

Ask God to bless your family. Pray this prayer now with your classmates.

Leader Lord God, show your wonderful love to all of our families.

Leader Bless our parents and grandparents,

All **we ask you, Lord.**

Leader Bless (*say other names silently in your heart*),

All **we ask you, Lord. Amen.**

With My Family

This Week . . .

In chapter 12, "The Catholic Family," your child learned:

- God invites a man and a woman to share their love for him and for one another forever in marriage.
- Matrimony is the sacrament that Catholics celebrate when they marry.
- The Christian family is the "Church of the home," or the "family Church."
- Families are signs of God's love. Families are the primary place where parents and children experience and grow in faith, hope, and love.
- Fidelity helps children and parents grow stronger as a family in their love for God and for one another. Fidelity helps us live the Fourth Commandment.

For more about related teachings of the Church, see the *Catechism of the Catholic Church*, 1601–1658 and 2197–2233, and the *United States Catholic Catechism for Adults*, pages 279–287.

Sharing God's Word

Read together Luke 2:41–52, the finding of the twelve-year-old Jesus in the Temple. Emphasize that in the Holy Family, Jesus grew in love for God and for his family. Talk about the things that your family does to help one another grow in love for God and for one another.

We Live as Disciples

The Christian home and family is a school of discipleship. Choose one or more of the following activities to do as a family, or design a similar activity of your own.

- When we pray as a family, we show that our family loves God. Make an extra effort this week to pray together as a family at least once a day.
- Talk about the many ways in which your family is a sign of God's love; for example, when you do kind things for one another, when you pray together, and so on. Encourage one another to continue doing these things.

Our Spiritual Journey

A blessing is a sacramental. Sacramentals are sacred signs given to us by the Church. We use blessings to dedicate things or special occasions or people to God. Read and pray the blessing prayer on page 113 together as a family.

For more ideas on ways your family can live as disciples of Jesus, visit **www.BeMyDisciples.com**

Unit 3 **Review**

Name ______________________________

A. Choose the Best Word

Complete the sentences. Color the circle next to the best choice.

1. Advent, Christmas, Lent and ____________ are all seasons of the Church year.

○ Winter ○ Easter

2. Jesus gave the Church seven ____________ to help us grow closer to God.

○ Commandments ○ Sacraments

3. The first ____________ of Jesus told everyone the Good News of God's love.

○ family ○ disciples

4. When we listen to the ____________, we come to know Jesus better and grow in faith.

○ Gospel ○ music

5. Christian families are ____________ of Jesus' love for his followers.

○ signs ○ homes

B. Show What You Know

Use purple, green, or gold to color the circle next to each season or time of the Church year. Use the correct color for each season or time.

○ Advent ○ Ordinary Time

○ Easter ○ Christmas

○ Lent

C. Connect with Scripture

What was your favorite story about Jesus in this unit? Draw something that happened in the story. Tell your class about it.

D. Be a Disciple

1. *What saint or holy person did you enjoy hearing about in this unit? Write the name here. Tell your class what this person did to follow Jesus.*

2. *What can you do to be a good disciple of Jesus?*

We Worship

Part Two

UNIT 4

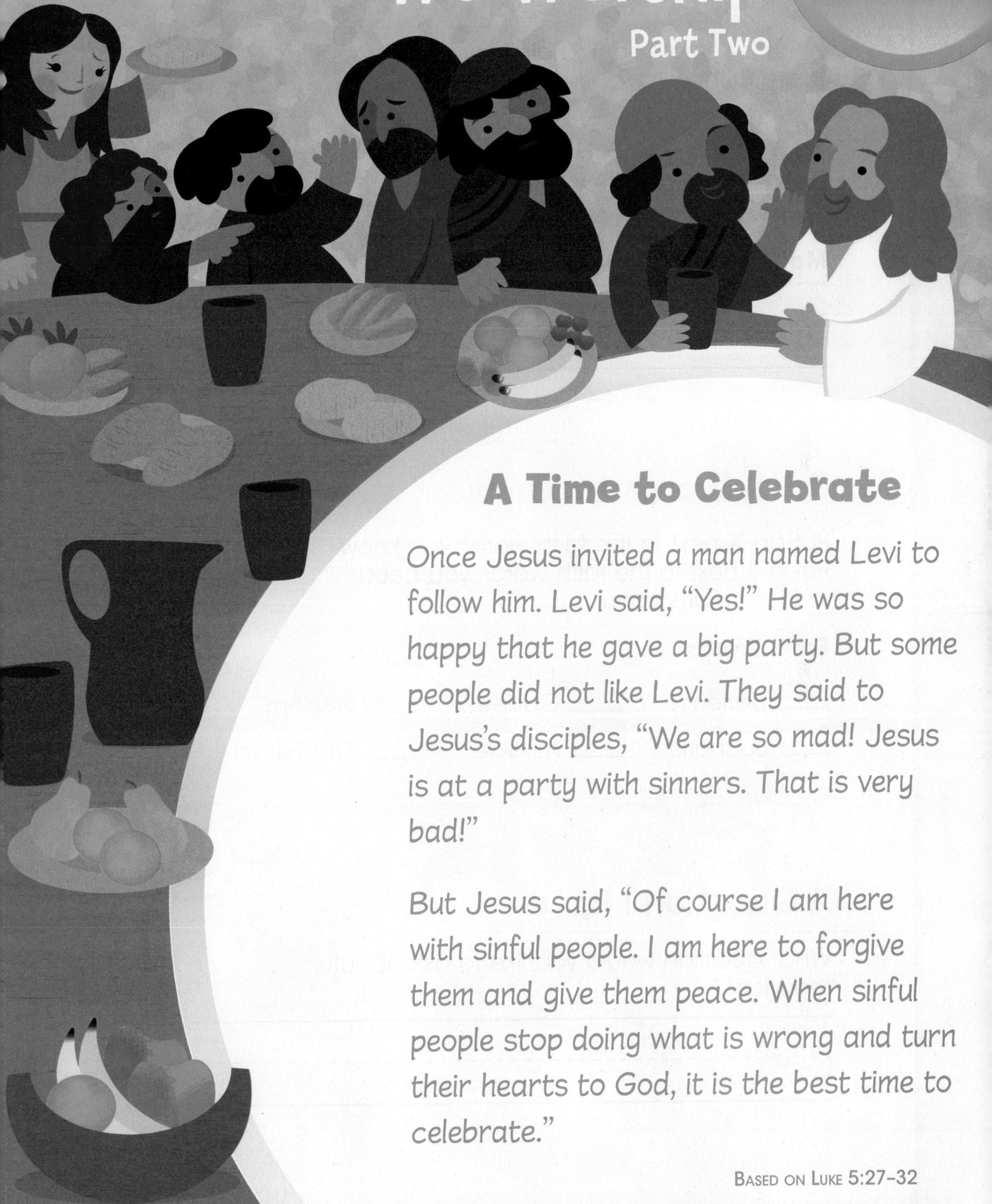

A Time to Celebrate

Once Jesus invited a man named Levi to follow him. Levi said, "Yes!" He was so happy that he gave a big party. But some people did not like Levi. They said to Jesus's disciples, "We are so mad! Jesus is at a party with sinners. That is very bad!"

But Jesus said, "Of course I am here with sinful people. I am here to forgive them and give them peace. When sinful people stop doing what is wrong and turn their hearts to God, it is the best time to celebrate."

Based on Luke 5:27–32

What I Have Learned

What is something you already know about these faith words?

prayer

Mass

Faith Words to Know

Put an **X** next to the faith words you know. Put a **?** next to the faith words you need to learn more about.

Faith Words

____ patience	____ Galilee	____ wisdom
____ Eucharist	____ miracle	____ Our Father

A Question I Have

What question would you like to ask about the Mass?

CHAPTER 13

We Pray

What are some of your favorite prayers?

Sometimes we pray for ourselves. Other times we pray for our family and friends. Jesus taught his followers how to pray. Listen to what Jesus said.

> You do not need to use many words when you talk to God. Talk to God from your heart. BASED ON MATTHEW 6:7

What prayers do you pray with your Church family?

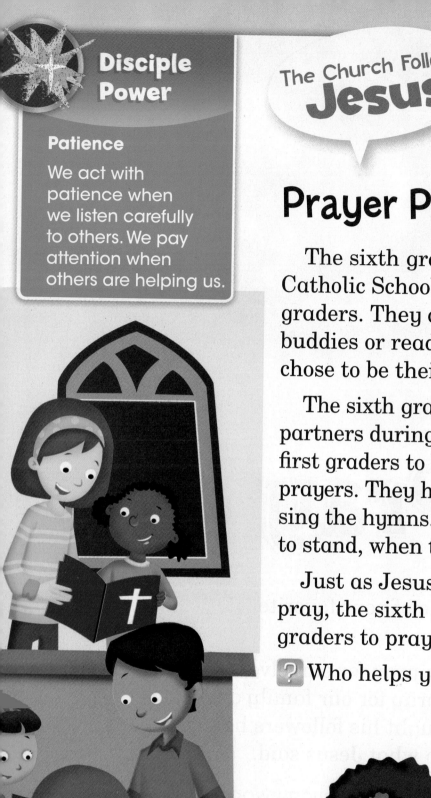

Disciple Power

Patience

We act with patience when we listen carefully to others. We pay attention when others are helping us.

The Church Follows Jesus

Prayer Partners

Read to Me

The sixth graders at Holy Nativity Catholic School wanted to help the first graders. They considered being lunch buddies or reading friends. Finally, they chose to be their prayer partners.

The sixth graders sat with their prayer partners during Mass. They helped the first graders to learn the words to the prayers. They helped the first graders to sing the hymns. They taught them when to stand, when to sit, and when to kneel.

Just as Jesus taught his disciples to pray, the sixth graders helped the first graders to pray.

? Who helps you learn how to pray?

God Hears Our Prayers

Friends and family members listen and talk to each other. We can share what is on our minds and in our hearts.

Prayer is listening and talking to God. We share with God what is on our minds and in our hearts.

We can pray anywhere and anytime. The Holy Spirit helps us to pray. When we pray, we grow in our love for God.

Faith Focus
Why is it important to pray?

Faith Word
prayer
Prayer is listening and talking to God.

Activity

Trace this message. Tell a partner how you pray.

I talk

to God.

Faith-Filled People

Saint Thérèse, the Little Flower

Saint Thérèse of Lisieux is also called the Little Flower. Thérèse found a favorite place to pray. When she was young, she would pray in the space between her bed and the wall. The Church celebrates the feast day of Saint Thérèse, the Little Flower, on October 1.

Jesus Shows Us How to Pray

Jesus prayed all during his life. Jesus prayed alone. Sometimes he prayed with his family. Sometimes Jesus prayed with his friends. Other times Jesus prayed with his neighbors.

Sometimes we pray alone. Sometimes we pray with others. We pray with our family. We pray with our friends. We pray with our Church family.

? Where is your favorite place to pray? Why is it your favorite place?

Thank you, God,
for loving me.
Bless my family and
friends. Amen.
9:02

God Always Listens

Jesus told us that God is our Father. God the Father invites us to talk with him in prayer. He wants us to share with him what is on our minds and in our hearts.

We do what Jesus taught us. We tell God the Father we love him. We thank him for his blessings.

We ask God to take care of us and our families. We ask God to help other people. We ask God to forgive us and to help us live as his children.

Catholics Believe

Meal Prayers

We pray before and after meals. We ask God to bless us and the food we eat. We ask God to help people who do not have enough food.

Activity

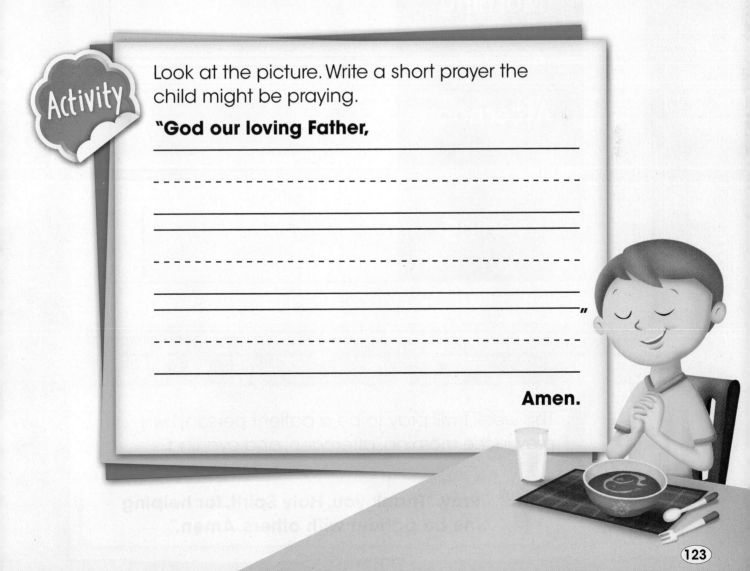

Look at the picture. Write a short prayer the child might be praying.

"God our loving Father,

"

Amen.

I Follow Jesus

The Holy Spirit teaches you to pray. You can talk to God about anything. You can pray anywhere and anytime. Patience helps you pray. It helps you spend time with God.

Activity

Times to Pray

Fill in the chart. Name something or someone you can pray for at different times during the day.

Morning	
Afternoon	
Evening	

My Faith Choice

This week I will pray to be a patient person. I will pray in the morning, afternoon, and evening.

Pray, "Thank you, Holy Spirit, for helping me be patient with others. Amen."

Chapter Review

Color the circle ● if the sentence is true.

Color the circle ● if the sentence is not true.

	T	F
1. We can talk to God about anything.	○	○
2. We can talk to God anywhere.	○	○
3. We can pray only by ourselves.	○	○
4. Jesus prayed often.	○	○
5. God cannot hear our prayers.	○	○

TO HELP YOU REMEMBER

1. Prayer is listening and talking to God.
2. When we pray, we grow in our love for God.
3. God always listens to our prayers.

Hail, Mary

Mary, the mother of Jesus, prays for us. She helps us to pray. Learn these words from the Hail Mary prayer. Pray them often. Pray them alone and with your family.

Hail, Mary, full of grace,
the Lord is with thee.
Blessed art thou among women
and blessed is the fruit of thy
womb, Jesus.

With My Family

This Week . . .

In chapter 13, "We Pray," your child learned:

- Prayer is listening and talking to God.
- Jesus is our example for how we are to pray.
- We can pray anywhere and anytime. We can share with God everything and anything that is on our minds and in our hearts.
- We demonstrate patience when we listen carefully to people and pay attention when they are helping us. Patience helps us to spend time with God in prayer, even when we want to do something else.

For more about related teachings of the Church, see the *Catechism of the Catholic Church*, 2558–2619; and the *United States Catholic Catechism for Adults*, pages 466–468 and 476–477.

Sharing God's Word

Read Matthew 7:7–11 together. Emphasize that prayer is listening and talking to God. We can pray anywhere and anytime. God knows what we need before we ask him.

We Live as Disciples

The Christian home and family is a school of discipleship. The first place where children should learn to live as disciples of Jesus. Choose one of the following activities to do as a family, or design a similar activity of your own.

- Go for a walk together. Thank God for everything you see and hear.
- **Family prayer time** helps us be aware that God is always with us. Evaluate your family prayer time. Do what it takes to integrate time for prayer into your family's daily activities and schedule.

Our Spiritual Journey

Mary is the first disciple of her son, Jesus. She is the model of what it means to be a disciple of Jesus. Devotion to Mary is beneficial to the life of Catholics. Incorporate frequent conversations with Mary. Seek direction for your life by meditating on the mysteries of the Rosary, the prayer devotion that Mary gave us.

For more ideas on ways your family can live as disciples of Jesus, visit **www.BeMyDisciples.com**

CHAPTER 14

We Are Peacemakers

Who has forgiven you?
Whom have you forgiven?

Forgiving others shows our love for one another. Christians forgive one another. Jesus tells us to forgive people who have hurt us. Jesus said,

> *"Ask God to forgive your sins and to help you forgive those who have hurt you."*
>
> BASED ON MATTHEW 6:12

What did Jesus teach about forgiveness?

Disciple Power

Peace

We live as peacemakers when we forgive those who have hurt us. We ask for forgiveness when we have hurt others. These actions bring peace to us and to others.

The Church Follows **Jesus**

Read to Me

The Pope Makes Peace

Pope John Paul II was riding in the back of his car. He was greeting and waving to people.

A man came out of the crowd and shot at the pope. The pope was hurt but soon got better.

Later, Pope John Paul II went to the prison. He visited the man who shot him. He put his arms around the man and forgave him. The pope made peace with him. He showed us what Jesus wants us to do.

? What are some of the ways people forgive others?

Making Peace

We can use words and actions to help others. Other times, we can choose to use our words and actions to hurt others. When we hurt others, we do not obey God. We **sin.**

Sin is choosing to do or say something that we know is against God's laws. When we sin, we turn away from God's love.

Sin is choosing not to love others as Jesus taught us. Sin hurts our friendship with God and with other people.

Faith Focus
Why is it important to say "I am sorry" when we choose to do or say something that is wrong?

Faith Word
sin
Sin is choosing to do or say something that we know is against God's laws.

Activity

Look at these pictures. Pretend you are making a movie. With a partner, act out a picture showing a good choice.

Faith-Filled People

Saint John Vianney

Father John Vianney heped people who were sorry for their sins. Railroad tracks were built to his town because so many people wanted to come to him to receive forgiveness for their sins. The Church celebrates the feast day of Saint John Vianney on August 4.

Asking for Forgiveness

We feel sorry when we sin. We want to be forgiven. We want to make up for our sin.

We need to say that we are sorry to people when we hurt them. We need to ask for forgiveness. We want everything to be right again.

We also need to tell God that we are sorry when we sin. We can tell God that we are sorry. We ask for forgiveness because we love God. God will always forgive us because he loves us.

Activity

Circle the words that show you are sorry. Also circle the words that show you forgive somone.

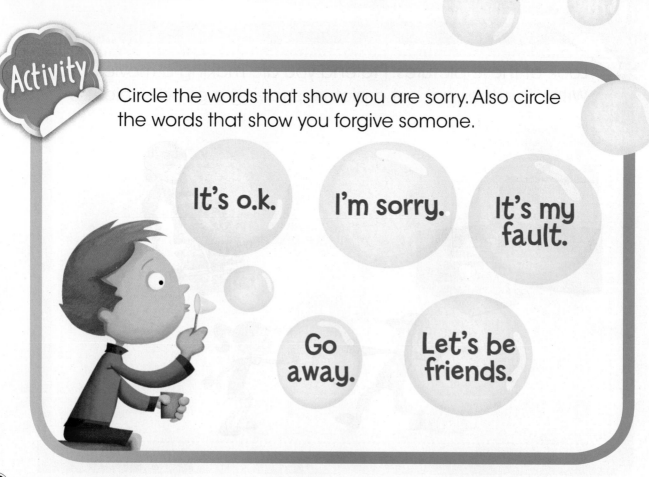

It's o.k.
I'm sorry.
It's my fault.
Go away.
Let's be friends.

Forgiving Others

Jesus tells us to forgive people who have hurt us. He tells us to forgive them over and over again.

Sometimes it is not easy to do what Jesus wants us to do. Sometimes we do not feel like forgiving people who have hurt us. The Holy Spirit can help us forgive others.

We open our hearts with love when we forgive others. We show our love for God and for one another. We are peacemakers.

? How do you show forgiveness? How do other people show you forgiveness?

Catholics Believe

Sign of Peace

Each Sunday at Mass, we shake hands or share another sign of peace with one another. This shows that we want to forgive those who have hurt us. We want to live together as the one family of God.

I Follow Jesus

The Holy Spirit teaches you and helps you forgive people. He also helps you to ask for forgiveness. When you forgive someone, you are a peacemaker.

Activity

A Forgiving Tree

In the leaves, write words or draw actions that show forgiveness.

My Faith Choice

This week I will use the forgiving words or actions that I wrote or drew in the activity. I will bring peace to others.

Pray, "Thank you, Holy Spirit, for teaching me and helping me to live as a peacemaker. Amen."

Chapter Review

Choose the best word and write it in the space in each sentence.

forgive	peace	sin

1. When we ________________, we turn away from God's love.

2. Jesus asks us to ________________ others.

3. We bring ________________ when we show our love for others.

TO HELP YOU REMEMBER

1. Sin hurts our friendship with God and others.
2. When we say that we are sorry, we show that we love God and others.
3. When we say that we are sorry, we ask for forgiveness from others and from God.

Prayer of Mercy

At the beginning of Mass, we ask God for his mercy. The word mercy reminds us that forgiveness is a gift of God's love. Pray this prayer together.

Leader: Lord, have mercy.

All: **Lord, have mercy.**

Leader: Christ, have mercy.

All: **Christ, have mercy.**

Leader: Lord, have mercy.

All: **Lord, have mercy. Amen.**

With My Family

This Week . . .

In chapter 14, "We Are Peacemakers," your child learned:

- People make choices that help others or hurt others. We can choose to follow or reject God's laws.
- People can sin. Sin always hurts our relationship with God and with others. When we sin, we need to say that we are sorry both to God and to those whom we have hurt. We need to ask for forgiveness. We need to reconcile our relationships with God and with people.
- We live as peacemakers when we are honest in our relationships and with God. When we forgive others, we are peacemakers.

For more about related teachings of the Church, see the *Catechism of the Catholic Church*, 1420–1484 and 1846–1869; and the *United States Catholic Catechism for Adults*, pages 235–236.

Sharing God's Word

Read together the Bible story in Matthew 18:21–35 about the Parable of the Unforgiving Servant. Emphasize that Jesus teaches us that we are to forgive others over and over again, as God always forgives us when we are truly sorry for our sins.

We Live as Disciples

The Christian home and family is a school of discipleship. Choose one of the following activities to do as a family, or design a similar activity of your own.

- When you participate in Mass this week, pay close attention to the prayer of mercy that we pray at the beginning. Remember that the word *mercy* reminds us that God's forgiveness is a gift of his love.
- Name ways that people show that they are sorry. Talk about ways that members of your family can show both forgiveness to one another and accept forgiveness from one another.

Our Spiritual Journey

In this chapter, your child prayed a prayer of mercy. This is one of the three forms of prayer that the Church uses for the Penitential Act in the Introductory Rites of the Mass. Through this prayer, we are reconciled with God and one another. We enter into the celebration of the Eucharist in the right relationship with God and one another, as peacemakers. Read and pray together the prayer on page 133.

For more ideas on ways your family can live as disciples of Jesus, visit **www.BeMyDisciples.com**

CHAPTER 15

We Go to Mass

When do you say thank you to others?

We can thank people in many ways. The Church thanks God in a special way at Mass. Listen to what the Bible tells us about giving thanks to God.

It is good to give thanks to God.

BASED ON PSALMS 92:2

When do you say thank you to God?

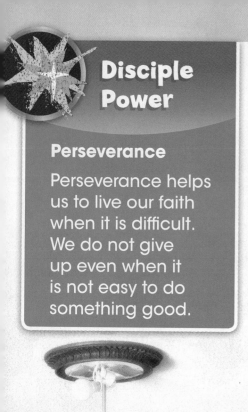

Disciple Power

Perseverance

Perseverance helps us to live our faith when it is difficult. We do not give up even when it is not easy to do something good.

The Church Follows **Jesus**

Sharing God's Love

Read to Me

First Holy Communion was a special day for Peyton and the other children of St. Mary's Church. At the end of Mass, they heard Father Julio say, "Go in peace, glorifying the Lord by your life."

At the next religion class, Peyton asked Mrs. Hensle, "What did Father Julio tell us to do? I don't understand."

Mrs. Hensle explained, "Father Julio said we need to show we are children of God by what we say and do." The children talked about what they could do.

They visited elderly people living in a retirement home. They played a board game together. This was their way of saying thank you for the gift of First Holy Communion. They shared God's love with other people.

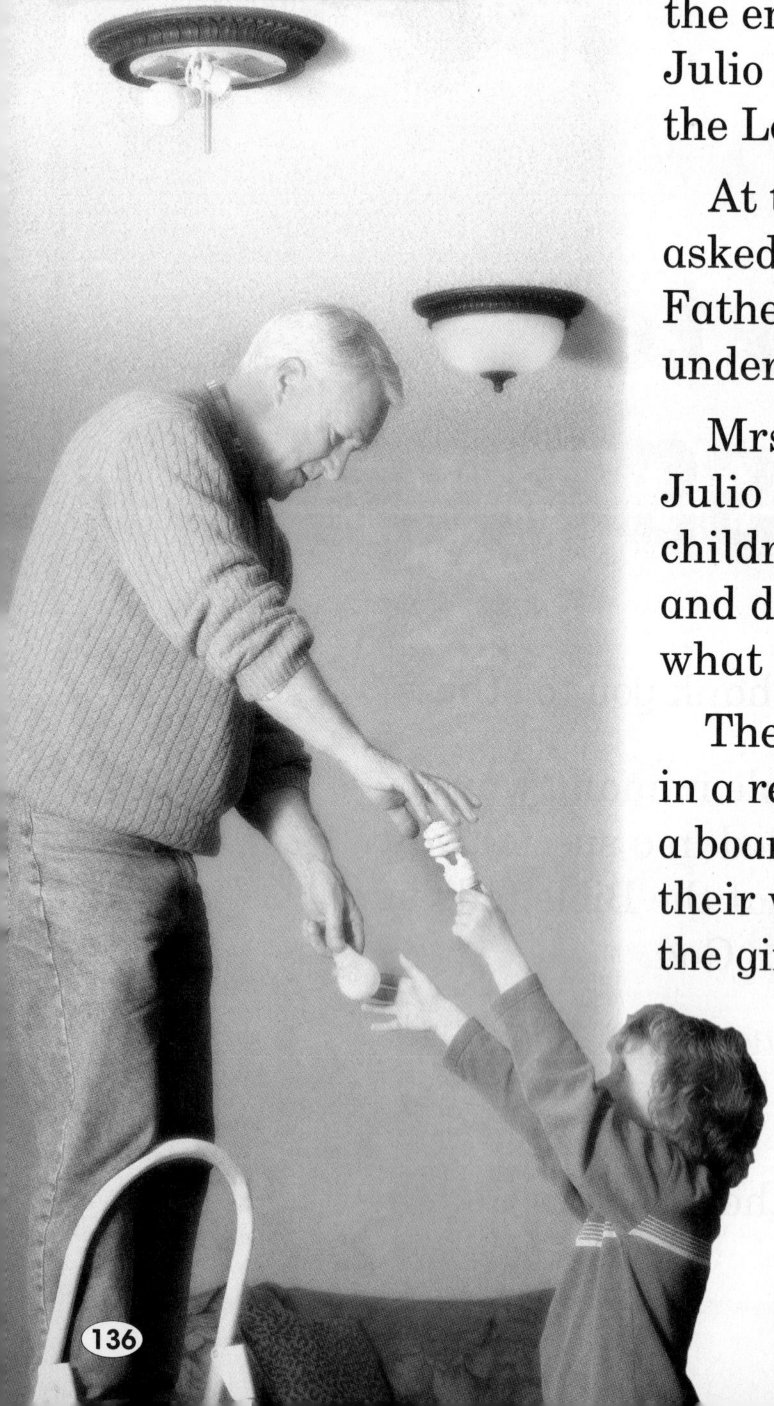

? How can you share God's love with people?

We Gather at Mass

The **Mass** is the most important celebration of the Church. We gather as the People of God. We give glory to God. We show that we love and honor God.

We listen to God's Word. We celebrate and share in the **Eucharist.** The Eucharist is the sacrament in which we receive the Body and Blood of Christ.

We begin the Mass by praying the Sign of the Cross. This reminds us of our Baptism. We remember that we belong to Jesus and are members of the Church.

Faith Focus
Why does our Church family gather to celebrate Mass?

Faith Words

Mass
The Mass is the most important celebration of the Church.

Eucharist
The Eucharist is the sacrament in which we receive the Body and Blood of Christ.

Activity

Draw you and your family at Mass.

Faith-Filled People

Priests

Priests are the Bishop's co-workers. They lead us in the celebration of the Mass. They teach us what Jesus taught. They help us to live as followers of Jesus.

We Listen to God's Word

We listen to readings from the Bible at every Mass. God tells us about his love for us. On Sunday we listen to three readings. The third reading is from the Gospel.

After the Gospel is read, the priest or deacon helps us to understand God's Word. We come to know and love God more. We learn ways to live as Jesus taught.

Next, we tell God that we have listened to his Word. We tell God we believe in him. Then we pray for other people and for ourselves.

Activity

Learn what we say after the First Reading at Mass. Trace these words.

Thanks

be to

God.

We Give Thanks to God

At Mass we celebrate the Eucharist. The word eucharist means "to give thanks."

In the celebration of the Eucharist, we give thanks to God. We remember and do what Jesus did at the Last Supper.

The Last Supper is the meal that Jesus ate with his disciples on the night before he died.

At the Last Supper Jesus took bread and wine. He took the bread and said, "This is my body." He took the cup of wine and said, "This is my blood."

Based on Matthew 26:26–28

Catholics Believe

Peacemakers

At the end of Mass we are told, "Go in peace." Peacemakers share God's love with people. Jesus taught that God blesses peacemakers. Jesus said, "Blessed are the peacemakers, for they will be called children of God" (Matthew 5:9).

Activity

Color the letters. Tell others what we do at Mass.

WE GIVE THANKS TO GOD.

I Follow Jesus

At Mass you can listen to God's Word. You learn ways to live as a follower of Jesus. In the celebration of the Eucharist, you give thanks to God. You can try to pay attention at Mass, even when it is hard.

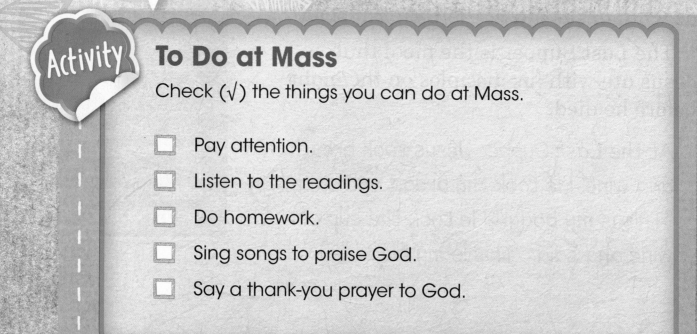

Activity

To Do at Mass

Check (√) the things you can do at Mass.

- ☐ Pay attention.
- ☐ Listen to the readings.
- ☐ Do homework.
- ☐ Sing songs to praise God.
- ☐ Say a thank-you prayer to God.

My Faith Choice

This week I will try hard to show my love for God and others. I will

____

____

____

Pray, "Thank you, Holy Spirit, for helping me listen to God's Word. Amen."

Chapter Review

Draw a line to match the words on the left with their meanings.

WORDS	MEANINGS
1. Last Supper	The celebration in which we listen to the Word of God. We say thank you to God.
2. Mass	The sacrament in which we receive the Body and Blood of Christ.
3. Eucharist	The meal Jesus ate with his disciples on the night before he died.

TO HELP YOU REMEMBER

1. At Mass, we worship God.
2. At Mass we listen to readings from the Bible.
3. At Mass we celebrate and share the Eucharist.

Thank You, God

We can pray quietly in our hearts, and we can pray aloud.

Leader Let us remember Jesus. Think about Jesus. *(Pause.)*

All **Thank you, God.**

Leader Think about what Jesus told us about God. *(Pause.)*

All **Thank you, God.**

Leader Think about people who share God's love with you. *(Pause.)*

All **Thank you, God. Amen.**

With My Family

This Week . . .

In chapter 15, "We Go to Mass," your child learned:

- ▶ The Mass is the most important celebration of the Church.
- ▶ During the Liturgy of the Word, we listen to the readings from the Bible.
- ▶ In the Liturgy of the Eucharist, we remember and do what Jesus did at the Last Supper. The bread and wine become the Body and Blood of Jesus.
- ▶ We receive the Body and Blood of Christ in Holy Communion.

For more about related teachings of the Church, see the *Catechism of the Catholic Church*, 1322–1405, and the *United States Catholic Catechism for Adults*, pages 215–227.

Sharing God's Word

Read together Matthew 26:26–29 the account of the Last Supper. Or read the adaptation of the story on page 139. Emphasize that at Mass the bread and wine become the Body and Blood of Christ. Discuss the importance of participating during Mass.

We Live as Disciples

The Christian home and family is a school of discipleship. Choose one of the following activities to do as a family, or design a similar activity of your own.

- ▶ At the end of Mass we are dismissed with these or similar words, "Go in peace, glorifying the Lord by your life." Choose one thing your family can do together to love and serve the Lord this week.
- ▶ Your child has learned that perseverance means that we try our best even when it is hard. Discuss times that this might be hard to do and that the Holy Spirit will help them.

Our Spiritual Journey

The Psalms are prayers that represent an array of faith experiences. They are both personal and communal. They heighten our memory of God's loving plan of Creation and Salvation. Pray them at times of joy and sadness, lament, and thanksgiving. Choose and memorize a variety of verses from the Psalms. Pause throughout the day to pray them.

For more ideas on ways your family can live as disciples of Jesus, visit **www.BeMyDisciples.com**

CHAPTER 16

Jesus Shows God's Love

Which foods are your family's favorites?

Healthy foods help us to grow. Jesus shared food with many people. Listen to part of one of those stories.

At the Last Supper, Jesus said to his disciples, "When you eat this bread and drink this wine, remember me." BASED ON LUKE 22:19–20

How do these words from the Bible remind you of the Mass?

Disciple Power

Wisdom

Wisdom helps us to know what God wants us to do. It helps us to live a holy life.

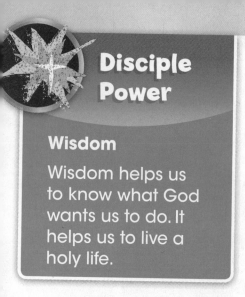

The Church Follows **Jesus**

CRS' Operation Rice Bowl

Read to Me

The Bible has many stories of Jesus sharing a meal with people. Like Jesus, we can share food with people too.

Each year during Lent, many Catholics participate in Operation Rice Bowl. Catholics put a small box on the table where they eat. Each family member puts money in the box.

At the end of Lent, Catholic families bring the box with the money to church. All the boxes are collected.

The Church uses the money to provide food and clean drinking water to people who need it. When we help people in need, we are sharing God's love too.

? What is one way that you can share God's love with people?

CRS OPERATION RICE BOWL Catholic Relief Services

www.orb.crs.org

Jesus Shares God's Love

Many of the first disciples of Jesus lived in **Galilee**. Galilee was the place where Jesus did much of his teaching. He also helped many people who lived there.

Some people in Galilee were fishermen. They fished in the Sea of Galilee for their food. Other people were farmers. They grew fruit and barley. They made bread from the barley.

Faith Focus
Why did Jesus share food with others?

Faith Words

Galilee
Galilee was one of the main places where Jesus taught and helped people.

miracle
A miracle is something only God can do. It is a sign of God's love.

Activity

Find the names on the map. Write the words in the right places in the sentences.

- Jesus was born in ______________________.
- Jesus lived in ______________________.
- Jesus ate fish from the Sea of ______________________.

Find these places on the map. Circle the words on the map.

Faith-Filled People

Saint Maria de Jesus

Sister Maria spent much of her life serving people who were poor and ill. Sister Maria loved going to Mass. Receiving Holy Communion helped her care for people as Jesus did. On May 21, 2000, Pope John Paul II named her Mexico's first woman saint.

Jesus Feeds the People

One time Jesus was teaching near the Sea of Galilee. A very large crowd of people gathered to hear Jesus. This is what happened.

It became late and the people were hungry. But Jesus' followers had only five loaves of bread and two fish. Jesus took the bread and the fish and prayed. His followers gave the food to the people. Everyone ate until they were full.

Based on Matthew 14:15–16, 19–20

Activity

Use the picture to tell a partner the story of Jesus feeding the people.

Jesus Cares for People

The story of Jesus sharing the bread and fish tells about a **miracle**. A miracle is something only God can do. It is a special sign of God's love.

The story of Jesus feeding the people shows how Jesus shared God's love with the people. Jesus took care of the people.

Jesus asks us to take care of one another too. This is one way we share God's love with people.

? How do you share God's love with other people?

Catholics Believe

Grace

The word *grace* means "gift." Grace is the gift of God's life and love. Grace helps us to share God's love with people. It helps us to live as children of God.

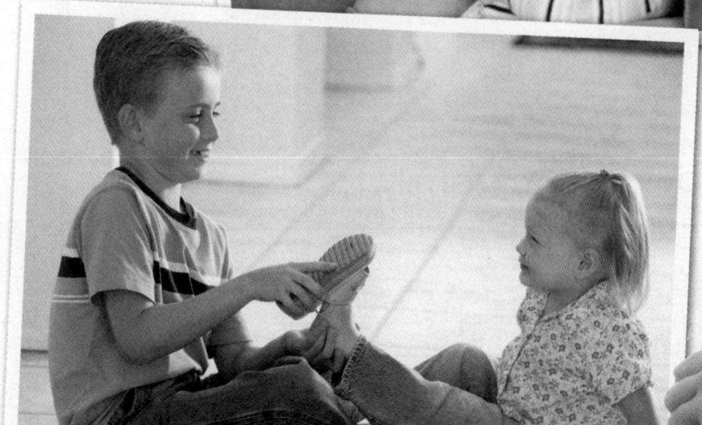

I Follow Jesus

The Holy Spirit helps you to show God's love and care for others. The Holy Spirit's gift of wisdom can help you to make good choices for living a holy life. When you share God's love, you are living a holy life.

Activity

Sharing God's Love

In the space write about or draw yourself sharing God's love with others.

My Faith Choice

I will share God's love with my family. I will

____

- -

____.

Pray, "Thank you, Jesus, for teaching me how to share God's love. Amen."

Chapter Review

Read again the story of Jesus feeding the people. Number the sentences in the order they happen in the story.

_______ Everyone ate until they were full.

_______ Jesus took the five loaves and two fish and prayed.

_______ A large crowd was listening to Jesus. It was evening and they were hungry.

TO HELP YOU REMEMBER

1. Jesus saw that the people were hungry and gave them all enough to eat.
2. Jesus showed people that God cares for them.
3. Jesus teaches us to care for people.

A Blessing Prayer

Blessing prayers tell God we know that all good things come from him. Pray this blessing prayer together.

Leader: Father, you care for everyone.

All: **Blessed be God.**

Leader: Jesus, you showed us how to care for people.

All: **Blessed be God.**

Leader: Holy Spirit, you help us to care for our families.

All: **Blessed be God. Amen.**

With My Family

This Week . . .

In chapter 16, "Jesus Shows God's Love," your child learned:

- Jesus fed a large crowd with only five loaves of bread and two fishes (Matthew 14:15–20).
- This story tells that Jesus took care of people to remind them of God's love for them.
- This story is one of the miracle stories in the Gospels and reveals God's loving care for people and all creation.
- Wisdom is a gift of the Holy Spirit. It helps us to know God's will for us and to make good choices.

For more about related teachings of the Church, see the *Catechism of the Catholic Church*, 302–308 and 547–550, and the *United States Catholic Catechism for Adults*, pages 79–80, 215–216, and 222–223.

Sharing God's Word

Read together Matthew 14:15–20 the account of Jesus feeding the crowd. Or read the adaptation of the story on page 146. Emphasize that everyone ate until they were full. Discuss that this is a sign of God's caring love for all people.

We Live as Disciples

The Christian home and family is a school of discipleship. Choose one of the following activities to do as a family, or design a similar activity of your own.

- Jesus fed the hungry people to show them that God loves and cares for them. Choose to do one thing this week to show people that God loves and cares for them.
- When you go grocery shopping this week, purchase food to donate to the local food pantry. Join with others to be a sign of God's loving care for all people.

Our Spiritual Journey

A blessing prayer is an expression of God's generosity and love. Our life can be a blessing prayer. The best way to bless God is to share our material and spiritual blessings with others, especially people in need. Pray the blessing prayer on page 149 together as a family.

For more ideas on ways your family can live as disciples of Jesus, visit **www.BeMyDisciples.com**

Unit 4 Review

Name ______________________

A. Choose the Best Word

Complete the sentences. Color the circle next to the best choice.

1. Every Sunday the people of the Church gather to celebrate __________.

○ Confirmation ○ Mass

2. At the __________, Jesus said, "This is my body" and "This is my blood."

○ Last Supper ○ First Easter

3. Prayer is listening and talking to __________.

○ our parents ○ God

4. We need to ask for __________ when we have hurt someone.

○ forgiveness ○ punishment

5. Jesus fed a crowd with two fish and __________ loaves of bread.

○ two ○ five

B. Show What You Know

Circle the number next to your favorite prayer. Tell your class when you can pray it.

1. The Sign of the Cross

2. The Our Father

3. The Glory Be Prayer

4. The Hail Mary

5. Grace Before Meals

C. Connect with Scripture

What was your favorite story about Jesus in this unit? Draw something that happened in the story. Tell your class about it.

D. Be a Disciple

1. *What saint or holy person did you enjoy hearing about in this unit? Write the name here. Tell your class what this person did to follow Jesus.*

2. *What can you do to be a good disciple of Jesus?*

We Live

Part One

UNIT 5

The Way to Heaven

One day a young man asked Jesus, "What must I do to live forever with God in heaven?" he asked. Jesus said, "Keep God's commandments."

The young man said, "I have always kept them." "Wonderful!" Jesus said. "Now sell all your things, give to the poor, and follow me."

The young man turned away. He did not want to give away his things. Jesus felt sad. "If a person loves their things more than they love God, it will be hard to get to heaven."

BASED ON MARK 19:16–23

What I Have Learned

What is something you already know about these faith words?

Christians

The Ten Commandments

Faith Words to Know

Put an **X** next to the faith words you know. Put a **?** next to the faith words you need to learn more about.

Faith Words

____ The Great Commandment

____ respect

____ community

____ worship

____ honor

A Question I Have

What question would you like to ask about The Ten Commandments?

CHAPTER 17

The First Christians

? What are some things that make you and your family smile?

The first followers of Jesus were like a family. Listen to what the Bible tells us about Jesus' followers:

Jesus' followers were filled with joy and the Holy Spirit BASED ON ACTS 13:52

? What do these words from the Bible tell you about Jesus' followers?

Disciple Power

Understanding

God the Holy Spirit gives us the gift of understanding. Stories in the Bible help us understand God's love for us. Stories in the Bible help us understand what Jesus taught us.

The Church Follows Jesus

Saint Martin de Porres

Read to Me

Martin de Porres loved God. The Holy Spirit helped Brother Martin to live as a disciple of Jesus. Brother Martin served people with a joyful heart.

Brother Martin worked with poor people. He opened a home for children whose parents had died or could not care for them. He opened a hospital and schools. He also took care of animals that were sick or hungry.

The Church named Brother Martin a saint. Today many people follow the good example of Saint Martin de Porres.

? Who is someone you know who helps people as Saint Martin de Porres did?

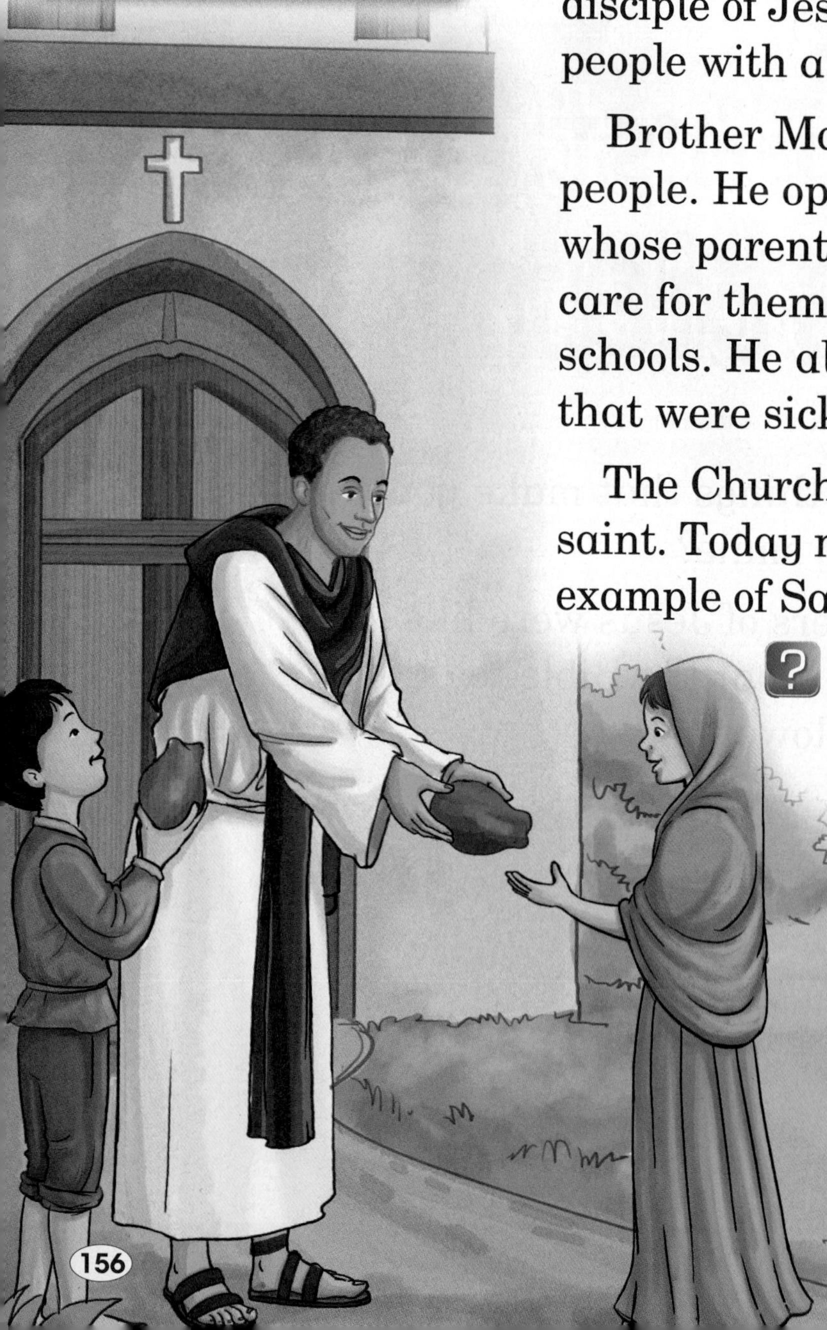

Christians Share Stories

The Church tells stories that help us understand God's love for us. The Church tells stories about Jesus. These stories help us to know what it means to be a Christian. **Christians** are followers of Jesus Christ.

Our Church also shares stories about what Christians did a long time ago. We can read many stories about the first Christians in the New Testament.

Faith Focus
How did the people of the Church live when the Church began?

Faith Word
Christians
Christians are followers of Jesus Christ. They believe in Jesus Christ and live as he taught.

Activity

On each of the road signs, write or draw one thing that followers of Jesus do today.

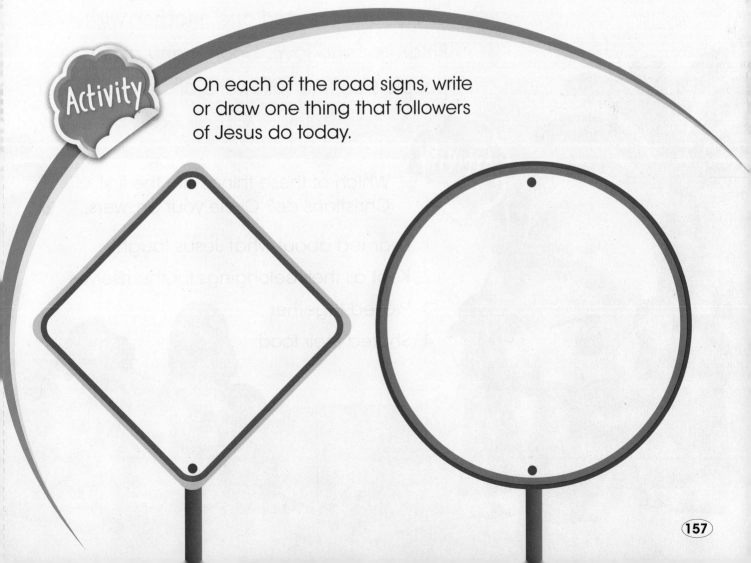

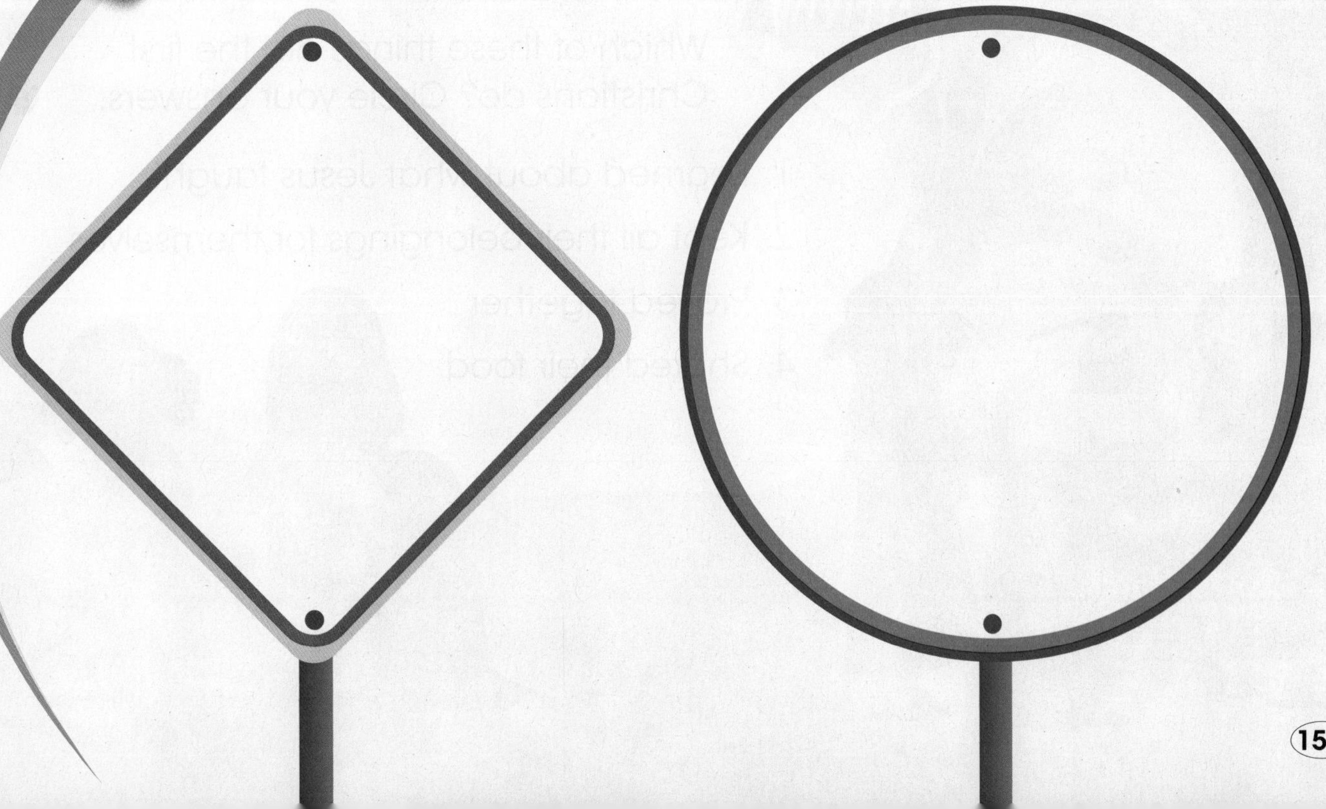

Faith-Filled People

Saint Paul the Apostle

Saint Paul became an apostle after Jesus' Resurrection. Paul traveled by land and by sea to teach people about Jesus. He invited them to believe in Jesus. The Church celebrates the feast day of Saint Paul the Apostle on June 29.

Christians Love One Another

This is a story about how the first Christians lived. It is a story that the Church has shared from her very beginning.

The first Christians spent time learning what Jesus taught. They shared their money and belongings with one another. They prayed together. They broke and shared bread together. Together they praised God.

Many people saw how the first Christians treated one another with kindness and love. Soon many other people became followers of Jesus.

BASED ON ACTS OF THE APOSTLES 2:42, 45–47

Activity

Which of these things did the first Christians do? Circle your answers.

1. Learned about what Jesus taught
2. Kept all their belongings for themselves
3. Prayed together
4. Shared their food

We Live as Jesus Taught

The first Christians did what Jesus did. They did what Jesus taught. Jesus taught us to love God and to love one another.

The stories about the first Christians teach us how to live as children of God. The first Christians showed their love for God. They prayed and shared the Eucharist. They thanked God for everything.

The first Christians showed their love for one another. They shared what they had with each other. They helped people in need.

Catholics Believe

Patron Saints

The saints help us remember how to love as Jesus taught. The Church names some saints to be patron saints. Patron saints help us live as followers of Jesus in a certain way. Saint Martin de Porres is the patron saint of African-Americans and health-care workers.

Activity

Think of a follower of Jesus that you know. Check one thing that this person does. Act it out for your class.

Person I Know

- ☐ Prays with others
- ☐ Cares for others
- ☐ Tells me about Jesus
- ☐ Shares with me

I Follow Jesus

Each day you can try your best to live as Jesus taught. The Holy Spirit helps you to understand how Jesus wants you to treat people. You can treat people with kindness. This brings people joy. It also fills your heart with joy.

Activity

Write a √ mark in the box next to each thing that you can do this week to live as Jesus taught.

☐ Say my prayers.	☐ Play fairly.
☐ Hurt someone.	☐ Speak unkind words.
☐ Learn about Jesus.	☐ Help at home.
☐ Share my toys.	☐ Pay attention at school.
☐ Obey my parents.	☐ Be a litter-bug.

My Faith Choice

This week I will do one of the Christian acts that I have checked. I will

____

____

____.

Pray, "Thank you, Holy Spirit, for helping me to act as Christians do."

Chapter Review

*Read each sentence. Circle **Yes** if the sentence is true. Circle **No** if it is not true.*

1. The first Christians shared stories about Jesus.

Yes No

2. The first Christians prayed together.

Yes No

3. The first Christians shared their belongings with one another.

Yes No

TO HELP YOU REMEMBER

1. Christians believe in Jesus Christ and do what he taught.
2. The first Christians gathered together and loved God and one another.
3. Christians today show their love for one another just as the first Christians did.

Praise the Lord

A sign of peace shows that we want to live as Jesus taught. We share a sign of peace at Mass.

Leader We thank you, God, for the Church.

All **Praise the Lord, for he is good!**

Leader Let us share a sign of peace with one another.

All **(Share a handshake or other sign of peace and friendship.)**

With My Family

This Week . . .

In chapter 17, "The First Christians," your child learned:

- The first Christians gathered to express their faith and belief in Jesus.
- The first Christians listened to the teachings of the Apostles. They shared all that they had with one another, especially with people in need. They gathered to pray and share the Eucharist.
- Christians today do the same things that the first Christians did. Every member of the Church is called to cooperate with the grace of the Holy Spirit and work together to live as Jesus taught.
- The Holy Spirit helps us to understand ways to live as followers of Jesus.

For more about related teachings of the Church, see the *Catechism of the Catholic Church*, 849–852, 1397, and 2030–2046, and the *United States Catholic Catechism for Adults*, pages 118–119.

Sharing God's Word

Read together the Acts of the Apostles 2:42–47, an account of the life of the first Christians. Or read the adaptation of the story on page 158. Emphasize that the first Christians shared with people in need and were known for their love for one another.

We Live as Disciples

The Christian home and family is a school of discipleship. Choose one of the following activities to do as a family, or design a similar activity of your own.

- Identify ways that your family lives as the first Christians did. Talk about ways you pray, learn about Jesus, and share things as a family. Invite each family member to choose one thing that they can do to help your family live as a Christian family.
- Decide one way in which your family can share your time and possessions with other people in your parish or neighborhood. For example, write get well cards to those who are sick or homebound.

Our Spiritual Journey

The peace that comes from living in communion with God, others, and all of Creation is the ultimate destination of our spiritual journey. You have received the gift of understanding to help you begin to achieve that peace. Pray the prayer on page 161 at mealtime and include a sign of peace.

For more ideas on ways your family can live as disciples of Jesus, visit **www.BeMyDisciples.com**

CHAPTER 18

We Love God

What are some good rules for a family?

God's rules are called commandments. Listen to what the Bible tells us about them:

God said, "I will show love to those who love me and keep my commandments."

Based on Exodus 20:6

What is one of God's rules that you know?

Disciple Power

Knowledge

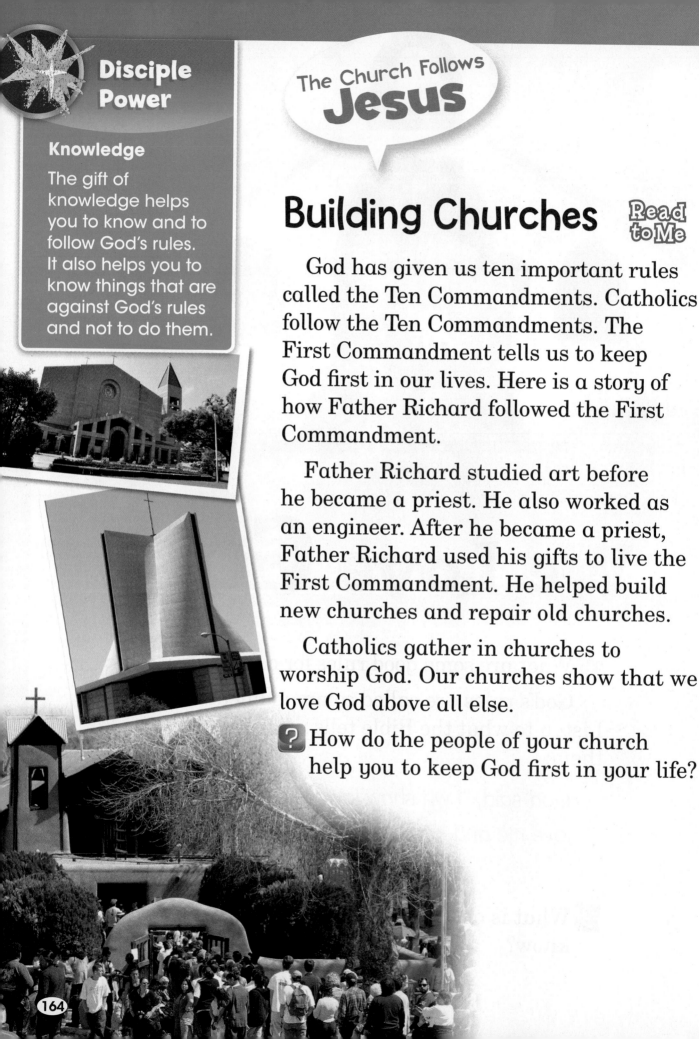

The gift of knowledge helps you to know and to follow God's rules. It also helps you to know things that are against God's rules and not to do them.

Building Churches

Read to Me

God has given us ten important rules called the Ten Commandments. Catholics follow the Ten Commandments. The First Commandment tells us to keep God first in our lives. Here is a story of how Father Richard followed the First Commandment.

Father Richard studied art before he became a priest. He also worked as an engineer. After he became a priest, Father Richard used his gifts to live the First Commandment. He helped build new churches and repair old churches.

Catholics gather in churches to worship God. Our churches show that we love God above all else.

? How do the people of your church help you to keep God first in your life?

God's Commandments

We have rules at home, at school, and in our community. Good rules help us to live together in peace.

The Bible tells us that God gave us ten very special rules. These rules are the **Ten Commandments**. He gave us the Ten Commandments because he loves us. The Ten Commandments are the laws that God has given us to live as children of God.

The Ten Commandments tell us how we are to love God and other people. They tell us to care for ourselves and for all creation.

Faith Focus
Why did God give us the Ten Commandments?

Faith Words

Ten Commandments
The Ten Commandments are the ten laws that God has given us to help us live as children of God.

worship
We worship God when we love and honor God more than anyone and anything else.

Activity

With your classmates, act out one way you can show your love for God. Act out one way you can show your love for your family.

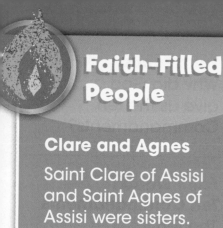

Faith-Filled People

Clare and Agnes

Saint Clare of Assisi and Saint Agnes of Assisi were sisters. They gave up everything to show their love for God. Saint Clare's feast day is August 11. Saint Agnes's feast day is November 16.

Jesus Teaches Us

Jesus taught us that we are to live the Ten Commandments. Jesus showed us how to love God. Jesus prayed to his Father. He always did what God the Father asked him to do.

Jesus showed us how to love one another. He was kind to everyone. Jesus told us to treat people as he did. He said,

> *"I give you this new commandment. You are to love one another as I have loved you."*
>
> Based on John 13:34

? How are the people in the pictures showing their love for God? How are they showing their love for people?

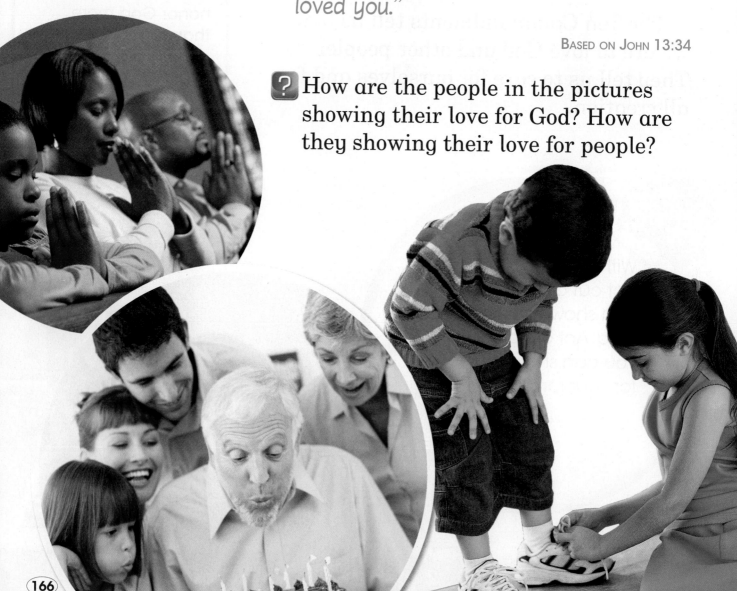

We Love God

The First, Second, and Third Commandments tell us ways to show our love for God.

The First Commandment tells us that we are to **worship** only God. We are to love God more than anything and anyone else.

The Second Commandment tells us to honor God. We are to speak God's name with love and respect.

The Third Commandment tells us to keep Sunday as a holy day. Every Sunday, we gather with our Church family for Mass. We give thanks and praise to God for all that he has done for us.

Catholics Believe

Cathedrals

There are many kinds of churches. The cathedral is the bishop's church. The name of the pope's cathedral is Saint John Lateran.

Activity

Thank God for all that he has done for you. Use words and pictures to make a Thank You card.

Thank You, God, for...

I Follow Jesus

You are learning about the Ten Commandments. Your family and the Church will help you. The Holy Spirit will always help you to know the Ten Commandments and to live them. The Holy Spirit's gift of knowledge helps you to worship and honor God.

Activity

A Letter to God

Write a letter. Tell God how you will show your love for him this week.

Dear God,

____

____

____

____

____

______________________________.

My Faith Choice

Read the letter you wrote to God. Write one way you will show your love for God this week.

____

____

____

Pray, "Thank you, Holy Spirit, for helping me to show my love for you."

Chapter Review

Draw lines to match the words in column A with their meanings in column B.

Column A	Column B
1. First Commandment	**a.** Keep Sunday holy.
2. Second Commandment	**b.** Worship only God.
3. Third Commandment	**c.** Speak God's name with respect.

TO HELP YOU REMEMBER

1. The Ten Commandments teach us to worship God.
2. The Ten Commandments teach us to speak God's name with love and respect.
3. The Ten Commandments teach us to take part in Mass on Sundays.

An Act of Love

When we pray an act of love, we tell God that we love him with all our hearts. Pray these words at home with your family. Pray them now with your class:

O my God, you created me.
You share your love with me.
You are all good.
I love you with my whole heart.
Amen.

With My Family

This Week . . .

In chapter 18, "We Love God," your child learned:

- God gave us the Ten Commandments.
- The Ten Commandments tell us ways to live as children of God.
- The First, Second, and Third Commandments tell us to love, honor, and worship God above all else.
- Jesus taught us to live the Ten Commandments. We are to love God and people as Jesus taught.
- Knowledge is one of the seven Gifts of the Holy Spirit. It helps us to know how we are to live and to follow God's will.

For more about related teachings of the Church, see the *Catechism of the Catholic Church*, 2052–2074, and the *United States Catholic Catechism for Adults*, pages 341–369.

Sharing God's Word

Read John 13:34–35 together, about Jesus giving his disciples the New Commandment, or you can read an adaptation of the story on page 166. Emphasize that Jesus by his example showed us how to love God and one another. He showed us how to live the Ten Commandments.

We Live as Disciples

The Christian home and family is a school of discipleship. Choose one of the following activities to do as a family, or design a similar activity of your own.

- Attend Mass together as a family on Sundays. Plan an activity for after Mass that the family can enjoy together.
- When you gather for dinner this week, invite family members to share one thing that they did that day to show their love for God.

Our Spiritual Journey

Knowing your way as you journey though life is vital. Knowing where you want to go and how to get there is essential. We can never reach our ultimate goal alone. Pray the prayer on page 169 with your family to help keep your priorities in life in the right order.

For more ideas on ways your family can live as disciples of Jesus, visit **www.BeMyDisciples.com**

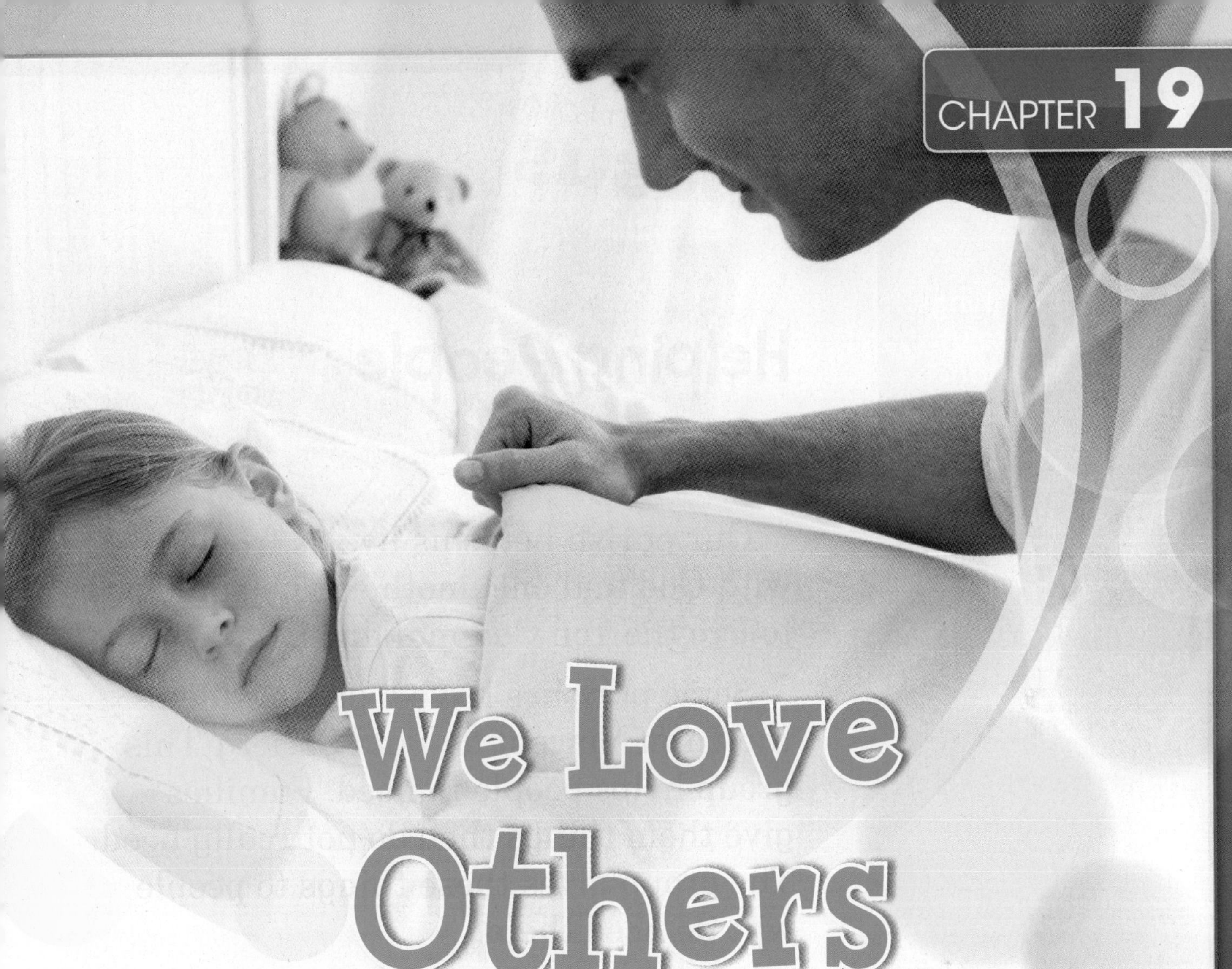

CHAPTER 19

We Love Others

What are some of the ways that you show your love for your family?

Listen to what the Bible tells us about God's Commandments:

God gave us the commandments so that we may live as friends with him and other people. Based on Deuteronomy 6:20–25

What do the Commandments help us to do?

Disciple Power

Temperance

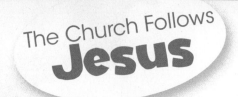

Having more things does not make us happy. Temperance helps us to know the difference between what we need and what we just want to have. It is important to know what will really make us happy.

The Church Follows **Jesus**

Helping People in Need

Read to Me

Our parish helps us live as friends with God and one another. They help us to live the Ten Commandments.

Some parishes have a group called the Saint Vincent de Paul Society. This group helps people in need. Families give them things they do not really need. The group gives these things to people who do need them.

The group also helps people to visit doctors and dentists. They run camps in the summertime for children. The Saint Vincent de Paul Society shows people how much God loves them.

? What are some of the ways that you see people being kind to one another?

We Respect People

The rest of the Ten Commandments tell us that we are to **respect** other people and ourselves. We show respect when we treat and **honor** other people and ourselves as children of God. Showing respect is a way to show love.

We show respect to people in many ways. We listen carefully to one another. We are polite and kind. We are fair to one another.

We show respect to ourselves in many ways. We take care of our bodies. We act safely.

Faith Focus
What do the Ten Commandments teach about respect?

Faith Words

respect
We show people respect when we love them because they are children of God.

honor
We honor people when we treat them with great respect.

Activity

With a partner, act out one way you can show respect to each other.

Faith-Filled People

Vincent de Paul

Saint Vincent de Paul showed us how to live the Ten Commandments. He treated all people with respect. He cared for people who were lonely. The Church celebrates the feast day of Saint Vincent de Paul on September 27.

We Care for Things

The Ten Commandments tell us to respect what belongs to us. It teaches us that we are to respect what belongs to other people, too. We are to take good care of the things that we have.

We show respect for what belongs to others. We are to ask before we borrow their things. We are to return the things that we borrow. We do not steal.

The Ten Commandments also teach us that we are to share our things with others.

? What is one way that you can care and share with others?

We Tell the Truth

The Ten Commandments teach us that we are to be honest. We are honest when we tell the truth. We are not being honest when we lie.

It is important to tell the truth. When we tell the truth, we show respect for ourselves and other people. Lying shows that we do not respect ourselves and other people. When we tell the truth, people trust us.

Catholics Believe

Collection at Mass

At Mass on Sunday there is a collection. The collection takes place at the beginning of the Liturgy of the Eucharist. The people of the parish share their money to help the Church.

Activity

Look at the pictures. What could the children say to be honest?

____

____

I Follow Jesus

When you are kind and fair, you treat people with respect. You love people as Jesus taught. When you tell the truth, you are a disciple of Jesus.

Activity

Fill in the blanks in the story.

I ask a friend, "May I please borrow your markers? ______________________________

I will take good ______________________________

of them." When I am finished, I will return the markers

"______________________________"

and say, ______________________________.

My Faith Choice

Check (√) ways you will show respect for other people. I will

- ☐ tell the truth.
- ☐ share my things.
- ☐ say kind words.
- ☐ play safely.

Pray, "Thank you, God, for teaching me to show respect for other people. Amen."

Chapter Review

*Read each sentence. Circle **Yes** if the sentence is true. Circle **No** if it is not true.*

1. Respecting others is a way to show love.	Yes	No
2. Listening to one another shows respect.	Yes	No
3. Taking care of what belongs to others shows respect.	Yes	No
4. Telling lies shows respect.	Yes	No

TO HELP YOU REMEMBER

1. We are to treat ourselves and others as children of God.
2. We are to show respect for other people.
3. We are to tell the truth.

Lord, Hear Our Prayer

Pray together:

Leader God, you love us. For people who are hungry, we pray,

All **Lord, hear our prayer.**

Leader For people who are sick, we pray,

All **Lord, hear our prayer.**

Leader Everyone pray quietly for someone. *(Pause.)*

All **Lord, hear our prayer.**

With My Family

This Week . . .

In chapter 19, "We Love Others," your child learned:

- The Fourth through Tenth Commandments tell us to love and respect other people, ourselves, and all God's creation.
- The last seven of the Ten Commandments name the ways that we are to live the second part of the Great Commandment and truly live as children of God.
- Temperance is a virtue that helps us to know the difference between what we need and what we simply want to have.

For more about related teachings of the Church, see the *Catechism of the Catholic Church*, 2052–2074, and the *United States Catholic Catechism for Adults*, pages 375–455.

Sharing God's Word

Read together Acts of the Apostles 2:42–47. Emphasize that this story tells about the first Christians living the Commandments as Jesus taught. Name the things that the first Christians did to show how they lived the Ten Commandments.

We Live as Disciples

The Christian home and family is a school of discipleship. Choose one of the following activities to do as a family, or design a similar activity of your own.

- Read together a children's book about treating people with respect. Discuss why showing respect is at the heart of our love for others.
- Have each family member create two lists of their possessions. In one list, write things that you really need. In the second list, write things you like but do not truly need. Choose items you might give away to a charitable group.

Our Spiritual Journey

We are a pilgrim people. We make our earthly journey together. In this chapter, your child prayed for others. This is called a prayer of intercession. Intercessory prayer is a prayer that we offer on the behalf of others. Read and pray together the prayer on page 177.

For more ideas on ways your family can live as disciples of Jesus, visit **www.BeMyDisciples.com**

CHAPTER 20

We Live as a Community

Who do you know in your neighborhood?

Jesus taught us how to live together.

Jesus said, "Love God with all your heart, with all your soul, and with all your mind. This is the first and greatest commandment. The second is like this one: Love others as much as you love yourself." BASED ON MATTHEW 22:34–39

How do good neighbors and friends treat each other?

Disciple Power

Justice

We practice justice when we treat people fairly. People who are just live as Jesus taught.

The Church Follows **Jesus**

Saint Peter Claver

Read to Me

Peter Claver did what Jesus taught. He helped people who were being treated unjustly. They had been taken from their homes in Africa. They were being sold as slaves.

The guards tried to stop Peter from going on the ships, but he would not be stopped. God gave him strength. Peter was a very brave man.

The ship was hot and dirty. The people had no water to drink or to wash with. Many of them were very sick. Peter Claver cared for them. Many became followers of Jesus.

? How did Saint Peter Claver follow Jesus?

God's Family

A **community** is a group of people who respect and care for one another. People in a community help one another.

We all belong to the community of God's people. God makes each of us special. God blesses each of us with gifts and talents. We respect each other's gifts. God asks us to use our gifts to live as a community.

Faith Focus
What does it mean to live the Great Commandment?

Faith Words

community
A community is a group of people who respect and care for one another.

Great Commandment
The Great Commandment is to love God above all else and to love others as we love ourselves.

Activity

How are the children in the pictures sharing their gifts? Tell a partner.

Faith-Filled People

Saint Louis

Louis the Ninth was the ninth king of France. He wrote many good laws. These laws helped people to treat one another fairly.

Good Rules Help Us

Good rules help us to live together in a community. They help us to respect one another. The rules that a community makes are called laws.

Good laws help a community to live together in peace. Good laws help people to have the things that they need to live. Good laws help us to live God's laws.

God's laws help us to know right and wrong. They help us to make good choices. They help us to respect one another and to care for one another. God's laws help us to live as children of God.

Activity

Follow the road in the picture. Circle a favorite place. Tell one way that you can live as a child of God there.

The Great Commandment

God wants all people to love him with their whole hearts. He wants all people to love and respect others as they love themselves. We call this the **Great Commandment**. It is also called the Great Law of God.

The Great Commandment helps us to live as the community of God's people. Jesus showed us how to live the Great Commandment.

Jesus gave us the gift of the Church. The Church helps us to live the Great Commandment.

Catholics Believe

Religious Communities

Members of religious communities show us how to live the Great Commandment. These men and women make special promises or vows. Their vows help them to love God more than all else. Their promises help them to love people and themselves.

Activity

Circle **G** next to ways that you can show love for God. Circle **P** next to ways that you can show love for yourself and other people.

Living the Great Commandment

G	P	Pray.
G	P	Say kind words.
G	P	Act fairly.
G	P	Take part in Mass.
G	P	Forgive others.

I Follow Jesus

The Holy Spirit helps you to live the Great Commandment. He helps you to love God above all else. He helps you to treat others with justice and with respect.

Activity

Living God's Laws

Finish each sentence. Write what you can do to live as a good member of your family, school, or parish.

1. I can share my ______________________.

2. I can help by ______________________.

3. I can pray for ______________________.

My Faith Choice

Look at what you wrote in the activity. Which one will you do this week? I will

______________________.

Pray, "Thank you, Jesus, for teaching me the Great Commandment. Thank you, Holy Spirit, for helping me to live the Great Commandment. Amen."

Chapter Review

Find and circle the three words hidden in the puzzle. Share with a partner what each word tells about the Great Commandment.

GOD **LOVE** **PEOPLE**

L	H	L	O	V	E	T	Y
Q	L	P	G	O	D	M	U
P	E	O	P	L	E	B	D

TO HELP YOU REMEMBER

1. The Great Commandment teaches us that we are to love God and to love other people as we love ourselves.
2. The Great Commandment helps us to follow Jesus.
3. The Great Commandment helps us to live as good members of our community.

Teach Me, Lord

The Bible has many prayers. This prayer is part of a psalm. Learn the words of this prayer by heart. Pray them each day. Ask God to help you live the Great Commandment.

Lord God, teach me your ways.
You are my God and Savior.

BASED ON PSALM 25:4–5

With My Family

This Week . . .

In chapter 20, "We Live as a Community," your child learned:

- Communities make laws to help people live together in peace. Good laws help us to live God's laws. God gives us laws to help us show our love for God, for ourselves, and for other people.
- The Great Commandment is the summary of all God's laws. The Church helps us to live God's Law and the good laws that communities make.
- The virtue of justice helps us to treat everyone fairly. When we treat people fairly, we live the Great Commandment.

For more about related teachings of the Church, see the *Catechism of the Catholic Church*, 1877–1942, 1949–1974, and 2234–2246, and the *United States Catholic Catechism for Adults*, pages 307–309.

Sharing God's Word

Read Matthew 22: 34-40 about Jesus teaching the Great Commandment. Emphasize that the Great Commandment has two parts. We are to love God, and we are to love all people as we love ourselves.

We Live as Disciples

The Christian home and family is a school of discipleship. It is the first place where children should learn to live as disciples of Jesus. Choose one of the following activities to do as a family or design a similar activity of your own.

- Good rules help us to live together. Talk about your family's good rules and how they help you to live together.
- Choose an activity to do this week to live the Great Commandment.

Our Spiritual Journey

The Ten Commandments are written on the heart of each person. They guide us toward living as God created us to live. They are the pulse of living the righteous life described in the Bible—that is, of our living in "right order" with God, with other people, and with all of creation. This week, pray as a family the psalm verse on page 185.

For more ideas on ways your family can live as disciples of Jesus, visit **www.BeMyDisciples.com**

Unit 5 Review

Name ______________________

A. Choose the Best Word

Complete the sentences. Color the circle next to the best choice.

1. The Great __________ is to love God with our whole heart and to love others as we love ourselves.

○ Commandment ○ Prayer

2. We show people __________ when we treat them as children of God.

○ respect ○ fear

3. The first __________ gathered together and showed how much they loved God and one another.

○ family ○ Christians

4. The Great Commandment helps us to live as good members of our __________.

○ community ○ team

5. The Ten Commandments teach us to __________ God.

○ respect ○ disobey

B. Show What You Know

Match the two columns. Draw a line from the words in column A to their meanings in column B.

Column A	Column B
1. community	**a.** Give praise and honor to God
2. worship	**b.** To treat people with great respect
3. honor	**c.** People who care for one another

C. Connect with Scripture

What was your favorite story about Jesus in this unit? Draw something that happened in the story. Tell your class about it.

D. Be a Disciple

1. *What saint or holy person did you enjoy hearing about in this unit? Write the name here. Tell your class what this person did to follow Jesus.*

2. *What can you do to be a good disciple of Jesus?*

We Live UNIT 6
Part Two

The Good Shepherd

Shepherds take care of sheep. A good shepherd knows each sheep by name. The sheep come when they hear the shepherd's voice.

One day Jesus said to his friends, "I am the Good Shepherd. I know you all by name. I care for you. I love you so much I will give up my life for you. You are mine. You belong to me."

BASED ON JOHN 10:2–14

What I Have Learned

What is something you already know about these faith words?

Heaven

parables

Faith Words to Know

Put an **X** next to the faith words you know. Put a **?** next to the faith words you need to learn more about.

Faith Words

____ Kingdom of God	____ glory	____ sin
____ joy	____ charity	____ peace

A Question I Have

What question would you like to ask about saints?

CHAPTER 21

Jesus and the Children

How do people show that they are friends?

Jesus invites everyone to be his friends. Listen to what the Bible tells us:

Jesus said, "I call you my friends. I taught you everything that I learned from God, my Father." BASED ON JOHN 15:15

How do we know that Jesus wants us to be his friends?

Disciple Power

Joy

We live with joy when we recognize that happiness does not come from money or possessions. True happiness comes from knowing and following Jesus.

The Church Follows **Jesus**

Read to Me

Children Helping Children

One cold day, a ten-year-old boy from France named Charles was out walking. He came upon another boy who was selling roasted chestnuts. Charles noticed that the boy was not wearing shoes. Charles took off his own shoes and gave them to the boy.

Charles grew up and became a priest and then a bishop. In 1839, Bishop Charles traveled from France to the United States and saw many poor children. Later, he asked children in France to help poor children in the United States. This was the beginning of the Holy Childhood Association.

Today children all over the world pray and help children in need.

? What are some ways that you can help other children?

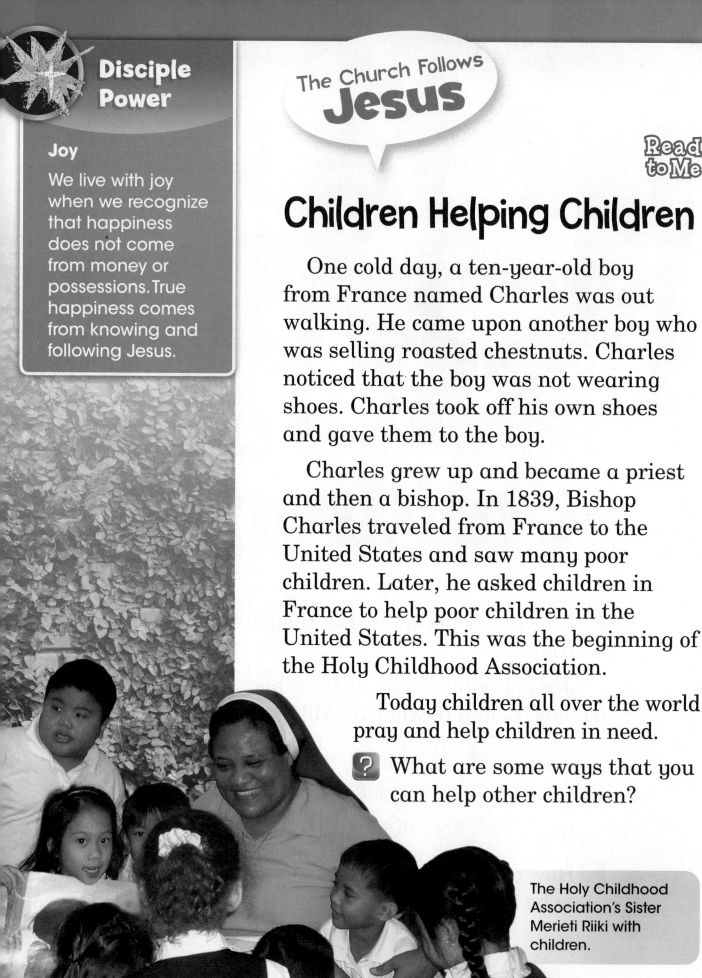

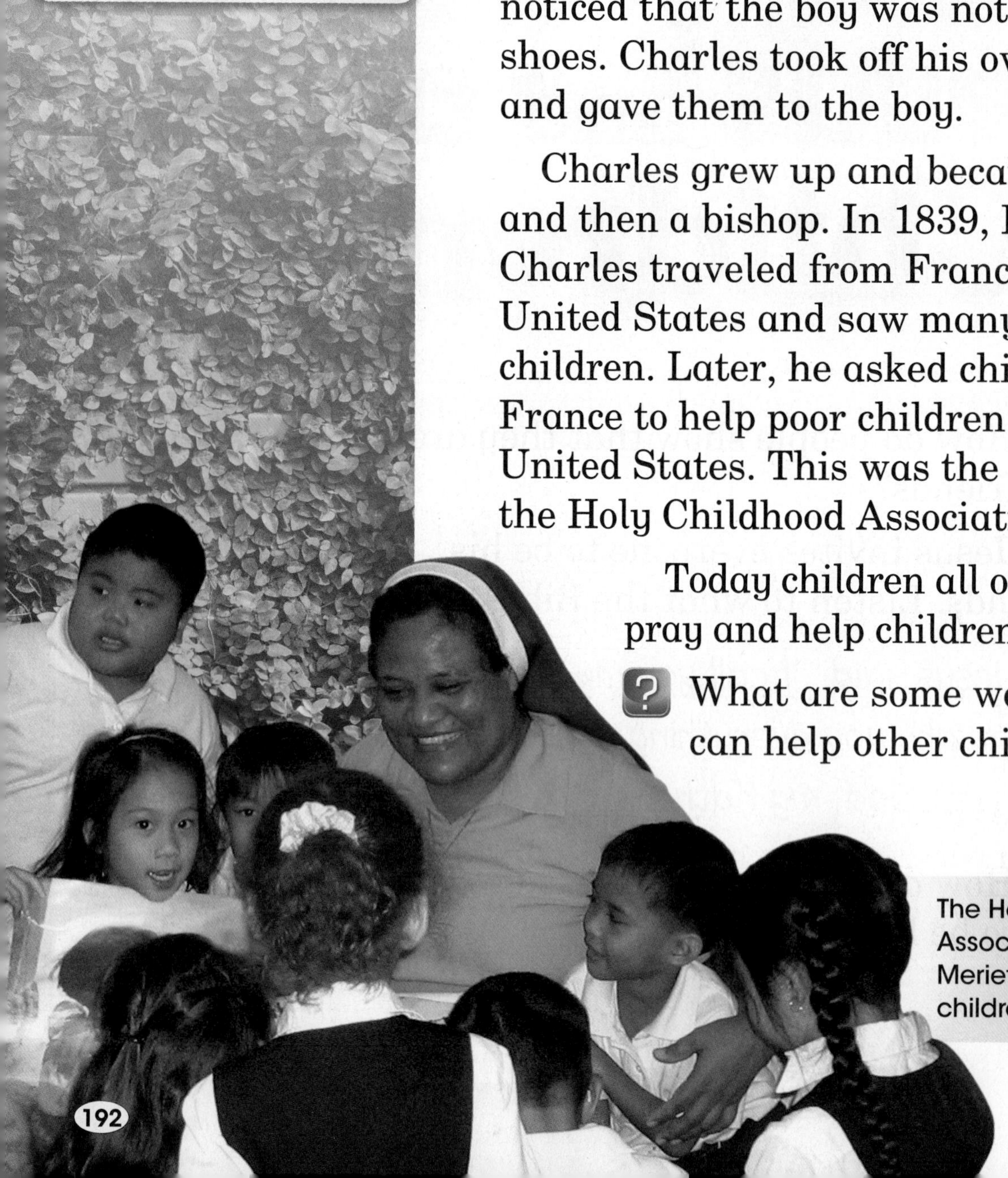

The Holy Childhood Association's Sister Merieti Riiki with children.

Children of God

Jesus showed that every person is special. He showed that God loves everyone. Jesus treated everyone as a child of God.

Jesus cured people who were sick. He was kind to people whom others did not like. Jesus forgave people who sinned. He loved those who wanted to hurt him.

Faith Focus
Why are all children so special to Jesus?

Faith Word
Kingdom of God
The Kingdom of God is Heaven. Heaven is happiness with God forever.

Activity

Look at the pictures of Jesus. With a partner, talk about what Jesus is doing or saying. How is he treating people as children of God?

"Love your enemies and do good to them."

Faith-Filled People

Saint Nicholas

Saint Nicholas tried to help children without being noticed. Many children leave their shoes out on December 6, the day that we celebrate the feast of Saint Nicholas. They hope that Saint Nicholas will fill their shoes with treats.

Jesus Welcomes Children

Here is a story from the Bible. It tells about Jesus inviting children to come to him.

People brought their children to Jesus.
But the disciples told them to go away.
Jesus said,
"Let the children come to me."
Then he blessed the children.

Based on Mark 10:13–14,16

Jesus invites all people to come to him. All people are invited to live in the **Kingdom of God**. The Kingdom of God is Heaven. Heaven is happiness with God forever.

Activity

Draw yourself in the picture.
Ask Jesus to bless you.

We Are Children of God

In the Bible story, Jesus taught that all children are special to God. Some children have big, bright eyes. Others have a happy smile. Some are very quiet. Others talk all the time. All children are very different. Our differences show how special we are.

We treat all people as children of God. We do our best to live as children of God. We trust and love God with our whole heart.

? Why are all the children in the pictures special to God? Tell a partner.

Catholics Believe

Children's Choir

Children help out in their parishes. One way they help is by singing in the children's choir. This shows that children have a special work to do in the Church.

I Follow Jesus

Jesus loves all children. Jesus loves you. The Holy Spirit helps you to share Jesus' love for others. True happiness comes from living as a child of God.

Activity

Use words and pictures to make an "I Care" button.

My Faith Choice

Underline one way that you will treat others as children of God. This week I will

1. invite a classmate to play with me.
2. tell my family I love them.
3. help out at home.

Pray, "Thank you, Jesus, for teaching me to treat others as children of God."

Chapter Review

Use one of these words to fill in the missing word in each sentence.

everyone	Heaven	invites

1. God loves ____________.
2. The Kingdom of God is ____________.
3. Jesus ____________ everyone to follow him.

TO HELP YOU REMEMBER

1. God loves all people.
2. God wants all people to come to him.
3. God wants us to live in Heaven.

Let the Children Come

Our imaginations can help us talk to Jesus and listen to him.

1. Sit quietly in a comfortable position.
2. Imagine that you are going with your family to see Jesus.
3. Imagine that you are talking and listening to Jesus.
4. Spend a minute quietly listening to what Jesus might be saying to you.

With My Family

This Week . . .

In chapter 21, "Jesus and the Children," your child learned:

- God loves all people and wants them to live with him forever in Heaven.
- Children of God share God's love with one another.
- We live with joy when we recognize that happiness does not come from money or possessions. True happiness comes from knowing and following Jesus.

For more about related teachings of the Church, see the *Catechism of the Catholic Church*, 541–550 and 2816–2821; and the *United States Catholic Catechism for Adults*, pages 67, 68, and 310.

Sharing God's Word

Read together the Bible story in Mark 10:13–16 about Jesus blessing the children, or you can read the adaptation of the story on page 194. Emphasize that Jesus invited the children to come to him and blessed them.

We Live as Disciples

The Christian home and family is a school of discipleship. Choose one of the following activities to do as a family, or design a similar activity of your own.

- Jesus welcomed everyone. He showed people that God loves them. As a family, do one thing that will show people that God loves them.
- Discuss the ways in which your parish welcomes children. Name activities, events, and opportunities that are available for children in your parish. Make an effort to participate in one of them.

Our Spiritual Journey

Your child prayed a prayer of meditation in this chapter. This kind of prayer is also sometimes called *guided imagery*. Talk with your child about how our imaginations can help us to pray and to be with Jesus. It can help us talk and listen to Jesus. Provide a time and space for quiet prayer in your family. Visit a church together outside of Mass time and spend a few moments in quiet meditation.

For more ideas on how your family can live as disciples of Jesus, visit **www.BeMyDisciples.com**

CHAPTER 22

We Are Children of God

What are some ways that people are different from one another?

All people are unique. No two people are exactly alike. People have different personalities, and special talents. We have different skin, hair, and eye colors. We even speak different languages. The Bible, however, tells us that all people are the same in one important way.

God created people in his image.

BASED ON GENESIS 1:27

How are all people the same?

Disciple Power

Gentleness

Gentle people act calmly. They avoid actions that might lead others to anger or feeling hurt. They treat all people as children of God.

The Church Follows **Jesus**

The Sisters of the Blessed Sacrament

Read to Me

Katharine Drexel cared for all people. She treated everyone as a child of God.

Saint Katharine began the Sisters of the Blessed Sacrament. They work with African Americans and Native Americans. They work in schools and colleges. They work in cities and on the lands where Native Americans live.

The Sisters of the Blessed Sacrament treat all people as children of God. They teach others to treat all people with respect, fairness, and gentleness.

? What are some of the ways that you see people treating one another as children of God?

Children of God

God created all people out of love. God created people in his image and likeness. God created all people to know, love, and serve him.

All people are part of God's family. We are part of God's family. We are **children of God**. Children of God love God and love one another. Our kind words and actions show others that we love them and care about them.

Faith Focus
What does it mean to be a child of God?

Faith Words

children of God
All people are children of God. God created all people in his image.

glory
Glory is another word for praise.

Activity

Work with a partner to show what the word gentle means. Act it out without any words.

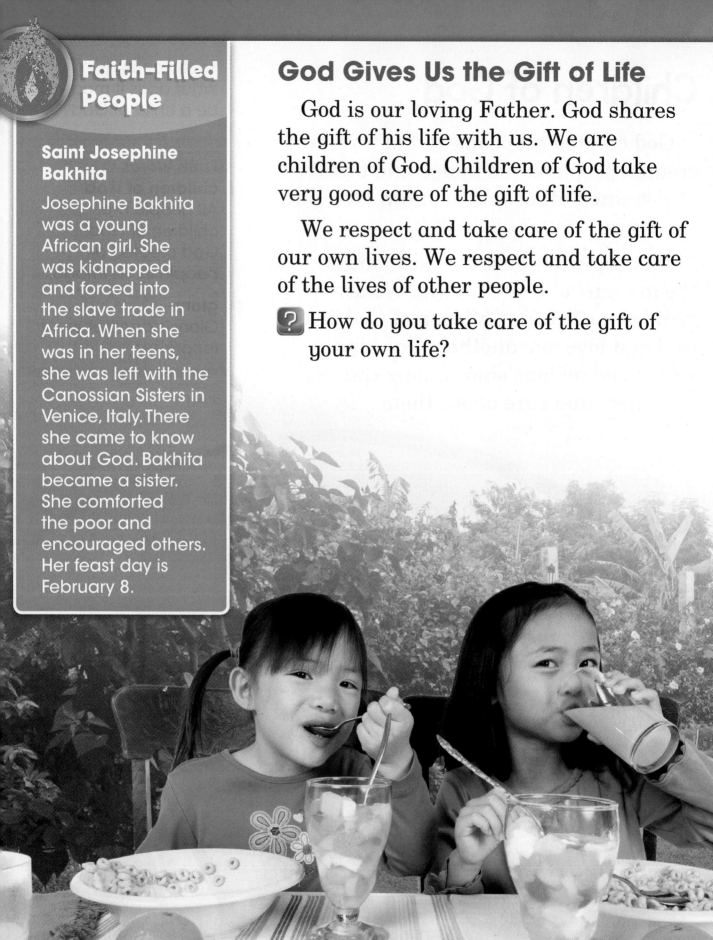

Faith-Filled People

Saint Josephine Bakhita

Josephine Bakhita was a young African girl. She was kidnapped and forced into the slave trade in Africa. When she was in her teens, she was left with the Canossian Sisters in Venice, Italy. There she came to know about God. Bakhita became a sister. She comforted the poor and encouraged others. Her feast day is February 8.

God Gives Us the Gift of Life

God is our loving Father. God shares the gift of his life with us. We are children of God. Children of God take very good care of the gift of life.

We respect and take care of the gift of our own lives. We respect and take care of the lives of other people.

? How do you take care of the gift of your own life?

We Show Our Love for God

God created us to know and to love him. Jesus taught about God's love. He showed people God's love with his actions. He spoke to people about God's love. Jesus taught that God wants us to be happy with him now and forever in Heaven.

Jesus showed us how to love God. We show our love for God when we help other people. We show our love for God when we pray. When we take care of creation, we are showing our love for God.

We give **glory** to God when we do these things. Children of God are to give glory to God in all they do and say.

Catholics Believe

Blessed Sacrament

The Blessed Sacrament is another name for the Eucharist. We keep the Blessed Sacrament in the tabernacle. We bring the Blessed Sacrament to people who are sick so that they can receive Holy Communion.

Activity

Write a sentence that tells about one way you show your love for God.

____

____

I Follow Jesus

God created you and all people to be children of God. The Holy Spirit helps you to treat all people as children of God.

Activity

Being Gentle

Draw yourself acting in a gentle way. Share your work with a partner.

My Faith Choice

Check (√) how you will live as a child of God.
This week I will

- ☐ be kind.
- ☐ pray.
- ☐ help my family.
- ☐ care for God's creation.

Pray, "Thank you, Holy Spirit, for helping me to treat all people as children of God."

Chapter Review

Use this number code. Find out the important message about ourselves.

A	C	D	E	G	H	I	L	N	O	R	S	W
1	2	3	4	5	6	7	8	9	10	11	12	13

___ ___   ___ ___ ___
13 4   1 11 4

___ ___ ___   ___ ___ ___ ___
1 8 8   5 10 3 12

___ ___ ___ ___ ___ ___ ___ ___
3 6 7 8 3 11 4 9

TO HELP YOU REMEMBER

1. God created all people in his image.
2. God gives us the gift of life.
3. We are to take care of the gift of life.

The Glory Prayer

All Christian prayer gives glory to God. Learn the Glory Be Prayer by heart. Pray it each day in English or Spanish.

Glory be to the Father
and to the Son
and to the Holy Spirit,
as it was in the beginning
is now, and ever shall be
world without end. Amen.

Here is the prayer in Spanish.

Gloria al Padre, al Hijo y al Espíritu Santo.
Como era en el principio, ahora y siempre,
por los siglos de los siglos. Amén.

With My Family

This Week . . .

In chapter 22, "We Are Children of God," your child learned:

- God created all people in his image and likeness. God created all people out of his infinite love.
- God calls all people to be responsible stewards of the gift of life. We are called to show our love for God, especially in the way that we treat other people.
- We are to care for and treat our own lives and the lives of all people with gentleness.

For more about related teachings of the Church, see the *Catechism of the Catholic Church*, 355–361 and 1699–1709; and the *United States Catholic Catechism for Adults*, pages 67–68.

Sharing God's Word

Read together 1 John 3:1. Emphasize that in Baptism we are joined to Jesus and become adopted children of God. We are to live as Jesus taught.

We Live as Disciples

The Christian home and family is a school of discipleship. Choose one of the following activities to do as a family, or design a similar activity of your own.

- All people have the dignity of being children of God. Children of God are to love God and one another. Talk together about how your family can live as children of God. What kinds of words and actions show others that we love and care about them?
- Look through a children's magazine or picture book with your child. Point out all the pictures that show people living as children of God.

Our Spiritual Journey

Giving glory and praise to God is so important that we are reminded to glorify God as we are sent forth at the end of Mass. In the Concluding Rites, the priest blesses us in the name of the Father, and of the Son, and of the Holy Spirit. We are then sent forth to do good works, praising and blessing the Lord with these words, "Go in peace, glorifying the Lord by your life." As a family, choose one thing you can do this week to glorify the Lord. Also, help your child to memorize the Glory Be Prayer on page 205 and pray it daily together.

For more ideas on ways your family can live as disciples of Jesus, visit **www.BeMyDisciples.com**

CHAPTER 23

Jesus Teaches about Love

Which stories can you think of that help you make good choices?

Jesus sometimes told stories to teach us how to live as his disciples.

Jesus asked, "Who was the good neighbor in the story?" Someone replied, "The traveler who helped the man lying on the road." Jesus said, "You are right. Now, you treat other people the same way." BASED ON LUKE 10:36–37

What do these words from the Bible tell you about Jesus?

Disciple Power

Charity

Charity is loving others as God loves us. We practice charity when we love our neighbor as Jesus taught us to.

The Church Follows Jesus

Saint Frances Cabrini

Read to Me

Frances Cabrini always kept trying. People told her she could not be a nun, but she did.

Frances and the other nuns did good work in the United States. They built homes and schools for children. They built hospitals for sick people. They built convents where other women learned how to serve God.

Sometimes life was hard for Frances, but she prayed to God and worked harder. All her life, Frances loved others as Jesus taught.

? How were Saint Frances and the other women good neighbors?

Jesus Teaches Love

Jesus' disciples called him "Teacher." In Jesus' times, this was a great honor and a sign of respect. As other teachers did, Jesus often used stories to teach.

One kind of story Jesus told is called a **parable**. In a parable, the teacher compares two things. The teacher uses one thing that his listeners know well to help them understand the main point of the story.

The parables that Jesus told helped his listeners to know and love God better. These parables also tell us how much God loves us. These parables show us how to live as good neighbors and children of God.

Faith Focus
What does the story of the Good Samaritan teach us?

Faith Word
parable
A parable is a story that compares two things. Jesus told parables to help people to know and love God better.

Activity

What words could you use to describe Jesus?

____

____

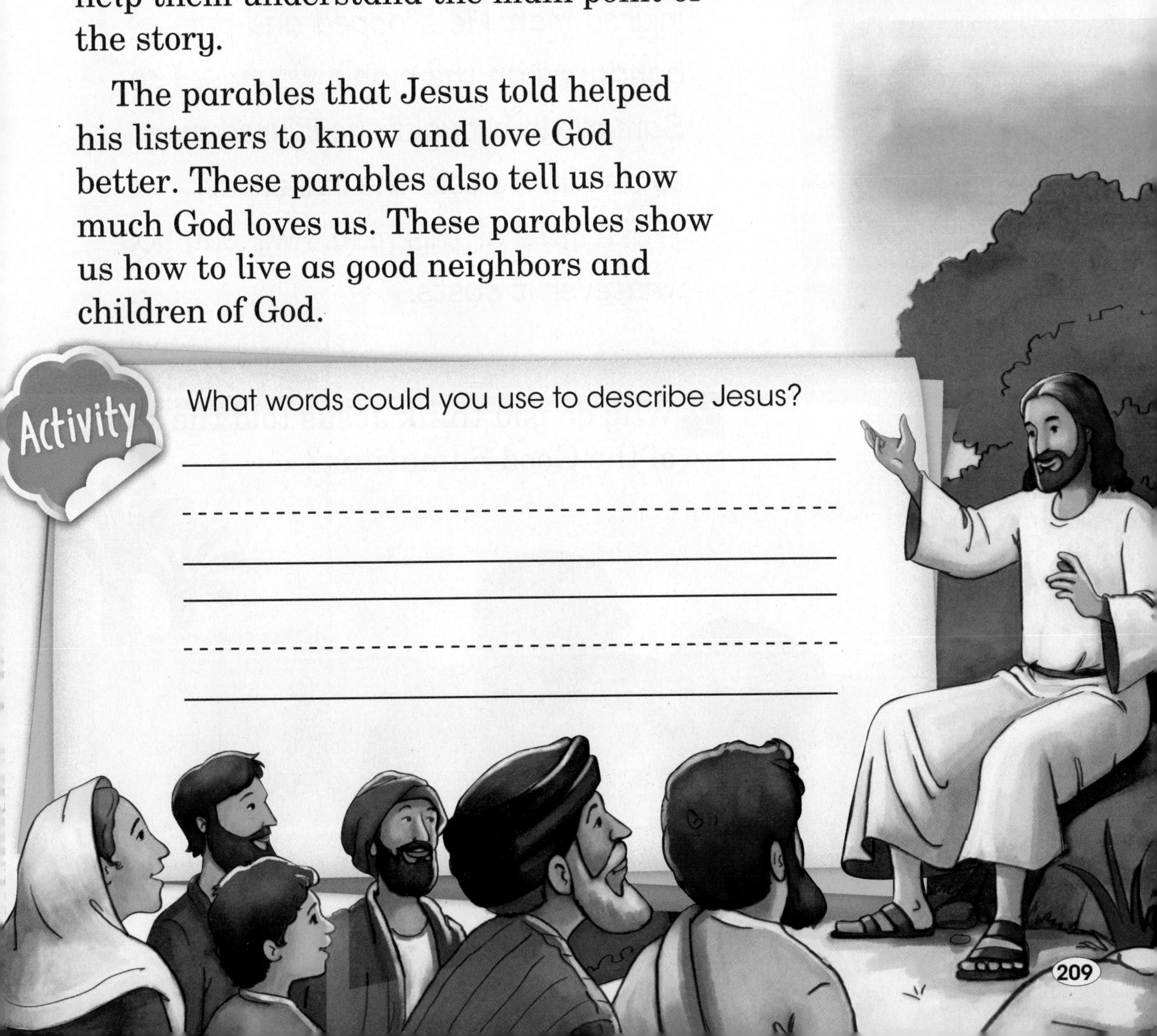

Faith-Filled People

Saint Isidore the Farmer

Isidore spent much of his life working on a farm in Spain. He and his wife, also a saint, showed their love for God by being kind to their neighbors. Although they were poor, Isidore and Maria shared their food with those poorer than they were. Isidore is the patron saint of farmers and migrant workers. The Church celebrates his feast day on May 15.

The Good Samaritan

The stories that Jesus told are in the Gospels. The Gospels are in the New Testament. Here is one story that Jesus told. He said,

One day robbers attacked a man on a road. They hurt the man and left him lying on the road.

A traveler from Samaria saw the injured man. He stopped and put bandages on the man's wounds. The Samaritan brought the injured man to an inn. He told the innkeeper, "Take care of this man. I will pay you whatever it costs."

BASED ON LUKE 10:30, 33–35

? Why do you think Jesus told the story of the Good Samaritan?

A Good Neighbor

The story of the Good Samaritan helps us to live as followers of Jesus. It teaches that God wants us to help one another. God wants us to help people even when we do not feel like helping. This story teaches us to be good neighbors to one another.

Catholics Believe

Prayer of the Faithful

We are good neighbors when we pray for one another. Each Sunday at Mass, we pray the Prayer of the Faithful. In this prayer we pray for the Church, people who are sick, and people who have died.

Activity

Finish the picture story. Draw or write about how the children can act as good Samaritans.

1.

2.

3.

I Follow Jesus

You can be a good neighbor. You can show people how much God loves them and cares about them.

Activity

Living as a Good Neighbor

Color a ☺ next to two ways that you can help someone this week as Jesus taught, and then write one other way that you can help.

☺ Say kind words to someone who is sad.

☺ Help to fold laundry at home.

☺ Give a get-well card to someone.

☺ ______________________________

My Faith Choice

I will do one of the things in the activity above. I will ______________________________.

Pray, "Thank you, Jesus, for teaching me to be a good neighbor."

Chapter Review

*Read each sentence. Circle **Yes** if the sentence is true. Circle **No** if it is not true.*

1. Jesus told stories called parables.

   Yes No

2. The Good Samaritan took care of the injured man.

   Yes No

3. Jesus told stories to teach us to help others.

   Yes No

TO HELP YOU REMEMBER

1. Jesus told the parable of the Good Samaritan to help us to live as his followers.
2. God wants us to care for others.
3. We show charity when we love our neighbors.

We Pray for Others

We pray the Prayer of the Faithful at Mass. We pray for other people.

Leader Dear God, help us show love. For the pope and Church leaders,

All **Lord, hear our prayer.**

Leader For our country's leaders,

All **Lord, hear our prayer.**

Leader Think of the people you wish to pray for. *(Pause.)*

All **Lord, hear our prayer.**

With My Family

This Week . . .

In chapter 23, "Jesus Teaches about Love," your child learned:

- Parables in the Bible help us come to know, love, and serve God.
- The parable of the Good Samaritan teaches us how we are to live as disciples of Jesus.
- We are to care about one another and to show our love by our actions as Jesus did.
- We practice charity when we love our neighbor as Jesus has taught us.

For more about related teachings of the Church, see the *Catechism of the Catholic Church*, 546; and the *United States Catholic Catechism for Adults*, pages 27–31, 79–80.

Sharing God's Word

Read together the parable of the Good Samaritan in Luke 10:29–37, or you can read the adaptation of the parable on page 210. Emphasize that the Samaritan was a good neighbor because he stopped and took the time to help the injured man.

We Live as Disciples

The Christian home and family is a school of discipleship. Choose one of the following activities to do as a family, or design a similar activity of your own.

- This week when you take part in the celebration of Mass, help your child pray the Prayer of the Faithful. After Mass, talk about the petitions that were used in the prayer.
- Talk about how your family can be good neighbors and show charity to one another this week. For example, help one another out without having to be asked.

Our Spiritual Journey

A prayer of the faithful is a prayer of intercession. Intercessory prayer is one of the Church's five main forms of prayer. In this chapter, your child prayed a prayer of the faithful. As the community of the faithful, we pray the Prayer of the Faithful at Mass or during the Liturgy of the Hours. Because these are the prayer intentions of the community, not individual people, appropriate subjects for the prayer have a communal nature. Subjects may include the Church and her ministers, civil leaders, the world and its people, those who are sick or dying, those who have died, those who are grieving, and anyone celebrating a sacrament. Read and pray together the prayer on page 213.

For more ideas on ways your family can live as disciples of Jesus, visit **www.BeMyDisciples.com**

CHAPTER 24

The Our Father

Who has helped you to learn something new?

Many people help us to learn new things. Jesus taught his disciples how to pray.

Jesus said, "Pray to God privately. God will see you and reward you. Speak from your heart and God will hear you. God knows what you need."

BASED ON MATTHEW 6:6–8

What do these words from the Bible tell you about prayer?

Disciple Power

Humility

Humility helps us know that all good things come from God.

The Church Follows Jesus

Blessed Teresa

Read to Me

Mother Teresa of Calcutta once said, "Prayer brings our heart closer to God. If our heart is close to God we can do very much."

Mother Teresa was very humble. She knew that praying often every day helped her to care for other people. Through a life of prayer and caring for others, she showed that God is everyone's Father.

Mother Teresa took care of people who had no one else. These people were very sick and very poor. They had no place to live. Mother Teresa fed them. She washed them.

The Church honors Mother Teresa as Blessed Teresa of Calcutta.

? What can you do to show that God is everyone's Father?

Jesus Prayed

Jesus prayed often. He talked to God his Father about everything. He listened to God the Father. He always did what his Father wanted him to do.

The followers of Jesus were with Jesus when he prayed. They saw him pray. They wanted to learn to pray as he prayed.

Faith Focus
What prayer did Jesus teach us to pray?

Faith Word
Our Father
The Our Father is the prayer Jesus taught his disciples.

Activity

In each picture frame draw a picture of someone who has helped you to pray. Write the person's name under the picture.

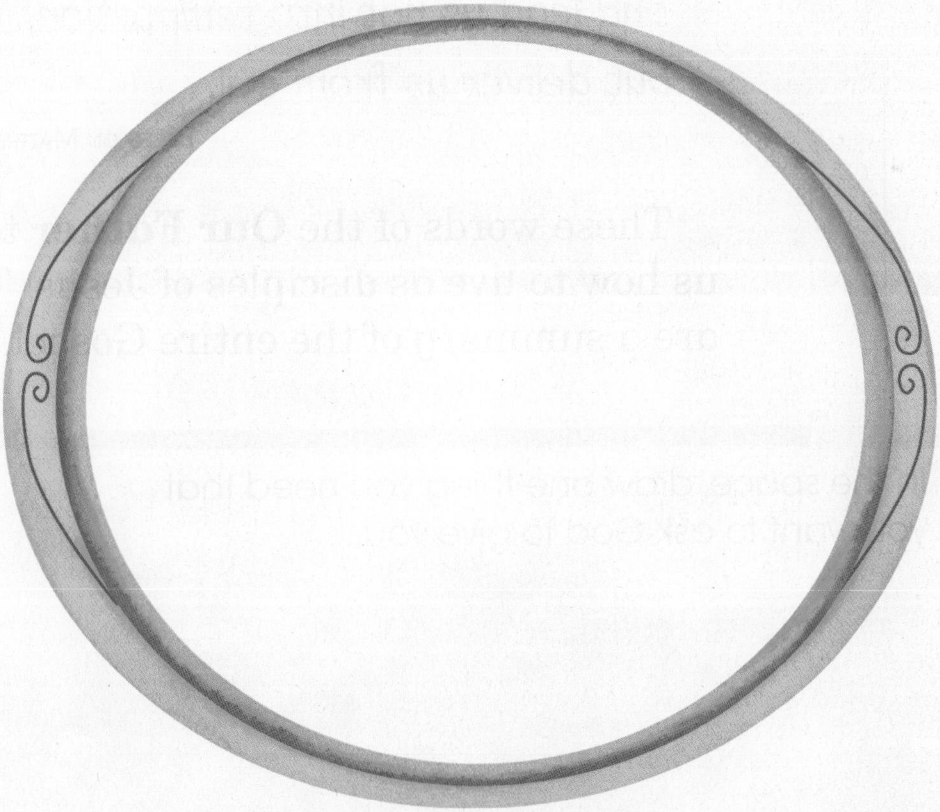

Faith-Filled People

Sister Thea Bowman

Sister Thea had the gift of singing. Singing was one way Sister Thea prayed. Everywhere she went, Sister Thea sang about God's love for everyone. She praised God in everything she did. She lived the words of the Our Father every day.

The Our Father

One day one of the disciples asked Jesus to teach them to pray. Jesus said,

"This is how you are to pray.
Our Father in heaven,
hallowed be your name,
your kingdom come,
your will be done,
on earth as in heaven.
Give us today our daily bread;
and forgive us our debts, as we forgive our debtors;
and lead us not into temptation,
but deliver us from evil."

BASED ON MATTHEW 6:9–13

These words of the **Our Father** teach us how to live as disciples of Jesus. They are a summary of the entire Gospel.

Activity

In the space, draw one thing you need that you want to ask God to give you.

Jesus Teaches Us to Pray

When we pray the Our Father, we tell God that we believe he is our Father. We honor the name of God. We trust him with all our hearts.

We ask God the Father to help us to live as his children. We ask for forgiveness. We tell God that we forgive those who hurt us. We ask him to help us to do what is good. We pray that we will live with him forever in Heaven.

Catholics Believe

The Lord's Prayer

The Lord's Prayer is another name for the Our Father. This is because Jesus our Lord gave us this prayer.

Activity

The Lord's Prayer

The children in the picture are bringing the Lord's Prayer to life. In the space draw what you could do to live this prayer.

I Follow Jesus

When you pray the Our Father, you show that you trust God. You show that everything good comes from God. You show that you believe that everyone is a child of God.

Activity

Check (√) where you can pray the Our Father.

- ☐ At Mass
- ☐ At Home
- ☐ On the school bus
- ☐ In the car
- ☐ In the park

My Faith Choice

This week I will choose to live as a child of God. I will bring the Lord's Prayer to life. I will

____

____

____.

Pray, "Thank you, Jesus, for teaching me to pray the Our Father."

Chapter Review

Find and circle the words in the puzzle. Use each word in a sentence. Tell a partner.

Jesus **forgive** **Father** **prayer**

F	O	R	G	I	V	E	T	P
M	C	J	E	S	U	S	W	Z
O	P	R	A	Y	E	R	K	H
L	P	R	F	A	T	H	E	R

TO HELP YOU REMEMBER

1. Jesus taught us to pray the Our Father.
2. The Our Father is a prayer for all God's children.
3. The Our Father is also called the Lord's Prayer.

The Our Father

Every day Christians all around the world pray the Our Father.

Leader Let us pray the Our Father together.
Our Father, who art in heaven,
hallowed be thy name;

All **thy kingdom come;**
thy will be done on earth
as it is in heaven.

Leader Give us this day our daily bread;
and forgive us our trespasses

All **as we forgive those who trespass**
against us;
and lead us not into temptation,
but deliver us from
evil. Amen.

With My Family

This Week . . .

In chapter 24, "The Our Father," your child learned that:

- Jesus gave the Our Father to his first disciples.
- Jesus gave this wonderful prayer to all Christians of all times.
- Praying the Our Father teaches us to pray. It is a summary of the entire message of the Gospel.
- Humility is a virtue that reminds us of our right place before God. It helps us know that all we have is a gift from God.

For more about related teachings of the Church, see the *Catechism of the Catholic Church*, 2759–2856; and the *United States Catholic Catechism for Adults*, pages 483–492.

Sharing God's Word

Read Matthew 6:9–13 together, the account of Jesus teaching the disciples to pray the Our Father. Or read the adaptation of the story on page 218. Emphasize that praying the Our Father honors God the Father and shows our trust in him.

We Live as Disciples

The Christian home and family is a school of discipleship. Choose one of the following activities to do as a family, or design a similar activity of your own.

- Practice saying the words of the Our Father with your child. When you take part in the celebration of Mass this week, help your child join in praying the Our Father.
- Use the Our Father as your mealtime prayer this week. Remember that the Our Father is the prayer of all God's children. Christians pray the Our Father every day all around the world.

Our Spiritual Journey

Saint Augustine called the Our Father the summary of the Gospel. Pray the Our Father as a prayer of meditation. Praying and living by the Our Father will create in you a pure and humble heart—a heart that keeps God and his love for you at the heart, or center, of your life. Make sure that your children know this prayer by heart.

For more ideas on how your family can live as disciples of Jesus, visit **www.BeMyDisciples.com**

Unit 6 Review

Name ______________________________

A. Choose the Best Word

Complete the sentences. Color the circle next to the best choice.

1. The Good Samaritan story teaches us that God wants us to ____________ one another.

○ respect ○ care for

2. God made all people ____________.

○ in his image ○ happy

3. God wants us to ____________ Heaven.

○ live in ○ remember

4. Mother Teresa was very ____________.

○ proud ○ humble

5. Jesus taught the disciples to pray the ____________.

○ Hail Mary ○ Our Father

B. Show What You Know

Circle the numbers next to the words that tell about the Bible story of the Good Neighbor.

1. parable

2. a man from Samaria

3. a camel

4. an innkeeper

5. care for one another

C. Connect with Scripture

What was your favorite story about Jesus in this unit? Draw something that happened in the story. Tell your class about it.

D. Be a Disciple

1. *What saint or holy person did you enjoy hearing about in this unit? Write the name here. Tell your class what this person did to follow Jesus.*

2. *What can you do to be a good disciple of Jesus?*

The Year of Grace

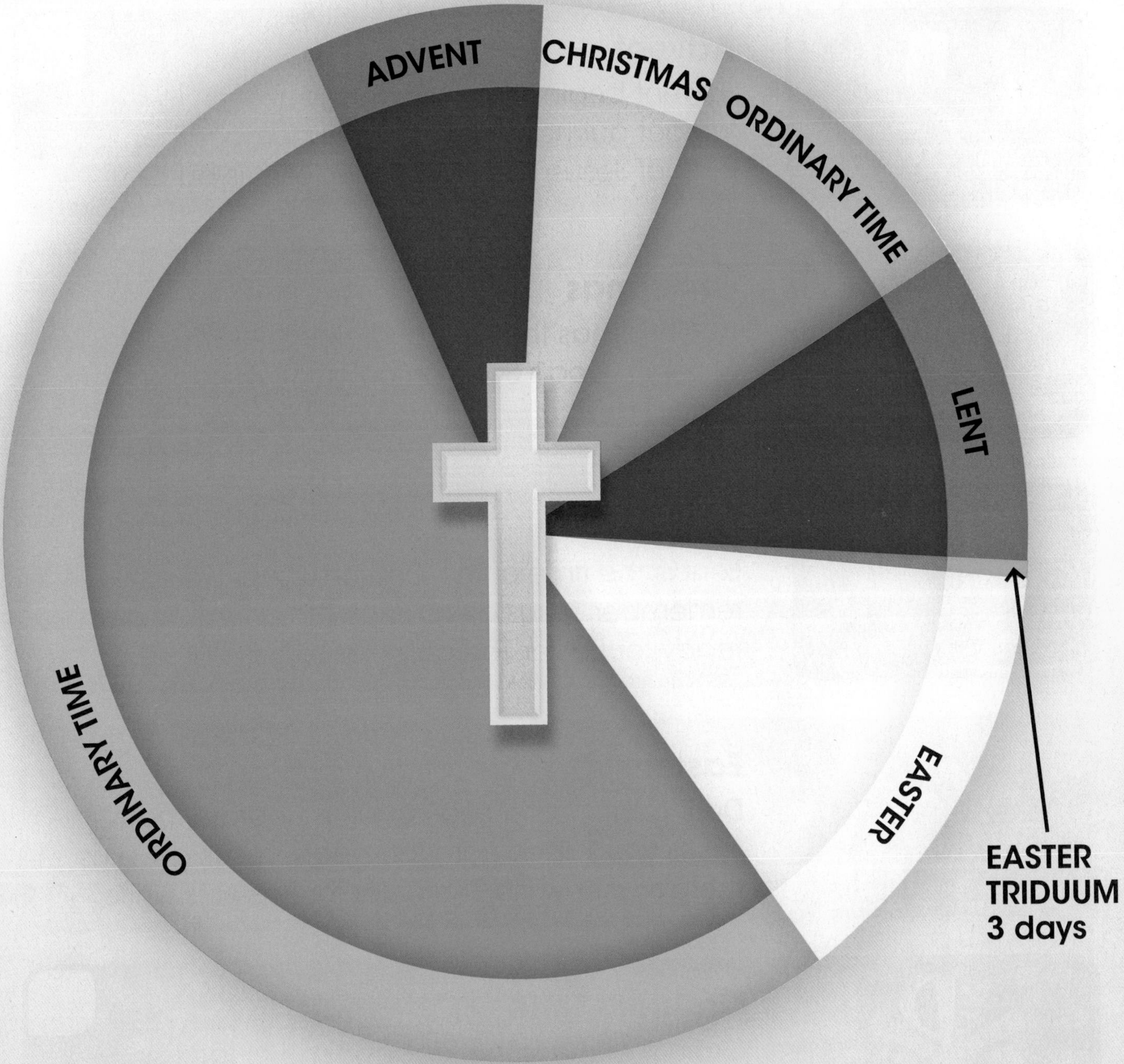

The Liturgical Year

The Church's year of prayer and worship is called the liturgical year.

Check (✓) your favorite season or time of the Church's year. Why is it your favorite?

Advent

Advent begins the Church's year. We get our hearts ready to remember the birth of Jesus. The color for Advent is purple.

Christmas

At Christmas the Church celebrates the birth of Jesus, God's Son. The color for Christmas is white.

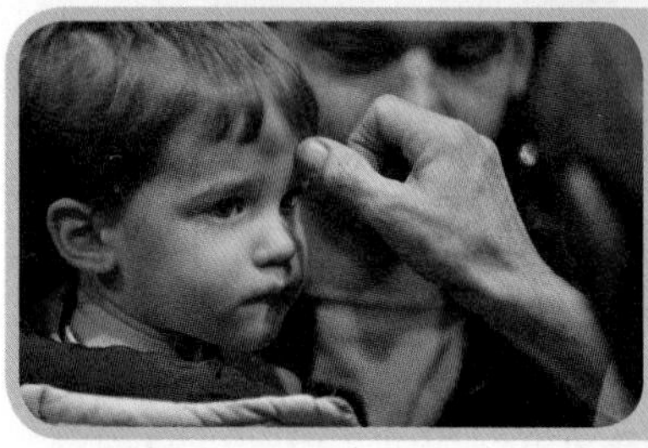

Lent

Lent is the time of the Church's year we remember Jesus died for us. It is a time to get ready for Easter. The color for Lent is purple.

Easter

During the Easter season we celebrate that Jesus was raised from the dead. Jesus gave us the gift of new life. The color for Easter is white.

Ordinary Time

Ordinary Time is the longest time of the Church's year. The color for Ordinary Time is green.

Faith Focus
Who are the saints?

Solemnity of All Saints

Saints are people who love God very much. They are holy people. They are members of our Church family who show us how to be good disciples of Jesus. Some saints are adults. Other saints are children. Saints come from all cultures and all nations. They live with Jesus in Heaven.

God wants each of us to become a saint. We pray to the saints to help us live as God's children. Mary, the mother of Jesus, is the greatest saint. We pray to Mary, too.

The saints hear our prayers and want us to be happy with God. The Church honors all saints on November 1 each year. This feast is the Solemnity of All Saints.

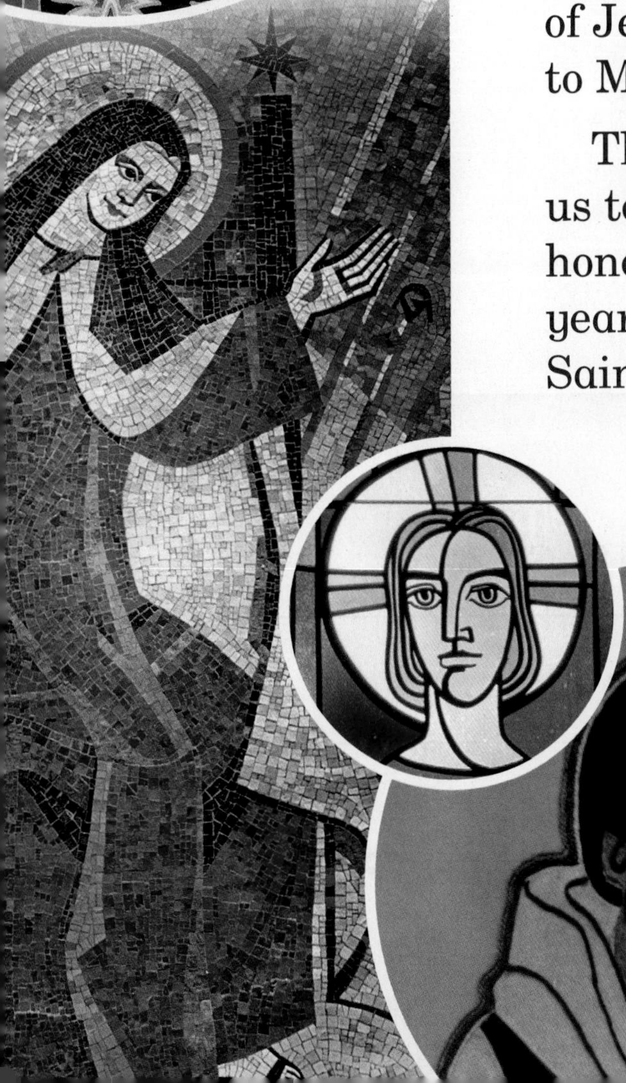

Mary, Saint Thérèse of Lisieux, Saint Andrew, and Saint Martin de Porres

The Greatest Saint

Draw a picture of Mary doing what God asks. Also draw a picture of yourself helping Mary.

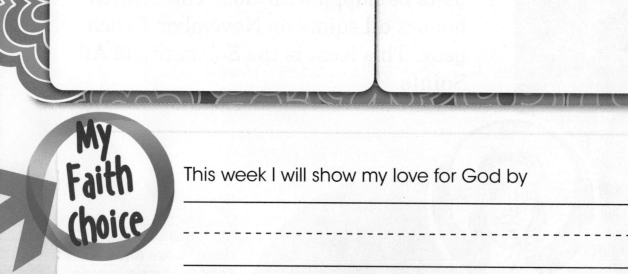

My Faith Choice

This week I will show my love for God by

____

____

____.

Pray, "Mary, help me to love God and follow Jesus. Amen."

Faith Focus
How does celebrating Advent help us prepare for Christmas?

Advent

The Church's season of Advent begins the Church's year. During Advent we prepare for Christmas. We light candles to chase away the winter darkness. These candles remind us that Jesus is the Light of the world.

Jesus asks us to be lights for the world too. During Advent we let our light shine. We help people. We make gifts. We do secret good deeds for each other.

We gather in church and prepare our hearts to welcome Jesus. We sing and pray together. We remember that Jesus is with us every day.

My Light Shines

Decide what you can do to get ready for Christmas. Color in the flames to show what you can do.

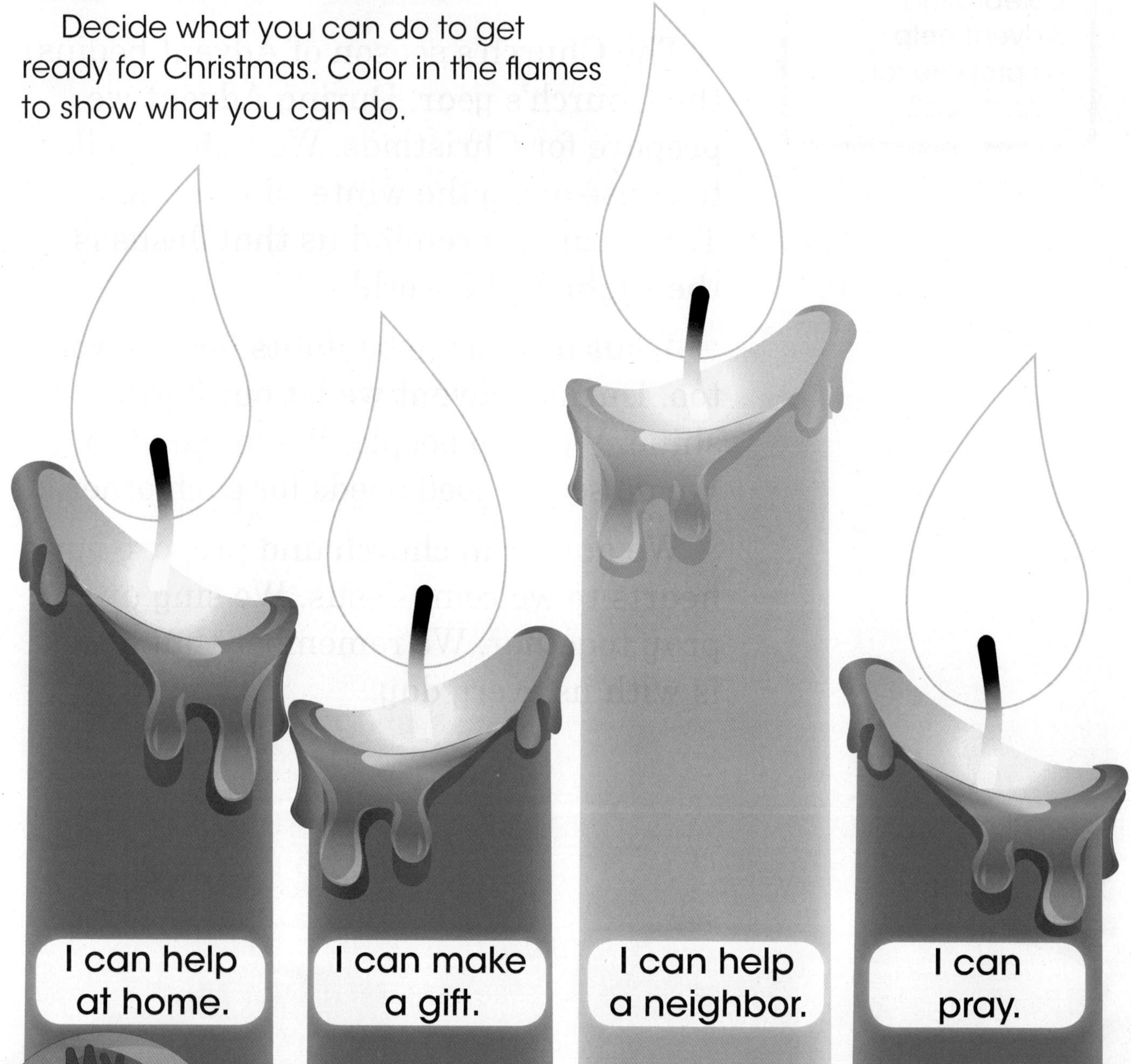
I can help at home.
I can make a gift.
I can help a neighbor.
I can pray.

My Faith Choice

During Advent I will share the light of Jesus. I will

____

____.

Pray, "Jesus, you are the Light of the world! Amen."

Faith Focus

Whom do we honor every year on December 8th?

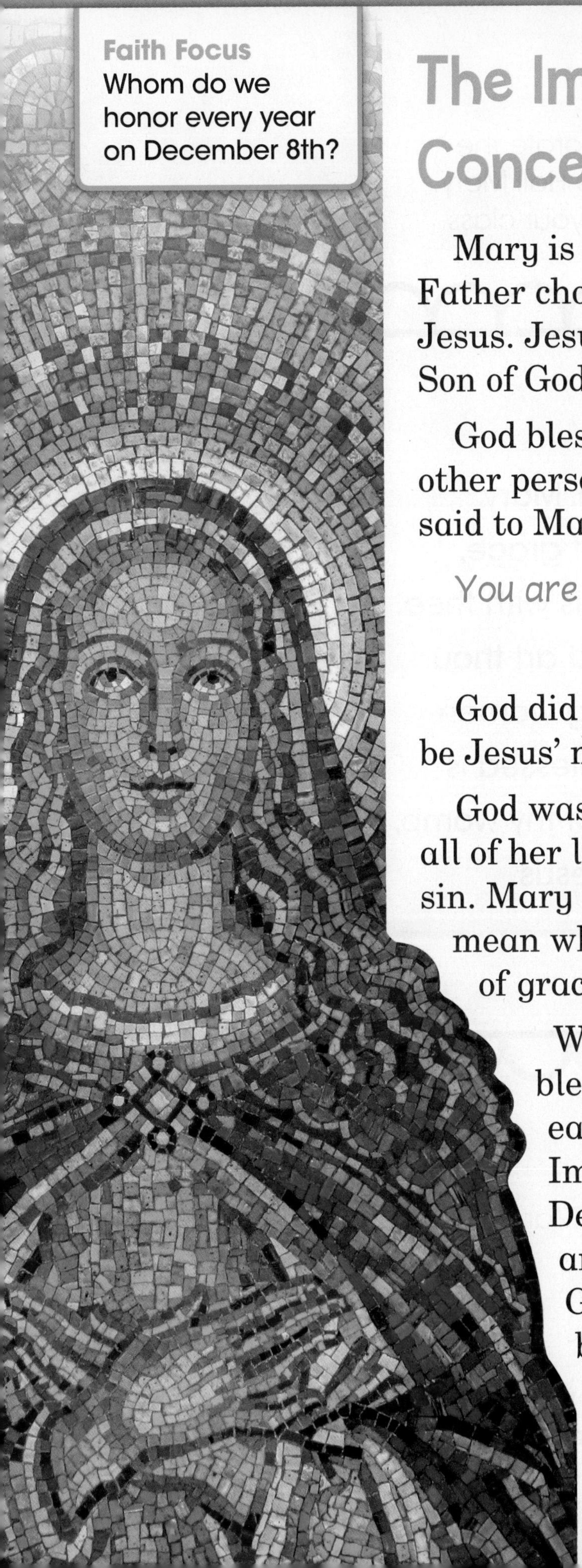

The Immaculate Conception

Mary is a very special mother. God the Father chose Mary to be the mother of Jesus. Jesus is the son of Mary and the Son of God.

God blessed Mary more than any other person. The Bible tells us that God said to Mary

You are blessed among all women.

BASED ON LUKE 1:42

God did this because he chose Mary to be Jesus' mother.

God was with Mary in a special way all of her life. Mary was born without sin. Mary never sinned. This is what we mean when we pray, "Hail Mary, full of grace, the Lord is with thee."

We celebrate this special blessing God gave Mary each year. We celebrate the Immaculate Conception on December 8th. We honor Mary, and we honor God. We thank God for the special way that he blessed Mary.

Hail Mary

Tell Mary how special she is. Decorate the space around these words from the Hail Mary. Pray this first part of the prayer with your class.

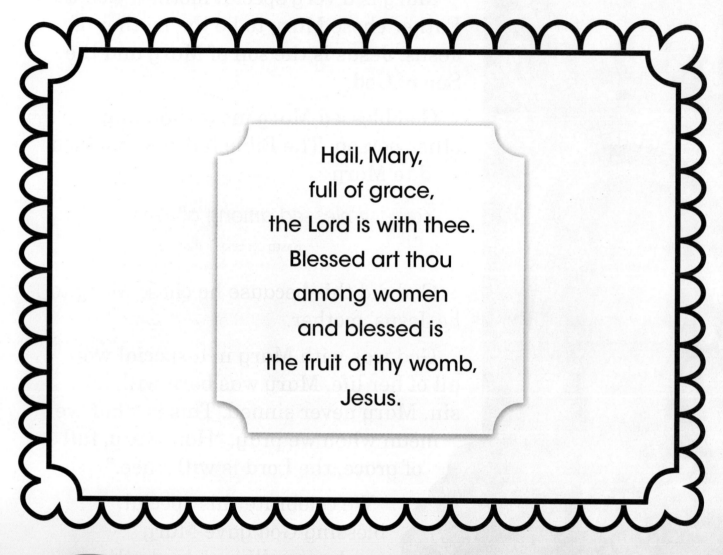

Hail, Mary,
full of grace,
the Lord is with thee.
Blessed art thou
among women
and blessed is
the fruit of thy womb,
Jesus.

My Faith Choice

This week I will honor Mary. I will learn to pray the Hail Mary by heart.

Pray, "Mary, God loves you. I love you too. Blessed are you! Amen."

Faith Focus

Who does our Blessed Mother Mary want us to love?

Our Lady of Guadalupe

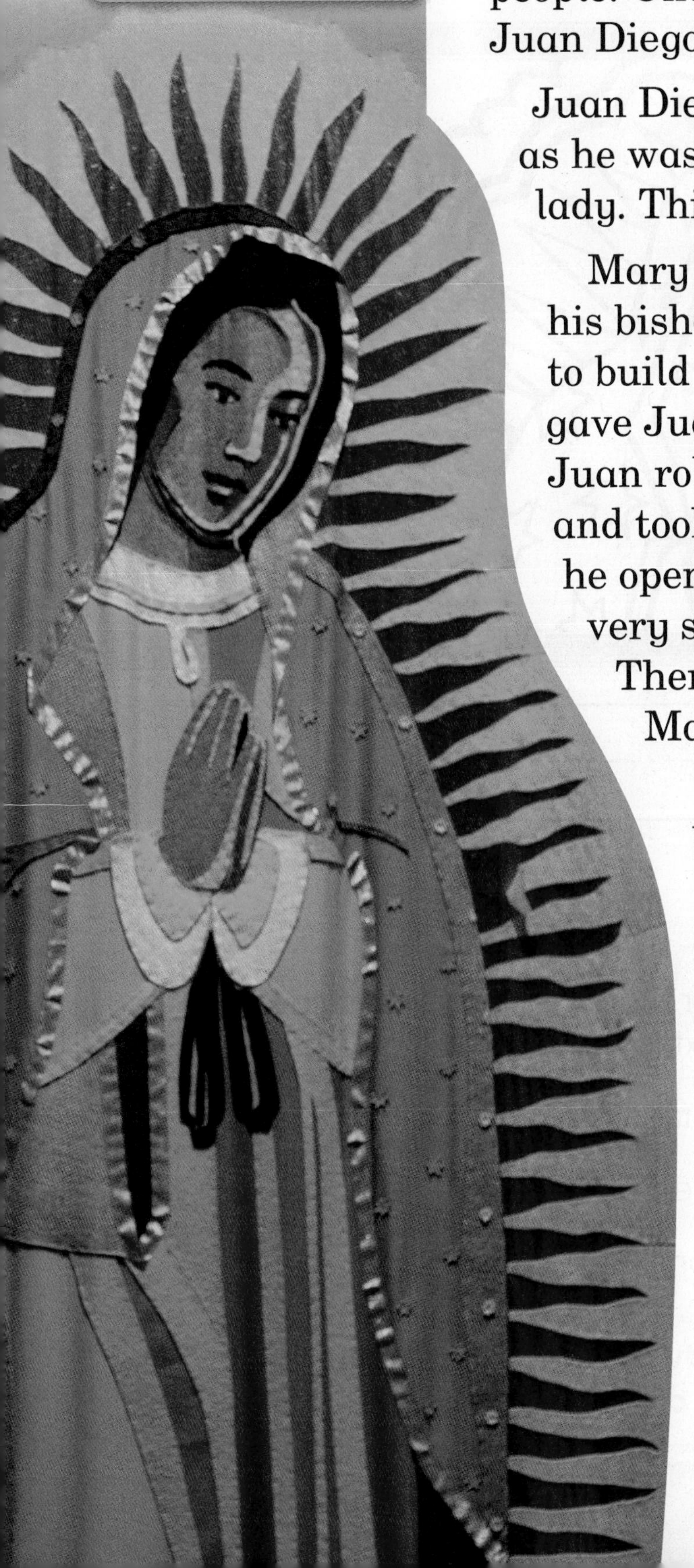

Our Blessed Mother Mary loves all people. One day Mary told a man named Juan Diego how much she loves us.

Juan Diego lived in Mexico. One day as he was walking to Mass, Juan saw a lady. This lady was Mary.

Mary gave Juan a message to give his bishop. She wanted the bishop to build a church in her name. Mary gave Juan roses to show the bishop. Juan rolled the roses up in his cloak and took them to the bishop. When he opened his cloak, everyone was very surprised at what they saw. There was a beautiful image of Mary on the cloth.

The Church that the bishop built is named Our Lady of Guadalupe. We celebrate the feast of Our Lady of Guadalupe on December 12.

Our Blessed Mother

Color the picture of Our Lady of Guadalupe. On the lines below the picture write, "I love you, Mary."

____

____

My Faith Choice

This week I will honor Mary. I will try my best to love others. Draw a ☺ next to the actions that you will do.

_______ Be kind to a friend.

_______ Help at home.

_______ Return crayons that I borrow.

Pray, "Mary, Our Lady of Guadalupe. Help me to love God as you do. Amen."

Faith Focus
Why did the angels visit the shepherds?

Christmas

We like good news. It makes us happy. On the night of Jesus' birth, some shepherds heard good news. Angels said to them,

> *"Today in Bethlehem the savior God promised to send you has been born."* BASED ON LUKE 2:11

The shepherds hurried to Bethlehem. They found Jesus there lying in a manger, just as the angels said. The shepherds were Jesus' first visitors. They told others all that happened.

We want to welcome Jesus just as the shepherds did. We thank God for bringing joy that will never end. We share the Good News with others.

Las Posadas

People in Mexico celebrate the journey of Mary and Joseph to the inn in Bethlehem. The words *las posadas* mean "the inns." You can perform this skit with your class.

Mary and Joseph In the name of God, can we stay here?

Innkeeper One We have no room for you. We are too crowded!

Mary and Joseph In the name of God, do you have room for us?

Innkeeper Two We have no room here.

Mary and Joseph In the name of God, do you have room for us?

Innkeeper Three My inn is full. There is a stable in the hills. It is warm there.

Reader *Read Luke 2:1–20.*

Leader God our Father, we rejoice in the birth of your Son. May we always welcome him when he comes. Amen.

My Faith Choice

This week I will treat others with love. I will

____

____.

Pray, "May Jesus' birth bring joy, peace, and love to all people. Amen."

Faith Focus

Who is the Mother of God?

Mary, the Mother of God

Gifts make us feel special. When someone gives us a gift we know they care about us. God gave us the best gift. God the Father gave us Jesus, his Son. On Christmas day we celebrate the birth of Jesus.

God the Father chose the Blessed Virgin Mary to be the mother of his Son, Jesus. Mary is the Mother of God. The Blessed Virgin Mary is our mother, too. She loves and cares for all the children of the world.

We honor Mary, the Mother of God, in a special way on January 1. On this day we go to Mass. We give thanks to God for the gift of our Blessed Mother. What a special way to start the New Year!

A Mother's Love

Our Blessed Mother did many things for her Son, Jesus. She does them for us too. Find the words in the border. Tell your class about times that mothers do these things. Then decorate the border.

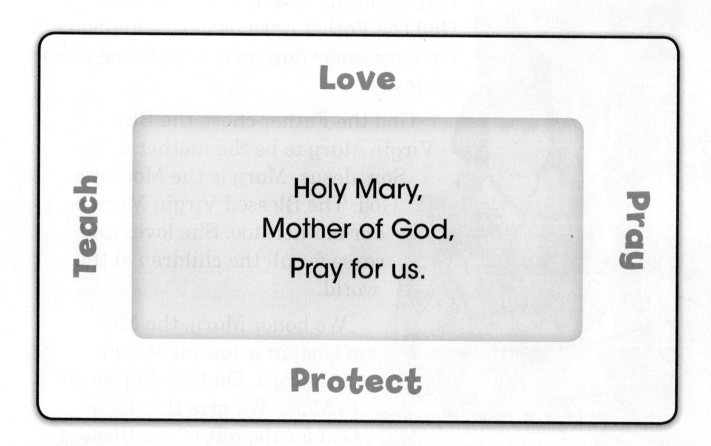

Love

Teach

Holy Mary,
Mother of God,
Pray for us.

Pray

Protect

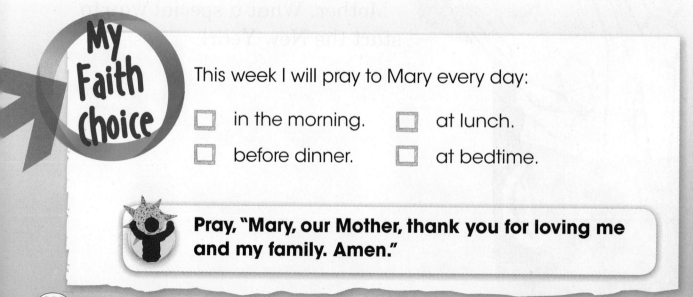

My Faith Choice

This week I will pray to Mary every day:

- ☐ in the morning.
- ☐ at lunch.
- ☐ before dinner.
- ☐ at bedtime.

Pray, "Mary, our Mother, thank you for loving me and my family. Amen."

Faith Focus
What do we celebrate on the Feast of Epiphany?

Epiphany

During Advent we waited and prepared for Christmas. We waited and prepared to welcome Jesus, the Son of God.

On Epiphany we hear the story of the Magi. These wise men traveled a long distance to find Jesus. They went to Bethlehem and honored Jesus.

We want the whole world to celebrate the birth of the newborn Savior. We want Heaven and nature to sing and rejoice. Jesus is the Savior of the world.

We Announce the Birth of the Savior

Make the cover for a Christmas card. Draw a picture and use words. Tell everyone that Jesus is the Savior of the world.

My Faith Choice

The Magi honored Jesus. I will honor Jesus by

____

____.

Pray, "Jesus, you are the Savior of the world! Amen."

Faith Focus
What is Ash Wednesday?

Ash Wednesday

Prepare! That's what we do whenever something important is going to happen. Parents prepare for a new baby. They visit the doctor and get everything ready at home. Students prepare for tests so they can learn as much as possible.

The most important time of the year for the Church is Easter. Lent is the time when we prepare for Easter. Ash Wednesday is the first day of Lent. It is the first day of our preparation for Easter.

On Ash Wednesday we go to church. The sign of the cross is made on our foreheads with ashes. We pray and ask God to help us to be more like Jesus. We ask God to help us celebrate Lent.

Being Like Jesus

Lent is a special time of prayer. In the spaces put words or pictures to complete your prayer.

Dear God,

I praise and thank you for

______________________________.

I ask you to watch over

______________________________.

Keep them in your care.

Amen.

My Faith Choice

This week I will remember to pray as Jesus did. I will

____

____

______________________________.

Pray, "Father in Heaven, thank you for helping me become more like your Son, Jesus. Amen."

Faith Focus
How does celebrating Lent help us to get ready for Easter?

Lent

Think about Spring. Remember how plants push their way up through the earth. Trees sprout leaves and buds. Birds sing their best songs.

During Spring we plant new seeds. We cut away dead twigs and stems. We prepare for new life.

Jesus talked about death and new life. He held up a seed and said,

> *"I say to you, unless a grain of wheat*
> *falls to the ground and dies,*
> *it remains just a grain of wheat;*
> *but if it dies, it produces much fruit."*
>
> John 12:24

During Lent we clear a place to plant seeds of faith and love. We work and pray. We grow in faith and love. We are getting ready for Easter.

New Life

Put this picture story in order. Number the pictures from 1 to 6. Share the story with a friend. Tell how the story helps us to understand Lent.

My Faith Choice

During Lent I can do good deeds and make sacrifices to get ready for Easter. I will

____

____.

Pray, "Thank you, Jesus, for helping us to change and grow during Lent. Amen."

Faith Focus
How do we begin our celebration of Holy Week?

Palm Sunday of the Lord's Passion

Sometimes important people come to our town or school. We go out and greet them. We cheer and rejoice!

Once Jesus came to the city of Jerusalem. He loved the people there. He wanted to gather them as a mother hen gathers her little chicks.

When Jesus came to the city, the people cheered. They waved branches from palm trees. They also spread their cloaks on the road to honor Jesus.

We remember this day at the beginning of Holy Week on Palm Sunday of the Lord's Passion. On this day we carry palm branches and honor Jesus too.

Honoring Jesus

These words are hidden in the puzzle. Find and circle the words. Use the words to tell a partner about Palm Sunday.

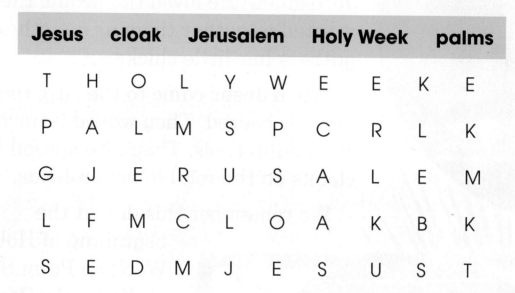

Jesus **cloak** **Jerusalem** **Holy Week** **palms**

T	H	O	L	Y	W	E	E	K	E
P	A	L	M	S	P	C	R	L	K
G	J	E	R	U	S	A	L	E	M
L	F	M	C	L	O	A	K	B	K
S	E	D	M	J	E	S	U	S	T

My Faith Choice

On Palm Sunday, the beginning of Holy Week, I can honor Jesus. I will

Pray, "Hosanna in the highest! We rejoice and honor you, Jesus. Amen."

Faith Focus
What does the Church celebrate on Holy Thursday?

Holy Thursday

Holy Thursday is one of the most important days for our Church. On this day we remember and celebrate the day on which Jesus gave us the Eucharist.

On the night before he died, Jesus celebrated a special meal with his disciples. We call this meal the Last Supper. At the Last Supper Jesus took bread and said to the disciples, "This is my body." He also took a cup of wine and said, "This is the cup of my blood." Then Jesus said to them, "Do this in memory of me" (based on Luke 22:14–19).

We celebrate the Eucharist every time we celebrate Mass. When we do, we are doing what Jesus asked.

Thank You, Jesus!

Use the code to color the stained-glass window. Use the stained-glass window to tell what happened at the Last Supper.

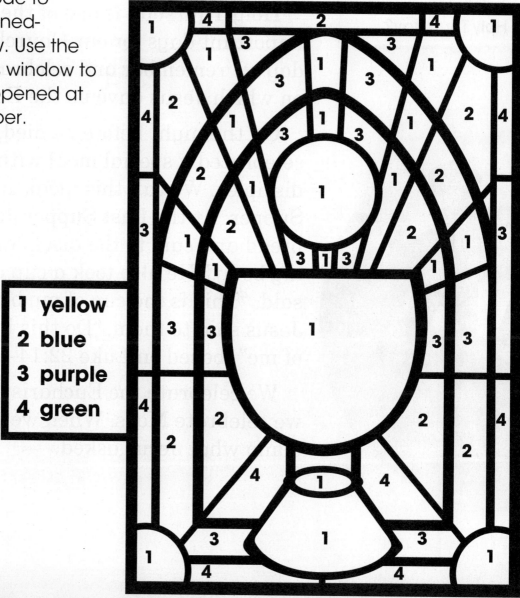

1 yellow
2 blue
3 purple
4 green

My Faith Choice

Jesus celebrated a special meal with his disciples and asked us to do the same. In Jesus' memory, I will

____

- -

____.

Pray, "Thank you, Jesus, for your gift of the Eucharist. Amen."

Faith Focus
What does the Church celebrate on Good Friday?

Good Friday

Sometimes we look at pictures or a gift that someone has given us. This helps us to remember and think about that person. What do you look at to help you remember someone?

The Friday of Holy Week is called Good Friday. It is a very special day for all Christians. It is the day we remember in a special way that Jesus suffered and died for us.

On Good Friday the deacon or priest holds up a cross for us to look at. Looking at the cross, we think about and remember how much Jesus loves us. One way we show our love for Jesus is by loving one another.

Showing Our Love for Others

Draw a ✝ next to the ways you can show your love for others. Write one more thing you will do.

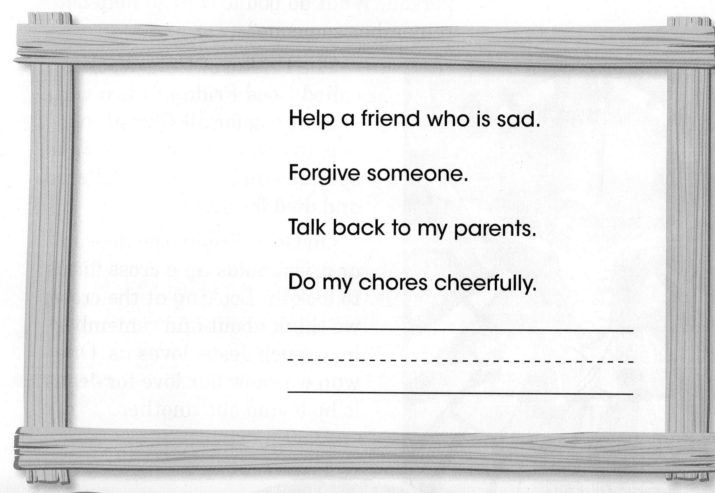

Help a friend who is sad.

Forgive someone.

Talk back to my parents.

Do my chores cheerfully.

____

____

____

My Faith Choice

I will show that I am thankful that Jesus died out of love for us. I will

____

____

______________________________.

Pray, "We adore you and thank you, Jesus, for suffering and dying on the cross for us. Amen."

Faith Focus
What do we celebrate at Easter?

Easter Sunday

At Easter we see signs of new life all around us. These signs remind us that Jesus was raised from the dead to new life. We call this the Resurrection of Jesus. On Easter Sunday Christians celebrate Jesus' Resurrection.

We are Easter people! Alleluia is our song! We sing Alleluia over and over during the fifty days of the Easter season. The word *Alleluia* means "Praise the Lord!" We praise God for raising Jesus from the dead to new life.

Every Sunday in the year is a little Easter. We sing. We rest. We enjoy one another. All year long we praise and thank God.

Praise the Lord

Decorate the Easter banner. Use colors and words about new life. Show your finished banner to your friends and your family. Tell them about the Resurrection of Jesus.

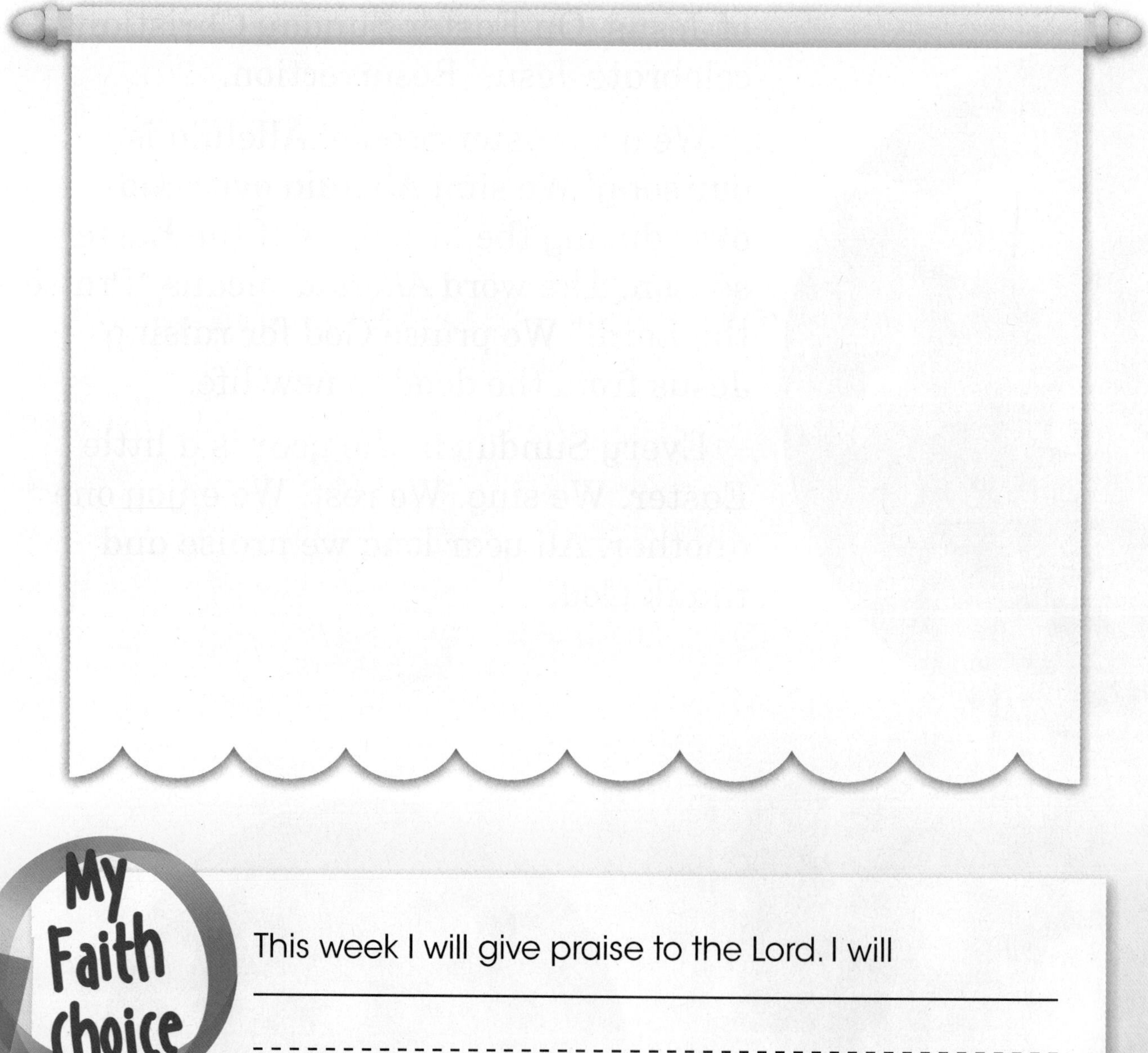

My Faith Choice

This week I will give praise to the Lord. I will

____

- -

____.

Pray, "Jesus, you are risen. Alleluia!"

Faith Focus

What is the Ascension of Jesus?

The Ascension

Forty days after Easter, Jesus led his disciples outside Jerusalem. He reminded them that he had suffered, died, and was raised to new life. Jesus said that we should share this good news with everyone.

Then he blessed the disciples and returned to his Father in Heaven. The Church celebrates the day Jesus returned to his Father. We call this day the Solemnity of the Ascension of the Lord.

Sing to Heaven

Sing this song. Use the melody to "Frère Jacques." Teach the song to your family and sing it together.

He is risen. He is risen.
Yes, he is. Yes, he is.
He will come in glory.
He will come in glory.
Yes, he will. Yes, he will.

Sound the trumpet.
Sound the trumpet.
He ascends. He ascends.
We await the Spirit.
We await the Spirit.
Yes, we do. Yes, we do.

Holy Spirit, Holy Spirit.
Come to us; come to us.
He will come and guide us.
He will come and guide us.
Yes, he will. Yes, he will.

My Faith Choice

Jesus asks us to share his good news with others. I will

____

____.

Pray, "Bless us always, Jesus, as we wait for you to come again in glory. Amen."

Faith Focus
When does the Holy Spirit help us to live as followers of Jesus?

Pentecost

Sometimes we receive a gift that we use to help others. We have received that kind of gift from Jesus.

After Jesus returned to his Father, the disciples received the gift of the Holy Spirit. The Spirit helped them to share the Good News about Jesus with others. He helped them to do good work in Jesus' name.

On Pentecost Sunday, we remember that the Holy Spirit came to the disciples. We too have received the gift of the Holy Spirit. The Holy Spirit helps us to do good. When we do good things in Jesus' name, we lead others to Jesus.

The Gift of the Holy Spirit

Work with a partner and follow this maze. At each place, stop to share the Good News about Jesus with each other.

This week I will honor the Holy Spirit. I will do good. I will

____

- -

____.

Pray, "Come, Holy Spirit, and fill my heart with your love. Amen."

Catholic Prayers and Practices

Sign of the Cross

In the name of the Father,
and of the Son,
and of the Holy Spirit. Amen.

Our Father

Our Father, who art in heaven,
hallowed be thy name;
thy kingdom come,
thy will be done
on earth as it is in heaven.
Give us this day our daily bread,
and forgive us our trespasses,
as we forgive those who trespass
against us;
and lead us not into temptation,
but deliver us from evil.
Amen.

Glory Be (Doxology)

Glory be to the Father
and to the Son
and to the Holy Spirit,
as it was in the beginning
is now, and ever shall be
world without end. Amen.

The Hail Mary

Hail, Mary, full of grace,
the Lord is with thee.
Blessed art thou among women
and blessed is the fruit
of thy womb, Jesus.
Holy Mary, Mother of God,
pray for us sinners,
now and at the hour of our death.
Amen.

Signum Crucis

In nómine Patris,
et Fílii,
et Spíritus Sancti. Amen.

Pater Noster

Pater noster, qui es in cælis:
sanctificétur nomen tuum;
advéniat regnum tuum;
fiat volúntas tua,
sicut in cælo, et in terra.
Panem nostrum cotidiánum
da nobis hódie;
et dimítte nobis débita nostra,
sicut et nos dimíttimus debitóribus
nostris;
et ne nos indúcas in tentatiónem;
sed líbera nos a malo. Amen.

Gloria Patri

Glória Patri
et Fílio
et Spirítui Sancto.
Sicut erat in princípio,
et nunc et semper
et in sæcula sæculórum. Amen.

Ave, Maria

Ave, María, grátia plena,
Dóminus tecum.
Benedícta tu in muliéribus,
et benedíctus fructus ventris tui,
Iesus.
Sancta María, Mater Dei,
ora pro nobis peccatóribus,
nunc et in hora mortis nostræ.
Amen.

Apostles' Creed

(from the Roman Missal)

I believe in God,
the Father almighty,
Creator of heaven and earth,
and in Jesus Christ, his only Son,
our Lord,

(At the words that follow, up to and including the Virgin Mary, *all bow.)*

who was conceived by the Holy Spirit,
born of the Virgin Mary,
suffered under Pontius Pilate,
was crucified, died and was buried;
he descended into hell;
on the third day he rose again from the dead;
he ascended into heaven,
and is seated at the right hand of God the Father almighty;
from there he will come to judge the living and the dead.
I believe in the Holy Spirit,
the holy catholic Church,
the communion of saints,
the forgiveness of sins,
the resurrection of the body,
and life everlasting. Amen.

Nicene Creed

(from the Roman Missal)

I believe in one God,
the Father almighty,
maker of heaven and earth,
of all things visible and invisible.

I believe in one Lord Jesus Christ,
the Only Begotten Son of God,
born of the Father before all ages.
God from God, Light from Light,
true God from true God,
begotten, not made, consubstantial with the Father;
through him all things were made.
For us men and for our salvation
he came down from heaven,

(At the words that follow, up to and including and became man, *all bow.)*

and by the Holy Spirit was incarnate of the Virgin Mary,
and became man.

For our sake he was crucified under Pontius Pilate,
he suffered death and was buried,
and rose again on the third day
in accordance with the Scriptures.
He ascended into heaven
and is seated at the right hand of the Father.
He will come again in glory
to judge the living and the dead
and his kingdom will have no end.

I believe in the Holy Spirit, the Lord, the giver of life,
who proceeds from the Father and the Son,
who with the Father and the Son is adored and glorified,
who has spoken through the prophets.

I believe in one, holy, catholic and apostolic Church.
I confess one Baptism for the forgiveness of sins
and I look forward to the resurrection of the dead
and the life of the world to come. Amen.

Morning Prayer

Dear God,
as I begin this day,
keep me in your love and care.
Help me to live as your child today.
Bless me, my family, and my friends
in all we do.
Keep us all close to you. Amen.

Grace Before Meals

Bless us, O Lord,
and these thy gifts,
which we are about to receive
from thy bounty,
through Christ our Lord.
Amen.

Grace After Meals

We give thee thanks,
for all thy benefits, almighty God,
who lives and reigns forever. Amen.

Evening Prayer

Dear God,
I thank you for today.
Keep me safe throughout the night.
Thank you for all the good I did today.
I am sorry for what I have chosen
to do wrong.
Bless my family and friends. Amen.

A Vocation Prayer

God, I know you will call me
for special work in my life.
Help me follow Jesus each day
and be ready to answer your call.
Amen.

Act of Contrition

My God,
I am sorry for my sins
with all my heart.
In choosing to do wrong
and failing to do good,
I have sinned against you,
whom I should love above all things.
I firmly intend, with your help,
to do penance,
to sin no more,
and to avoid whatever leads me
to sin.
Our Savior Jesus Christ
suffered and died for us.
In his name, my God, have mercy.
Amen.

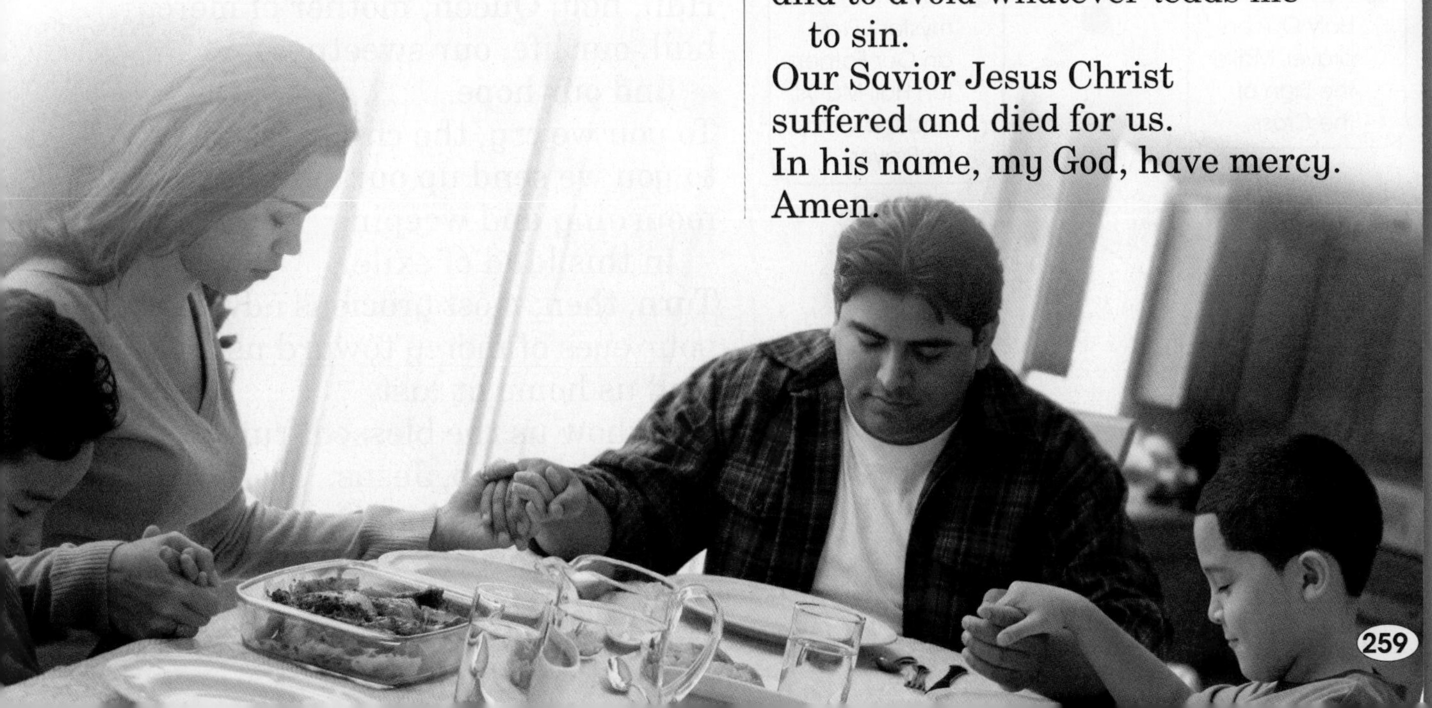

Rosary

Catholics pray the Rosary to honor Mary and remember the important events in the life of Jesus and Mary. There are twenty mysteries of the Rosary. Follow the steps from 1 to 5.

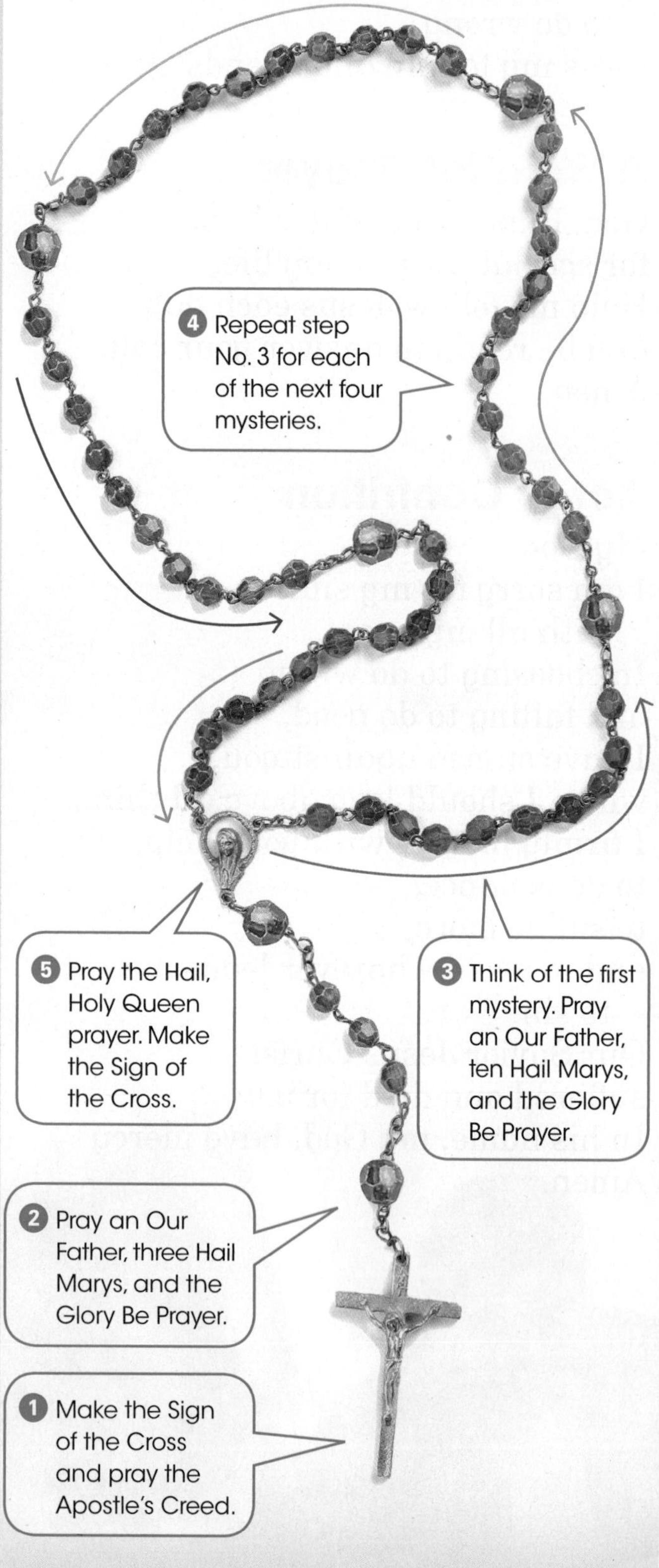

4 Repeat step No. 3 for each of the next four mysteries.
5 Pray the Hail, Holy Queen prayer. Make the Sign of the Cross.
3 Think of the first mystery. Pray an Our Father, ten Hail Marys, and the Glory Be Prayer.
2 Pray an Our Father, three Hail Marys, and the Glory Be Prayer.
1 Make the Sign of the Cross and pray the Apostle's Creed.

Joyful Mysteries

1. The Annunciation
2. The Visitation
3. The Nativity
4. The Presentation
5. The Finding of Jesus in the Temple

Mysteries of Light

1. The Baptism of Jesus in the Jordan River
2. The Miracle at the Wedding at Cana
3. The Proclamation of the Kingdom of God
4. The Transfiguration of Jesus
5. The Institution of the Eucharist

Sorrowful Mysteries

1. The Agony in the Garden
2. The Scourging at the Pillar
3. The Crowning with Thorns
4. The Carrying of the Cross
5. The Crucifixion

Glorious Mysteries

1. The Resurrection
2. The Ascension
3. The Coming of the Holy Spirit
4. The Assumption of Mary
5. The Coronation of Mary

Hail, Holy Queen

Hail, holy Queen, mother of mercy:
hail, our life, our sweetness,
and our hope.
To you we cry, the children of Eve;
to you we send up our sighs,
mourning and weeping
in this land of exile.
Turn, then, most gracious advocate,
your eyes of mercy toward us;
lead us home at last
and show us the blessed fruit
of your womb, Jesus.
O clement, O loving, O sweet
Virgin Mary.

The Ten Commandments

1. I am the LORD your God: you shall not have strange gods before me.
2. You shall not take the name of the LORD your God in vain.
3. Remember to keep holy the LORD's Day.
4. Honor your father and your mother.
5. You shall not kill.
6. You shall not commit adultery.
7. You shall not steal.
8. You shall not lie.
9. You shall not covet your neighbor's wife.
10. You shall not covet your neighbor's goods.

Based on Exodus 20:2–3, 7–17

Precepts of the Church

1. Participate in Mass on Sundays and holy days of obligation, and rest from unnecessary work.
2. Confess sins at least once a year.
3. Receive Holy Communion at least during the Easter season.
4. Observe the prescribed days of fasting and abstinence.
5. Provide for the material needs of the Church, according to one's abilities.

The Great Commandment

"You shall love the Lord, your God, with all your heart, with all your soul, and with all your mind. . . . You shall love your neighbor as yourself." Matthew 22:37, 39

The Law of Love

"This is my commandment: love one another as I love you."

John 15:12

The Seven Sacraments

Jesus gave the Church the Seven Sacraments. The Seven Sacraments are signs of God's love for us. When we celebrate the Sacraments, Jesus is really present with us. We share in the life of the Holy Trinity.

Baptism

We are joined to Christ. We become members of the Body of Christ, the Church.

Confirmation

The Holy Spirit strengthens us to live as children of God.

Eucharist

We receive the Body and Blood of Jesus.

Reconciliation

We receive God's gift of forgiveness and peace.

Anointing of the Sick

We receive God's healing strength when we are sick or dying, or weak because of old age.

Holy Orders

A baptized man is ordained to serve the Church as a bishop, priest, or deacon.

Matrimony

A baptized man and a baptized woman make a lifelong promise to love and respect each other as husband and wife. They promise to accept the gift of children from God.

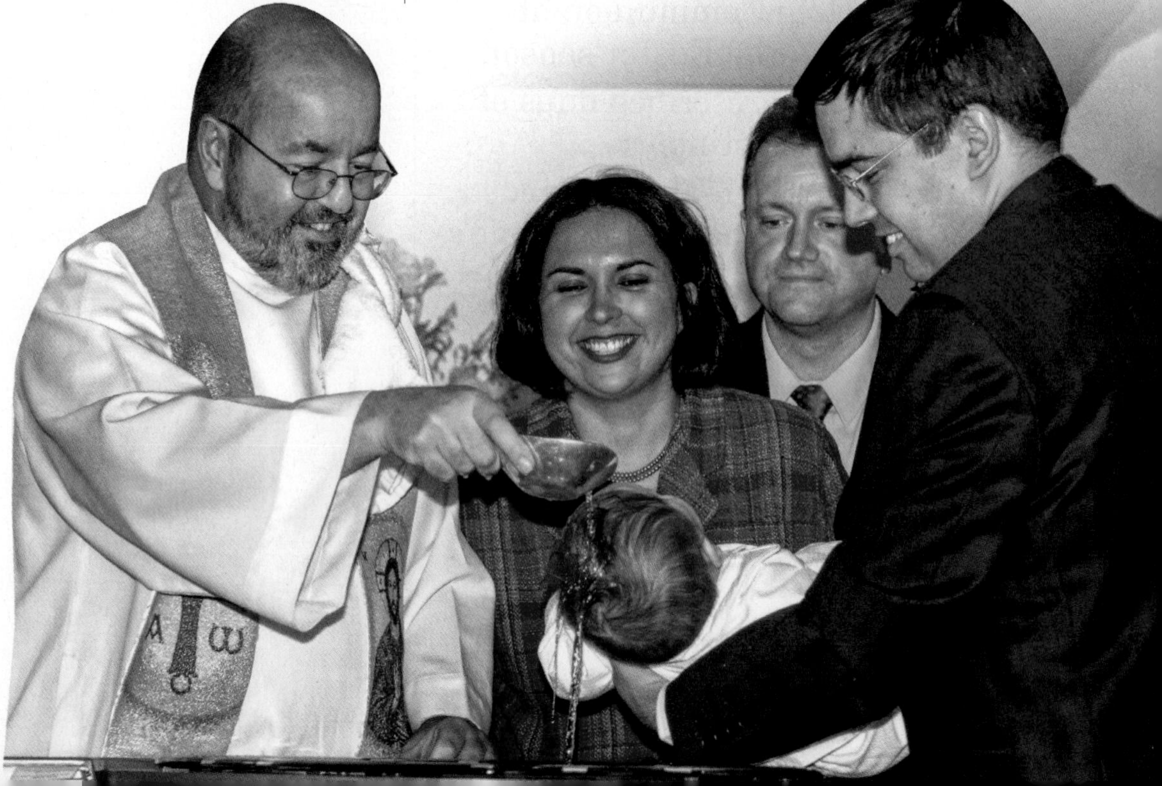

We Celebrate the Mass

The Introductory Rites

We remember that we are the community of the Church. We prepare to listen to the Word of God and to celebrate the Eucharist.

The Entrance

We stand as the priest, deacon, and other ministers enter the assembly. We sing a gathering song. The priest and deacon kiss the altar. The priest then goes to the chair, where he presides over the celebration.

Greeting of the Altar and of the People Gathered

The priest leads us in praying the Sign of the Cross. The priest greets us, and we say,

"And with your spirit."

The Penitential Act

We admit our wrongdoings. We bless God for his mercy.

The Gloria

We praise God for all the good that he has done for us.

The Collect

The priest leads us in praying the Collect, or the opening prayer. We respond, "Amen."

The Liturgy of the Word

God speaks to us today.
We listen and respond to God's Word.

The First Reading from Scripture

We sit and listen as the reader reads from the Old Testament or from the Acts of the Apostles. The reader concludes, "The Word of the Lord."
We respond,

"Thanks be to God."

The Responsorial Psalm

The song leader leads us in singing a psalm.

The Second Reading from Scripture

The reader reads from the New Testament, but not from the four Gospels. The reader concludes, "The Word of the Lord." We respond,

"Thanks be to God."

The Acclamation

We stand to honor Christ, present with us in the Gospel. The song leader leads us in singing "**Alleluia, Alleluia, Alleluia**," or another chant during Lent.

The Gospel

The deacon or priest proclaims, "A reading from the holy Gospel according to (name of Gospel writer)." We respond,

"Glory to you, O Lord."

He proclaims the Gospel. At the end he says, "The Gospel of the Lord." We respond,

"Praise to you,
Lord Jesus Christ."

The Homily

We sit. The priest or deacon preaches the homily. He helps the people gathered to understand the Word of God spoken to us in the readings.

The Profession of Faith

We stand and profess our faith. We pray the Nicene Creed together.

The Prayer of the Faithful

The priest leads us in praying for our Church and her leaders, for our country and its leaders, for ourselves and others, for those who are sick and those who have died. We can respond to each prayer in several ways. One way that we respond is,

"Lord, hear our prayer."

The Liturgy of the Eucharist

We join with Jesus and the Holy Spirit
to give thanks and praise to God the Father.

The Preparation of the Gifts

We sit as the altar table is prepared and the collection is taken up. We share our blessings with the community of the Church and especially with those in need. The song leader may lead us in singing a song. The gifts of bread and wine are brought to the altar.

The priest lifts up the bread and blesses God for all our gifts. He prays, "Blessed are you, Lord God of all creation . . ." We respond,

"Blessed be God for ever."

The priest lifts up the cup of wine and prays, "Blessed are you, Lord God of all creation . . . " We respond,

"Blessed be God for ever."

The priest invites us,

"Pray, brothers and sisters,
that my sacrifice and yours
may be acceptable to God,
the almighty Father."

We stand and respond,

"May the Lord accept the sacrifice at your hands for the praise and glory of his name, for our good, and the good of all his holy Church."

The Prayer over the Offerings

The priest leads us in praying the Prayer over the Offerings.
We respond, "**Amen**."

Preface

The priest invites us to join in praying the Church's great prayer of praise and thanksgiving to God the Father.

Priest: "The Lord be with you."

Assembly: "And with your spirit."

Priest: "Lift up your hearts."

Assembly: "We lift them up to the Lord."

Priest: "Let us give thanks to the Lord our God."

Assembly: "It is right and just."

After the priest sings or prays aloud the preface, we join in acclaiming,

"Holy, Holy, Holy Lord God of hosts.
Heaven and earth are full of your glory.
Hosanna in the highest.
Blessed is he who comes in the name of the Lord.
Hosanna in the highest."

The Eucharistic Prayer

The priest leads the assembly in praying the Eucharistic Prayer. We call on the Holy Spirit to make our gifts of bread and wine holy and that they become the Body and Blood of Jesus. We recall what happened at the Last Supper. The bread and wine become the Body and Blood of the Lord. Jesus is truly and really present under the appearances of bread and wine.

The priest sings or says aloud, "The mystery of faith." We respond using this or another acclamation used by the Church,

"We proclaim your Death, O Lord, and profess your Resurrection until you come again."

The priest then prays for the Church. He prays for the living and the dead.

Doxology

The priest concludes the praying of the Eucharistic Prayer. He sings or prays aloud,

"Through him, and with him, and in him,
O God, almighty Father,
in the unity of the Holy Spirit, all glory and honor is yours,
for ever and ever."

We respond by singing, "**Amen**."

The Communion Rite

The Lord's Prayer

We pray the Lord's Prayer together.

The Sign of Peace

The priest invites us to share a sign of peace, saying, "The peace of the Lord be with you always." We respond,

"And with your spirit."

We share a sign of peace.

The Fraction, or the Breaking of the Bread

The priest breaks the host, the consecrated bread. We sing or pray aloud,

"Lamb of God, you take away the sins of the world,
have mercy on us.
Lamb of God, you take away the sins of the world,
have mercy on us.
Lamb of God, you take away the sins of the world,
grant us peace."

Communion

The priest raises the host and says aloud,

"Behold the Lamb of God,
behold him who takes away the sins of the world.
Blessed are those called to the supper of the Lamb."

We join with him and say,

"Lord, I am not worthy that you should enter under my roof, but only say the word and my soul shall be healed."

The priest receives Communion. Next, the deacon and the extraordinary ministers of Holy Communion and the members of the assembly receive Communion.

The priest, deacon, or extraordinary minister of Holy Communion holds up the host. We bow, and the priest, deacon, or extraordinary minister of Holy Communion says, "The Body of Christ." We respond, "**Amen**." We then receive the consecrated host in our hands or on our tongues.

If we are to receive the Blood of Christ, the priest, deacon, or extraordinary minister of Holy Communion holds up the cup containing the consecrated wine. We bow, and the priest, deacon, or extraordinary minister of Holy Communion says, "The Blood of Christ." We respond, "**Amen**." We take the cup in our hands and drink from it.

The Prayer after Communion

We stand as the priest invites us to pray, saying, "Let us pray." He prays the Prayer after Communion. We respond,

"**Amen**."

The Concluding Rites

We are sent forth to do good works,
praising and blessing the Lord.

Greeting

We stand. The priest greets us as we prepare to leave. He says, "The Lord be with you." We respond,

"And with your spirit."

Final Blessing

The priest or deacon may invite us,

"Bow your heads and pray for God's blessing."

The priest blesses us, saying,

"May almighty God bless you: the Father, and the Son, and the Holy Spirit."

We respond, "**Amen**."

Dismissal of the People

The priest or deacon sends us forth, using these or similar words,

"Go in peace, glorifying the Lord by your life."

We respond,

"Thanks be to God."

We sing a hymn. The priest and the deacon kiss the altar. The priest, deacon, and other ministers bow to the altar and leave in procession.

The Sacrament of Reconciliation

Individual Rite

Greeting
Scripture Reading
Confession of Sins
and Acceptance of Penance
Act of Contrition
Absolution
Closing Prayer

Communal Rite

Greeting
Scripture Reading
Homily
Examination of Conscience, a
Litany of Contrition, and the
Lord's Prayer
Individual Confession and Absolution
Closing Prayer

Key Teachings of the Catholic Church

The Mystery of God

Divine Revelation

Who am I?

You are a person created by God. God wants you to live in friendship with him on Earth and forever in Heaven.

How do we know this about ourselves?

God knows and loves all people. God wants us to know and love him too. God tells us about ourselves. God also tells us about himself.

How did God tell us?

God tells us in many ways. First, all the things God has created tell us about him. We see God's goodness and beauty in creation. Second, God came to us and he told us about himself. He told us the most when he sent his Son, Jesus Christ. God's Son became one of us and lived among us. He showed us who God is.

What is faith?

Faith is a gift from God. It helps us to know and to believe in God.

What is a mystery of faith?

A mystery of faith can never be known completely. We cannot know everything about God. We only know who God is because he told us about himself.

What is Divine Revelation?

God wants us to know about him. Divine Revelation is how he freely makes himself known to us. God has told us about himself and his plan for us. He has done this so that we can live in friendship with him and with one another forever.

What is Sacred Tradition?

The word *tradition* means "to pass on." The Church's Sacred Tradition passes on what God has told us. The Holy Spirit guides the Church to tell us about God.

Sacred Scripture

What is Sacred Scripture?

Sacred Scripture means "holy writings." Sacred Scripture are writings that tell God's story.

What is the Bible?

The Bible is God's Word. It is a holy book. The stories in the Bible teach about God. The Bible tells the stories about Jesus. When you listen to the Bible, you are listening to God.

What does it mean to say that the Bible is inspired?

This means that the Holy Spirit helped people write about God. The Holy Spirit helped the writers tell what God wants us to know about him.

What is the Old Testament?

The Old Testament is the first part of the Bible. It has forty-six books. They were written before the birth of Jesus. The Old Testament tells the story of creation. It tells about Adam and Eve. It tells about the promise, or Covenant, between God and his people.

What is the Covenant?

The Covenant is the promise that God and his people freely made. It is God's promise always to love and be kind to his people.

What are the writings of the prophets?

God chose people to speak in his name. These people are called the prophets. We read the message of the prophets in the Bible. The prophets remind God's people that God is faithful. They remind God's people to be faithful to the Covenant.

What is the New Testament?

The New Testament is the second part of the Bible. It has twenty-seven books. These books were inspired by the Holy Spirit. They were written during the time of the Apostles. They are about Jesus Christ. They tell about his saving work.

What are the Gospels?

The Gospels are the four books at the beginning of the New Testament. They tell the story of Jesus and his teachings. The four Gospels are Matthew, Mark, Luke, and John.

What are the letters of Saint Paul?

The letters of Saint Paul are in the New Testament. The letters teach about the Church. They tell how to follow Jesus. Some of these letters were written before the Gospels.

The Holy Trinity

Who is the Mystery of the Holy Trinity?

The Holy Trinity is the mystery of one God in three Persons—God the Father, God the Son, and God the Holy Spirit.

Who is God the Father?

God the Father is the First Person of the Holy Trinity.

Who is God the Son?

God the Son is Jesus Christ. He is the Second Person of the Holy Trinity. God the Father sent his Son to be one of us and live with us.

Who is God the Holy Spirit?

The Holy Spirit is the Third Person of the Holy Trinity. God sends us the Holy Spirit to help us to know and love God better. The Holy Spirit helps us live as children of God.

Divine Work of Creation

What does it mean to call God the Creator?

God is the Creator. He has made everyone and everything out of love. He has created everyone and everything without any help.

Who are angels?

Angels are spiritual beings. They do not have bodies like we do. Angels give glory to God at all times. They sometimes serve God by bringing his message to people.

Why are human beings special?

God creates every human being in his image and likeness. God shares his life with us. God wants us to be happy with him, forever.

What is the soul?

The soul is the spiritual part of a person. The soul will never die. It is the part of us that lives forever. It bears the image of God.

What is free will?

Free will is the power God gives us to choose between good and evil. Free will gives us the power to turn toward God.

What is Original Sin?

Original Sin is the sin of Adam and Eve. They chose to disobey God. As a result of Original Sin, death, sin, and suffering came into the world.

Jesus Christ, Son of God, Son of Mary

What is the Annunciation?

At the Annunciation the angel Gabriel came to Mary. The angel had a message for her. God had chosen her to be the Mother of his Son, Jesus.

What is the Incarnation?

The Incarnation is the Son of God becoming a man and still being God. Jesus Christ is true God and true man.

What does it mean that Jesus is Lord?

The word *lord* means "master or ruler." When we call Jesus "Lord," we mean that he is truly God.

What is the Paschal Mystery?

The Paschal Mystery is the Passion, Death, Resurrection, and Ascension of Jesus Christ. Jesus passed over from death into new and glorious life.

What is Salvation?

The word *salvation* means "to save." It is the saving of all people from sin and death through Jesus Christ.

What is the Resurrection?

The Resurrection is God's raising Jesus from the dead to new life.

What is the Ascension?

The Ascension is the return of the Risen Jesus to his Father in Heaven.

What is the Second Coming of Christ?

Christ will come again in glory at the end of time. This is the Second Coming of Christ. He will judge the living and the dead. This is the fulfillment of God's plan.

What does it mean that Jesus is the Messiah?

The word *messiah* means "anointed one." He is the Messiah. God promised to send the Messiah to save all people. Jesus is the Savior of the world.

The Mystery of the Church

What is the Church?

The word *church* means "those who are called together." The Church is the Body of Christ. It is the new People of God.

What does the Church do?

The Church tells all people the Good News of Jesus Christ. The Church invites all people to know, love, and serve Jesus.

What is the Body of Christ?

The Church is the Body of Christ on Earth. Jesus Christ is the Head of the Church and all baptized people are its members.

Who are the People of God?

The Church is the People of God. God invites all people to belong to the People of God. The People of God live as one family in God.

What is the Communion of Saints?

The Communion of Saints is all of the holy people that make up the Church. It is the faithful followers of Jesus on Earth. It is those who have died who are still becoming holier. It is also those who have died and are happy forever with God in Heaven.

What are the Marks of the Church?

There are four main ways to describe the Church. We call these the Four Marks of the Church. The Church is one, holy, catholic, and apostolic.

Who are the Apostles?

The Apostles were the disciples who Jesus chose. He sent them to preach the Gospel to the whole world in his name. Some of their names are Peter, Andrew, James, and John.

What is Pentecost?

Pentecost is the day the Holy Spirit came to the disciples of Jesus. This happened fifty days after the Resurrection. The work of the Church began on this day.

Who are the clergy?

The clergy are bishops, priests, and deacons. They have received the Sacrament of Holy Orders. They serve the whole Church.

What is the work of the pope?

Jesus Christ is the true Head of the Church. The pope and the bishops lead the Church in his name. The pope is the bishop of Rome. He is the successor to Saint Peter the Apostle, the first pope. The pope brings the Church together. The Holy Spirit guides the pope when he speaks about faith and about what Catholics believe.

What is the work of the bishops?

The other bishops are the successors of the other Apostles. They teach and lead the Church in their dioceses. The Holy Spirit always guides the pope and all of the bishops. He guides them when they make important decisions.

What is religious life?

Some men and women want to follow Jesus in a special way. They choose the religious life. They promise not to marry. They dedicate their whole lives to doing Jesus' work. They promise to live holy lives. They promise to live simply. They share what they have with others. They live together in groups and they promise to obey the rules of their community. They may lead quiet lives of prayer, or teach, or take care of people who are sick or poor.

Who are lay people?

Many people do not receive the Sacrament of Holy Orders. Many are not members of a religious community. These are lay people. Lay people follow Christ every day by what they do and say.

The Blessed Virgin Mary

Who is Mary?

God chose Mary to be the mother of his only Son, Jesus. Mary is the Mother of God. She is the Mother of Jesus. She is the Mother of the Church. Mary is the greatest saint.

What is the Immaculate Conception?

From the first moment of her being, Mary was preserved from sin. This special grace from God continued throughout her whole life. We call this the Immaculate Conception.

What is the Assumption of Mary?

At the end of her life on Earth, the Blessed Virgin Mary was taken body and soul into Heaven. Mary hears our prayers. She tells her Son what we need. She reminds us of the life that we all hope to share when Christ, her Son, comes again in glory.

Life Everlasting

What is eternal life?
Eternal life is life after death. At death the soul leaves the body. It passes into eternal life.

What is Heaven?
Heaven is living with God and with Mary and all the saints in happiness forever after we die.

What is the Kingdom of God?
The Kingdom of God is also called the Kingdom of Heaven. It is all people and creation living in friendship with God.

What is Purgatory?
Purgatory is the chance to grow in love for God after we die so we can live forever in heaven.

What is Hell?
Hell is life away from God and the saints forever after death.

Celebration of the Christian Life and Mystery

Liturgy and Worship

What is worship?
Worship is the praise we give God. The Church worships God in the liturgy.

What is liturgy?
The liturgy is the Church's worship of God. It is the work of the Body of Christ. Christ is present by the power of the Holy Spirit.

What is the liturgical year?
The liturgical year is the name of the seasons and feasts that make up the Church's year of worship. The main seasons of the Church year are Advent, Christmas, Lent, and Easter. The Triduum is the three holy days just before Easter. The rest of the liturgical year is called Ordinary Time.

The Sacraments

What are the sacraments?
The sacraments are the seven signs of God's love for us that Jesus gave the Church. We share in God's love when we celebrate the sacraments.

What are the Sacraments of Christian Initiation?
The Sacraments of Christian Initiation are Baptism, Confirmation, and Eucharist.

What is the Sacrament of Baptism?
Baptism joins us to Christ. It makes us members of the Church. We receive the gift of the Holy Spirit. Original Sin and our personal sins are forgiven. Through Baptism, we belong to Christ.

What is the Sacrament of Confirmation?
At Confirmation we receive the gift of the Holy Spirit. The Holy Spirit strengthens us to live our Baptism.

What is the Sacrament of Eucharist?
In the Eucharist, we join with Christ. We give thanksgiving, honor, and glory to God the Father. Through the power of the Holy Spirit, the bread and wine become the Body and Blood of Jesus Christ.

Why do we have to participate at Sunday Mass?

Catholics participate in the Eucharist on Sundays and holy days of obligation. Sunday is the Lord's Day. Participating at the Mass, and receiving Holy Communion, the Body and Blood of Christ, when we are old enough, are necessary for Christians.

What is the Mass?

The Mass is the main celebration of the Church. At Mass we worship God. We listen to God's Word. We celebrate and share in the Eucharist.

What are the Sacraments of Healing?

The two Sacraments of Healing are the Sacrament of Penance and Reconciliation and the Sacrament of Anointing of the Sick.

What is confession?

Confession is telling our sins to a priest in the Sacrament of Penance. Confession is another name for the Sacrament of Penance.

What is contrition?

Contrition is being truly sorry for our sins. We want to make up for the hurt our sins have caused. We do not want to sin again.

What is penance?

A penance is a prayer or act of kindness. The penance we do shows that we are truly sorry for our sins. The priest gives us a penance to help repair the hurt caused by our sin.

What is absolution?

Absolution is the forgiveness of sins by God through the words and actions of the priest.

What is the Sacrament of Anointing of the Sick?

The Sacrament of Anointing of the Sick is one of the two Sacraments of Healing. We receive this sacrament when we are very sick, old, or dying. This sacrament helps make our faith and trust in God strong.

What are the Sacraments at the Service of Communion?

Holy Orders and Matrimony, or Marriage, are the two Sacraments at the Service of Communion. People who receive these sacraments serve God.

What is the Sacrament of Holy Orders?

In this sacrament, baptized men are consecrated as bishops, priests, or deacons. They serve the whole Church. They serve in the name and person of Christ.

Who is a bishop?

A bishop is a priest. He receives the fullness of the Sacrament of Holy Orders. He is a successor to the Apostles. He leads and serves in a diocese. He teaches and leads worship in the name of Jesus.

Who is a priest?

A priest is a baptized man who receives the Sacrament of Holy Orders. Priests work with their bishops. The priest teaches about the Catholic faith. He celebrates Mass. Priests help to guide the Church.

Who is a deacon?

A deacon is ordained to help bishops and priests. He is not a priest. He is ordained to serve the Church.

What is the Sacrament of Matrimony?

In the Sacrament of Matrimony, or Marriage, a baptized man and a baptized woman make a lifelong promise. They promise to serve the Church as a married couple. They promise to love each other. They show Christ's love to others.

What are the sacramentals of the Church?

Sacramentals are objects and blessings the Church uses. They help us worship God.

Life in the Spirit

The Moral Life

Why did God create us?

God created us to give honor and glory to him. God created us to live a life of blessing with him here on Earth and forever in Heaven.

What does it mean to live a moral life?

God wants us to be happy. He gives us the gift of his grace. When we accept God's gift by living the way Jesus taught us, we are being moral.

What is the Great Commandment?

Jesus taught us to love God above all else. He taught us to love our neighbor as ourselves. This is the path to happiness.

What are the Ten Commandments?

The Ten Commandments are the laws that God gave Moses. They teach us to live as God's people. They teach us to love God, others, and ourselves. The Commandments are written on the hearts of all people.

What are the Beatitudes?

The Beatitudes are teachings of Jesus. They tell us what real happiness is. The Beatitudes tell us about the Kingdom of God. They help us live as followers of Jesus. They help us keep God at the center of our lives.

What are the Works of Mercy?

God's love and kindness is at work in the world. This is what mercy is. Human works of mercy are acts of loving kindness. We reach out to people. We help them with what they need for their bodies and their spirits.

What are the Precepts of the Church?

The Precepts of the Church are five rules. These rules help us worship God and grow in love of God and our neighbor.

Holiness of Life and Grace

What is holiness?

Holiness is life with God. Holy people are in the right relationship with God, with people, and with all of creation.

What is grace?

Grace is the gift of God sharing of his life and love with us.

What is sanctifying grace?

Sanctifying grace is the grace we receive at Baptism. It is a free gift of God, given by the Holy Spirit.

What are the Gifts of the Holy Spirit?

The seven Gifts of the Holy Spirit help us to live our Baptism. They are wisdom, understanding, right judgment, courage, knowledge, reverence, and wonder and awe.

The Virtues

What are the virtues?

The virtues are spiritual powers or habits. The virtues help us to do what is good.

What are the most important virtues?

The most important virtues are the three virtues of faith, hope, and love. These virtues are gifts from God. They help us keep God at the center of our lives.

What is conscience?

Every person has a conscience. It is a gift God gives to every person. It helps us know and judge what is right and what is wrong. Our consciences move us to do good and avoid evil.

Evil and Sin

What is evil?

Evil is the harm we choose to do to one another and to God's creation.

What is temptation?

Temptations are feelings, people, and things that try to get us to turn away from God's love and not live a holy life.

What is sin?

Sin is freely choosing to do or say something that we know God does not want us to do or say.

What is mortal sin?

A mortal sin is doing or saying something on purpose that is very bad. A mortal sin is against what God wants us to do or say. When we commit a mortal sin, we lose sanctifying grace.

What are venial sins?

Venial sins are sins that are less serious than mortal sins. They weaken our love for God and for one another. They make us less holy.

Christian Prayer

What is prayer?

Prayer is talking to and listening to God. When we pray, we raise our minds and hearts to God the Father, Son, and Holy Spirit.

What is the Our Father?

The Lord's Prayer, or Our Father, is the prayer of all Christians. Jesus taught his disciples the Our Father. Jesus gave this prayer to the Church. When we pray the Our Father, we come closer to God and to his Son, Jesus Christ. The Our Father helps us become like Jesus.

What kinds of prayer are there?

Some kinds of prayer use words that we say aloud or quietly in our hearts. Some silent prayers use our imagination to bring us closer to God. Another silent prayer is simply being with God.

Glossary

angels
[page 49]

________________ are God's messengers and helpers.

Baptism
[page 93]

________________ is the first sacrament that we celebrate. In Baptism, we receive the gift of God's life and become members of the Church.

believe
[page 21]

To ________________ means to have faith in God.

Bible
[page 13]

The ________________ is the written Word of God.

Catholics
[page 73]

________________ are followers of Jesus and members of the Catholic Church.

charity
[page 208]

________________ is loving others as God loves us.

children of God
[page 201]

All people are ________________, created in God's image.

Christians
[page 157]

________________ believe in Jesus Christ and live as he taught.

Church
[page 73]

The ______________________ is the People of God who believe in Jesus and live as his followers.

Church's year
[page 85]

The ______________________ is made up of four main seasons. They are Advent, Christmas, Lent, and Easter.

community
[page 181]

A ______________________ is a group of people who respect and care for one another.

counsel
[page 64]

______________________ is another word for the help that a good teacher gives us. Counsel is a gift of the Holy Spirit.

courage
[page 48]

The virtue of ______________________ helps us to trust in God and live our faith.

Creator
[page 29]

God is the ______________________. He created out of love and without any help.

cross
[page 57]

Jesus died on a ______________________ so that we could live forever in Heaven.

disciples
[page 57]

______________________ are followers of Jesus.

Easter
[page 85]

________________ is the season when we celebrate that Jesus is risen.

Eucharist
[page 137]

The ________________ is the sacrament in which we receive the Body and Blood of Christ.

faith
[page 21]

________________ is a gift from God. It helps us to know God and to believe in him.

faithful
[page 12]

Good friends of Jesus are ________________ to him. They are loyal to him.

fidelity
[page 108]

Parents demonstrate ________________ when they love and care for their children.

Galilee
[page 145]

________________ was one of the main places where Jesus taught and helped people.

generosity
[page 20]

We share our things with others. We show ________________ to them.

gentleness
[page 200]

Gentle people act calmly. They treat all people with ________________.

glory
[page 201]

____

______________________ is another word for praise.

goodness
[page 100]

____

______________________ is a sign that we are living our Baptism. When we are good to people, we honor God.

Gospel
[page 101]

____

The ______________________ is the Good News that Jesus told us about God's love.

Great Commandment
[page 181]

____

The ______________________ is to love God above all else and to love others as we love ourselves.

Holy Family
[page 37]

____

The ______________________ is the family of Jesus, Mary, and Joseph.

Holy Spirit
[page 65]

____

The ______________________ is the Third Person of the Holy Trinity.

Holy Trinity
[page 65]

____

The ______________________ is one God in Three Divine Persons—God the Father, God the Son, and God the Holy Spirit.

honor
[page 173]

____

We ______________________ people when we treat them with great respect.

hope
[page 56]

The virtue of ____________ helps us to remember that one day we may live in happiness with God forever in Heaven.

hospitality
[page 92]

We demonstrate ____________ when we welcome others as God's children.

humility
[page 216]

____________ helps us know that all good things come from God.

image of God
[page 29]

We are created in the ____________ .

joy
[page 192]

We live with ____________ when we recognize that true happiness comes from knowing and following Jesus.

justice
[page 180]

We practice ____________ when we treat people fairly.

kindness
[page 36]

We live the virtue of ____________ by treating others as we want to be treated.

Kingdom of God
[page 193]

The ____________ is Heaven. Heaven is happiness with God forever.

knowledge
[page 164]

The gift of ______
helps you to know and to follow God's rules.

marriage
[page 109]

A ______ is the lifelong promise of love made by a man and a woman to live as a family.

Mass
[page 137]

The ______ is the most important celebration of the Church.

Matrimony
[page 109]

______ is the sacrament that Catholics celebrate when they marry.

miracle
[page 145]

A ______ is something only God can do. It is a sign of God's love.

Our Father
[page 217]

The ______
is the prayer Jesus taught his disciples.

parable
[page 209]

Jesus often told a ______
to help people to know and love God better.

patience
[page 120]

We act with ______
when we listen carefully to others.

peace
[page 128]

We live as ________ makers when we forgive those who have hurt us.

perseverance
[page 136]

________ helps us to live our faith when it is difficult.

prayer
[page 121]

________ is listening and talking to God.

prudence
[page 84]

________ helps us ask advice from others when making important decisions.

respect
[page 173]

We show people ________ when we love them because they are children of God.

Resurrection
[page 57]

God's raising Jesus from the dead to new life is called the ________.

reverence
[page 72]

We show ________ to others when we honor them and give them great respect.

Sacraments
[page 93]

The ________ are the seven signs and celebrations of God's love that Jesus gave the Church.

sin
[page 129]

__________ is choosing to do or say something that we know is against God's laws.

Son of God
[page 37]

Jesus is the __________.

temperance
[page 172]

__________ helps us to know the difference between what we need and what we just want to have.

Ten Commandments
[page 165]

The __________ are the ten laws that God has given us to help us live as children of God.

understanding
[page 156]

God the Holy Spirit gives us the gift of __________. Stories in the Bible help us understand God's love for us.

wisdom
[page 144]

__________ helps us to know what God wants us to do. It helps us to live a holy life.

wonder
[page 28]

__________ is a gift from God to help us know how good He is.

worship
[page 165]

We __________ God when we love and honor God more than anyone and anything else.

Index

Credits

Cover Illustration: Marcia Adams Ho

PHOTO CREDITS

Frontmatter: Page 6, © Laurence Monneret/Getty Images; 7, © Ladushka/Shutterstock.
Chapter 1: Page 11, © Andersen Ross; 17, © Ken Seet/Corbis; 18, © Ocean/Corbis.
Chapter 2: Page 19, © Fever Images/Jupiterimages; 26, © Asia Images Group Pte Ltd/Alamy.
Chapter 3: Page 30, © Dmitriy Shironosov/Shutterstock.com; 34, © Jupiterimages.
Chapter 4: Page 39, © Bounce/Getty Images; 39, © Design Pics Inc./Alamy; 39, © Fuse/Getty Images; 42, © Design Pics Inc./Alamy.
Chapter 5: Page 47, © Roger Cracknell 01/classic / Alamy; 50, © LWA/Jay Newman/Jupiterimages; 51, © Jupiterimages; 53, © Plush Studios/Jupiterimages; 54, © Design Pics/Kristy-Anne Glubish/Getty Images.
Chapter 6: Page 56, © AFP/Getty Images; 57, © Bill Wittman; 61, © Jose Luis Pelaez Inc/Jupiterimages; 62, © kali9/iStockphoto.
Chapter 7: Page 63, iofoto/Shutterstock; 64, © The Crosiers/Gene Plaisted, OSC; 66, © Tischenko Irina /Shutterstock; 67, © Fuse/Jupiterimages; 67, © Bruce Forster/Getty Images; 67, © Fuse/Jupiterimages; 69, © Bill Wittman; 70, © Blend Images/Alamy.
Chapter 8: Page 77, © Design Pics/Don Hammond/Jupiterimages; 78, © Blend Images/Alamy.
Chapter 9: Page 83, © Dorling Kindersley/Jupiterimages; 89, © Tetra Images/Jupiterimages; 90, © AFP/Getty Images.
Chapter 10: Page 93, © Dmitry Naumov / Shutterstock; 94, © Bill Wittman; 95, © Ted Foxx/Alamy; 98, © a la france/Alamy.
Chapter 11: Page 102, © Tim Graham/Getty Images; 106, © Michael Hitoshi/Jupiterimages.
Chapter 12: Page 107, © OJO Images/Jupiterimages; 108, © Jose Luis Pelaez Inc/Jupiterimages; 108, © Fancy/Veer/Corbis/Jupiterimages; 109, © Jupiterimages; 109, © Monkey Business Images/Shutterstock; 113, © Fuse/Jupiterimages; 114, © Purestock/Getty Images.
Chapter 13: Page 121, © Brand X Pictures/Jupiterimages; 125, © The Crosiers/Gene Plaisted, OSC; 126, © Fuse/Jupiterimages.
Chapter 14: Page 128, © AFP/Getty Images; 131, © Steve Gorton/Getty Images; 133, © Myrleen Ferguson Cate/Photo Edit; 134, © moodboard/Alamy.
Chapter 15: Page 135, © SW Productions/Jupiterimages; 136, © Design Pics Inc./Alamy; 141, © Stockbyte/Jupiterimages; 142, © Bill Wittman.
Chapter 16: Page 143, © Juice Images/Jupiterimages; 144, © CRS; 144, © Jim Stipe/CRS; 144, © Jim Stipe/CRS; 147, © Ariel Skelley/Jupiterimages; 147, © SW Productions/Jupiterimages; 147, © Andersen Ross/Jupiterimages; 149, © Design Pics Inc./Alamy; 150, © Juice Images/Alamy.
Chapter 17: Page 155, © Danita Delimont/Alamy; 161, © Inspirestock Inc./Alamy; 162, © Tony Freeman/Photo Edit.
Chapter 18: Page 164, © 501room/Shutterstock; 164, © Ritu Manoj Jethani/Shutterstock; 164, © Neil Jacobs/Getty Images; 165, © sonya etchison/Shutterstock; 166, © Colorblind/Jupiterimages; 166, © George Doyle/Jupiterimages; 166, © Daniel Pangbourne/Jupiterimages; 169, © Stockbyte/Jupiterimages; 170, © Jupiterimages.
Chapter 19: Page 171, ©Monkey Business Images/Shutterstock; 172, © Tony Freeman/Photo Edit; 173, © Karl Kost/Alamy; 173, © SW Productions/Getty Images; 177, © Paul Burns/Jupiterimages; 178, © BananaStock/Jupiterimages.
Chapter 20: Page 179, © Zigy Kaluzny/Getty Images; 181, © Fuse/Jupiterimages; 181, © Tom Merton/Jupiterimages; 181, © Colin Hawkins/Jupiterimages; 185, © Brand X Pictures/Jupiterimages; 186, © Ocean/Corbis.
Chapter 21: Page 191, © Stockbyte/Getty Images; 192, © Photograph courtesy of the Pontifical Mission Societies; 192, © taelove7/Shutterstock; 195, © Dmitriy Shironosov/Shutterstock; 195, © GlowImage/Alamy; 195, © Blend Images/Alamy; 197, © Digital Vision/Jupiterimages; 198, © Michael Newman / Photo Edit.
Chapter 22: Page 200, © William Thomas Cain/Getty Images; 202, © Thomas M Perkins/Shutterstock; 202, © laszlo a. lim/Shutterstock; 205, © Zdorov Kirill Vladimirovich/Shutterstock; 206, © Dave & Les Jacobs/Jupiterimages.
Chapter 23: Page 213, © Bill Wittman 214, © Design Pics/SW Productions/Jupiterimages.
Chapter 24: Page 215, © Image Source/Jupiterimages; 216, © Jean-Claude FRANCOLON/Gamma-Rapho via Getty Images; 216, © Luciano Mortula /Shutterstock; 219, © Peter Zander/Getty Images; 221, © ONOKY - Photononstop/Alamy; 222, © Jupiterimages.
Liturgical Seasons: Page 225, © Design Pics Inc./Alamy; 226, © Chris Salvo/Getty Images; 226, © sodapix sodapix/Getty Images; 226, © ATTILA KISBENEDEK/AFP/Getty Images; 226, © The Crosiers/Gene Plaisted, OSC; 226, © ClassicStock/Alamy; 227, © Stockbyte/Jupiterimages; 227, © The Crosiers/Gene Plaisted, OSC; 227, © The Crosiers/Gene Plaisted, OSC; 227, © The Crosiers/Gene Plaisted, OSC; 229, © McPHOTO / SHU; 231, © The Crosiers/Gene Plaisted, OSC; 233, © The Crosiers/Gene Plaisted, OSC; 235, © Serp/Shutterstock; 235, © Cathy Baxter/Private Collection/The Bridgeman Art Library; 236, © Spencer Grant / Photo Edit; 237, © The Crosiers/Gene Plaisted, OSC; 239, © The Crosiers/Gene Plaisted, OSC; 241, © AP Photo/The Miami Herald, Walter Michot; 241, © Amanda Brown/Star Ledger/Corbis; 243, © Dejan Ristovski /Getty Images; 245, © asharkyu/Shutterstock; 245, © Bill Wittman; 245, © The Crosiers/Gene Plaisted, OSC; 247, © Bill Wittman; 249, © The Crosiers/Gene Plaisted, OSC; 249, © Bill Wittman; 251, © Ocean/Corbis; 251, © Bill Wittman; 251, © Cornelia Doerr/Getty Images; 253, © The Crosiers/Gene Plaisted, OSC; 255, © The Crosiers/Gene Plaisted, OSC; 255, © Stephanie Neal Photography/Getty Images.
Backmatter: Page 257, © Fuse/Jupiterimages; 259, © Blend Images/Alamy; 262, © Bill Wittman; 263, © Bill Wittman; 264, © Bill Wittman; 265, © Bill Wittman; 266, © Bill Wittman; 267, © Bill Wittman; 269, © Bill Wittman.

ILLUSTRATION CREDITS

Listed from Top to Bottom; Left to Right

Frontmatter: Page 8,
Chapter 1: Page 12, Q2A Media; **14-15 Julia Woolf**
Chapter 2: Page 20, Q2A Media; 21, Julia Woolf; 22, Q2A Media; 23, Q2A Media; 25, Q2A Media.
Chapter 3: Page 27, Q2A Media; 28, Kristin Sorra; 29, Q2A Media; 32, Q2A Media; 33, Q2A Media.
Chapter 4: Page 35, Julia Woolf; 36, Q2A Media; 37, Julia Woolf; 38, Julia Woolf; 41, Q2A Media.
Chapter 5: Page 48, Q2A Media; 49, Julia Woolf.
Chapter 6: Page 55, Julia Woolf; 58-**59**, Julia Woolf.
Chapter 7: Page 64, Q2A Media; 66, Q2A Media.
Chapter 8: Page 71, Q2A Media; 72, Q2A Media; 73, Julia Woolf; 75, Q2A Media.
Chapter 9: Page 84, Q2A Media; 85, Q2A Media; 86, Q2A Media; 87, Q2A Media; 88, Q2A Media.
Chapter 10: Page 91, Julia Woolf; 92, Q2A Media; 94, Q2A Media.
Chapter 11: Page 99, Julia Woolf; 100, Q2A Media; 101, Julia Woolf; 103, Q2A Media; 104, Q2A Media; 105, Q2A Media.
Chapter 12: Page 110, Q2A Media; 111, Julia Woolf.
Chapter 13: Page 120, Q2A Media; 122, Q2A Media; 123, Q2A Media.
Chapter 14: Page 127, Sole Otero; 129, Q2A Media; 130, Q2A Media; 132, Q2A Media.
Chapter 15: Page 137, Q2A Media; 139, Julia Woolf.
Chapter 16: Page 145, Lyn Boyer; 146, Julia Woolf.
Chapter 17: Page 156, Q2A Media; 158-**9**, Julia Woolf.
Chapter 18: Page 163, Ivanke and Lola; 167, Q2A Media.
Chapter 19: Page 174, Q2A Media; 175, Rémy Simard.
Chapter 20: Page 180, Colleen Madden; 182, Q2A Media; 183, Q2A Media.
Chapter 21: Page 193, Kristin Sorra; 194, Julia Woolf.
Chapter 22: Page 199, Julia Woolf; 201, Q2A Media.
Chapter 23: Page 207, Julia Woolf; 208, Q2A Media; 209, Q2A Media; 210, Julia Woolf; 211, Rémy Simard.
Chapter 24: Page 217, Q2A Media.
Liturgical Seasons: Page 234, Pamela Becker; 244, Ivanke and Lola; 256, Ivanke and Lola.